AF703201

GENETICS AND BREEDING OF VEGETABLE CROPS

Kruger Brentt Publishers

GENETICS AND BREEDING OF VEGETABLE CROPS

Ricky Bates
Gustavo Rodriguez
Veronica Di Stilio

Kruger Brentt
Publishers

2025

Kruger Brentt Publishers UK. LTD.
Company Number 9728962

Regd. Office: 68 St Margarets Road, Edgware, Middlesex HA8 9UU

© 2025 AUTHORS

ISBN: 978-1-78715-392-9

Disclaimer:

Every possible effort has been made to ensure that the information contained in this book is accurate at the time of going to press, and the publisher and author cannot accept responsibility for any errors or omissions, however caused. No responsibility for loss or damage occasioned to any person acting, or refraining from action, as a result of the material in this publication can be accepted by the editor, the publisher or the author. The Publisher is not associated with any product or vendor mentioned in the book. The contents of this work are intended to further general scientific research, understanding and discussion only. Readers should consult with a specialist where appropriate.

Every effort has been made to trace the owners of copyright material used in this book, if any. The author and the publisher will be grateful for any omission brought to their notice for acknowledgement in the future editions of the book.

All Rights reserved under International Copyright Conventions. No part of this publication may be reproduced, stored in a retrieval system, or transmitted in any form or by any means, electronic, mechanical, photocopying, recording or otherwise without the prior written consent of the publisher and the copyright owner.

For information on all our publications visit our website at http://krugerbrentt.com/

PREFACE

"Genetics and Breeding of Vegetable Crops" offers a comprehensive exploration of the principles, methodologies, and applications of molecular breeding in the improvement of vegetable crops. This book serves as an essential resource for researchers, breeders, students, and practitioners involved in the development and enhancement of vegetable crops through modern biotechnological approaches.

Vegetable crops are vital components of global agriculture, providing essential nutrients, flavors, and culinary diversity to diets worldwide. The improvement of vegetable crops through traditional breeding methods has been limited by factors such as long breeding cycles, genetic variability, and complex traits. Molecular breeding, which encompasses a range of biotechnological tools and techniques, offers new opportunities to accelerate the breeding process, enhance genetic diversity, and develop improved vegetable varieties with desirable traits.

As editors of this volume, we have endeavored to compile a diverse array of perspectives, methodologies, and case studies from leading experts and practitioners in the field of molecular breeding of vegetable crops. Through a blend of theoretical foundations, practical insights, and real-world applications, this book covers a wide range of topics, including molecular markers, genetic mapping, genomic selection, gene editing, and biotechnological approaches for trait improvement in vegetable crops.

Our overarching goal in presenting this book is to provide readers with a comprehensive understanding of the principles and applications of molecular breeding in vegetable crop improvement. By offering insights into the latest advancements, best practices, and emerging trends in molecular breeding, we aim to inspire innovation, collaboration, and excellence in the development of improved vegetable varieties to meet the challenges of food security, climate change, and sustainability.

We extend our sincere gratitude to all the contributors who have generously shared their expertise, experiences, and insights in the field of molecular breeding of vegetable crops. It is our hope that this book will serve as a valuable reference and guide for researchers, breeders, students, and practitioners as they work towards the development of resilient, nutritious, and high-yielding vegetable crops to feed the growing global population.

Ricky Bates

Gustavo Rodriguez

Veronica Di Stilio

CONTENTS

1 Chapter

RECENT ADVANCES IN MOLECULAR BREEDING OF VEGETABLE CROPS

1. PROGRESS IN GENE EDITING IN THE CONTEXT OF MOLECULAR BREEDING OF VEGETABLES

Genome editing has emerged as a reliable and potent technique for accurately manipulating the genetic makeup of plants in recent times. This novel strategy addresses the limitations of traditional breeding methods, such as prolonged artificial selection and restricted genetic germplasm resources, in the context of molecular vegetable breeding applications [1,2]. Wan et al. conducted a comprehensive examination of the progress and utilization of CRISPR-Cas9 gene editing in vegetable crops. Presently, this approach has been employed to enhance the longevity of produce, the caliber of fruits, and the ability of main vegetable crops, such as tomato and cabbage, to withstand stress. Genome editing has achieved success in a restricted number of B. oleracea cultivars when it comes to broccoli [3]. While the use of genome editing is widespread, there is a need for further research and improvement in order to gain germplasm resources through gene editing of Cis-regulatory elements (CREs) and develop a universal regeneration system for vegetable crops [1,4].

2. ASSESSMENT OF GENETIC VARIATION IN VEGETABLE GERMPLASM FOR THE PURPOSE OF ENHANCING CROP QUALITY

Assessing genetic diversity in agronomic and qualitative attributes remains a crucial method and approach in the contemporary breeding process, serving as a foundation for the exploitation and exploration of germplasm. Uddin et al. conducted a study where they analyzed the characteristics and genetic diversity of 130 eggplant germplasms [5]. The inherent variation of several qualities was determined and relevant parents for eggplant improvement were selected based on an analysis of trait variance, correlation matrix, and MGIDI index. Simple sequence repeats (SSRs) are extensively employed as genetic markers in several crops for the purpose of studying genetic diversity. This is primarily because SSRs possess co-dominant features and exhibit a high degree of polymorphism. Zhong et al. utilized this sequencing approach in Capsicum frutescens to generate SSR molecular markers and assess the genetic diversity for pepper breeding purposes [6]. The analysis of the entire genome identified SSR markers, indicating that trinucleotides were the most prevalent repetitive motif. A total of 147 pepper cultivars were identified and classified into seven primary groups based on the investigation of their genetic diversity and phylogenetic relationships. A total of 103,056 SSR loci were identified in Cucurbita moschata using in silico PCR. Among these loci, di-nucleotide patterns were the most prevalent kind [7]. The examination of cross-species SSR markers revealed that the primary syntenic linkages across Cucurbita species have remained remarkably preserved throughout evolution.

3. THE GENETIC BASIS OF BIOTIC RESISTANCE IN VEGETABLE CROPS

Fungal infections continue to pose issues that hinder the sustainable development of vegetable production. While pesticides can effectively prevent and manage fungal illnesses, their excessive application has caused significant harm to both the environment and human health. Enhancing disease resistance has emerged as a crucial goal in breeding. The advent and implementation of molecular marker technology enables the direct and efficient selection of germplasm resources. Kim et al. [8] evaluated the use of RAPD-PCR to develop molecular markers that are linked to resistance against gray mold disease in onion (*Allium cepa L.*). The genetic link between the resistant and susceptible lines was determined and a SCAR marker was developed using RAPD analysis. Furthermore, we conducted RNA-seq analysis on both the gray mold-resistant and -susceptible onion lines in order to establish a selectable marker specifically for the resistant line. Phytophthora blight is a prevalent ailment that leads to reduced productivity and diminished quality in pepper plants (*Capsicum annuum L.*). Li et al. utilized SLAF-seq and QTL analysis to produce a detailed genetic map of pepper that is linked to resistance to Phytophthora capsici [9]. CQPc5.1 has been discovered as a significant quantitative trait locus (QTL) for P. capsici resistance. This QTL contains a total of 23 candidate genes that are located within the specified interval.

4. IDENTIFYING GENES ASSOCIATED WITH ABIOTIC STRESSES TO ENHANCE VEGETABLE TRAITS

Abiotic stressors in vegetable crops result in significant harm, restricting development and disrupting physiological metabolic systems. Hence, it is imperative to conduct gene screening for abiotic challenges in order to develop resistant types through breeding. Pepper exhibits a heightened sensitivity to elevated temperatures, resulting in the manifestation of severe symptoms, including pollination failure, growth abnormalities, and other related issues. Wang et al. conducted a transcriptomics investigation to identify genes that are expressed differently in pepper leaves of heat-tolerant and heat-sensitive types [10]. Heat shock proteins (HSPs) and heat shock transcription factors (HSFs) were found to be sensitive to heat stress or recovery. A tiny heat shock protein, CaHSP18.1a, was isolated and characterized from pepper [11]. Liu et al. established that CaHSP18.1a exhibited sensitivity to heat stress and exhibited elevated expression levels in the thermo-tolerant line. When CaHSP18.1a was silenced, there was an increase in MDA levels and a loss in resistance to heat, drought, and salt shocks. This suggests that CaHSP18.1a has a favorable role in regulating tolerance to abiotic stress. Gong et al. [12] conducted a study to identify Hsf and Hsp genes in eggplant that are affected by heat stress at the genome-wide level. The RNA-seq analysis revealed distinct expression levels of Hsf and Hsp genes in the thermotolerant line 05-4 and the thermosensitive line 05-1. This finding serves as a foundation for investigating the correlation between thermotolerance and heat-response genes. Currently, the excessive and illogical application of nitrogen fertilizer has led to unfavorable plant growth and decreased pepper production. The precise mechanism responsible for the genetic diversity in N-use efficiency (NUE) is not well understood. Wang et al. conducted a comparative transcriptome analysis to identify two genotypes of pepper with differing low-N tolerance. The purpose was to investigate the variation in nitrogen use efficiency (NUE) in these genotypes [13]. A multitude of differentially expressed genes (DEGs) related to nitrogen (N) metabolism and other physiological processes were discovered, offering potential candidate genes for enhancing nitrogen use in pepper plants.

5. INTERACTIONS BETWEEN ORGANIC COMPOUNDS IN VEGETABLES AND THE ENVIRONMENT

Flavonoids and volatile organic compounds play crucial roles in the establishment and development of vegetable crops. They are involved in attracting insect pollinators, preventing plant illnesses, and enhancing weed management. Anthocyanins in broccoli are responsible for its purple color and function as beneficial antioxidants that promote health. Liu et al. discovered significant genetic regions and potential genes that play a crucial role in the production of anthocyanin in broccoli [14]. Two quantitative trait loci (QTLs) located on chromosomes 7 were shown to have a strong correlation with the process of anthocyanin production, as determined by QTL-seq bulk segregant analysis. Subsequent detailed mapping revealed 14 possible genes that could serve as a molecular indicator for the development of new kinds with high levels of anthocyanins. Plant-derived volatile organic molecules are associated with the allelopathy phenomenon, which refers to the chemical interactions between plants. Xie et al. conducted a study on the recent progress in understanding the allelopathic effects of volatile organic compounds (VOCs) emitted by plants [15]. Volatile organic compounds (VOCs) exert several allelopathic effects on plants, including influencing enzyme activity, dormancy, disease resistance, scavenging reactive oxygen species (ROS), facilitating plant-to-plant communication, and impacting other areas [15]. The study indicated that the allelopathic effects of volatile organic compounds (VOCs) can be used to create cost-effective and efficient strategies for sustainable agriculture [15]. Scientists and breeders are collaborating closely to enhance breeding technology, such as gene editing, and create superior cultivars through activities including germplasm evaluation, gene isolation, and marker creation. Given the increasing availability of extensive genomic data and advanced technologies, it is highly beneficial to engage in additional collaboration to advance next-generation breeding technology. One example of such technology is genome-based breeding by design, which aims to develop vegetable cultivars that are both environmentally adaptive and have high yield and quality.

AT A GLANCE

In recent years, there has been significant advancement in the field of molecular vegetable breeding. Significant progress has been achieved in the assessment of germplasm, identification of genes, modification of plants, editing of genes, and breeding with the assistance of molecular markers. The objective of this Special Issue is to showcase cutting-edge research in the field of molecular vegetable breeding, ranging from fundamental to practical applications. The papers chosen for publication in the Special Issue of Horticulturae demonstrate a wide range of approaches in the field of molecular vegetable breeding. The studies mentioned in this editorial are particularly remarkable.

REFERENCES

1. Wan, L.; Wang, Z.; Tang, M.; Hong, D.; Sun, Y.; Ren, J.; Zhang, N.; Zeng, H. CRISPR-Cas9 Gene Editing for Fruit and Vegetable Crops: Strategies and Prospects. *Horticulturae* 2021, 7, 193.
2. Chen, Y.; Mao, W.; Liu, T.; Feng, Q.; Li, L.; Li, B. Genome Editing as A Versatile Tool to Improve Horticultural Crop Qualities. *Hortic. Plant J.* 2020, 6, 372.
3. Han, F.; Liu, Y.; Fang, Z.; Yang, L.; Zhuang, M.; Zhang, Y.; Lv, H.; Wang, Y.; Ji, J.; Li, Z. Advances in Genetics and Molecular Breeding of Broccoli. *Horticulturae* 2021, 7, 280.
4. Li, X.; Li, H.; Zhao, Y.; Zong, P.; Zhan, Z.; Piao, Z. Establishment of a Simple and Efficient Agrobacterium-mediated Genetic Transformation System to Chinese Cabbage (*Brassica rapa* L. ssp. *pekinensis*). *Hortic. Plant J.* 2021, 7, 117.
5. Uddin, M.; Billah, M.; Afroz, R.; Rahman, S.; Jahan, N.; Hossain, M.; Bagum, S.; Uddin, M.; Khaldun, A.; Azam, M.; et al. Evaluation of 130 Eggplant (*Solanum melongena* L.) Genotypes for Future Breeding Program Based on Qualitative and Quantitative Traits, and Various Genetic Parameters. *Horticulturae* 2021, 7, 376.

6. Zhong, Y.; Cheng, Y.; Ruan, M.; Ye, Q.; Wang, R.; Yao, Z.; Zhou, G.; Liu, J.; Yu, J.; Wan, H. High-Throughput SSR Marker Development and the Analysis of Genetic Diversity in Capsicum frutescens. *Horticulturae* 2021, 7, 187.

7. Zhu, L.; Zhu, H.; Li, Y.; Wang, Y.; Wu, X.; Li, J.; Zhang, Z.; Wang, Y.; Hu, J.; Yang, S.; et al. Genome Wide Characterization, Comparative and Genetic Diversity Analysis of Simple Sequence Repeats in Cucurbita Species. *Horticulturae* 2021, 7, 143.

8. Kim, S.; Park, J.; Park, T.; Lee, H.; Choi, J.; Park, Y. Development of Molecular Markers Associated with Resistance to Gray Mold Disease in Onion (*Allium cepa* L.) through RAPD-PCR and Transcriptome Analysis. *Horticulturae* 2021, 7, 436.

9. Li, Y.; Zhang, S.; Yang, X.; Wang, C.; Huang, Q.; Huang, R. Generation of a High-Density Genetic Map of Pepper (*Capsicum annuum* L.) by SLAF-seq and QTL Analysis of Phytophthora capsici Resistance. *Horticulturae* 2021, 7, 92.

10. Wang, F.; Yin, Y.; Yu, C.; Li, N.; Shen, S.; Liu, Y.; Gao, S.; Jiao, C.; Yao, M. Transcriptomics Analysis of Heat Stress-Induced Genes in Pepper (*Capsicum annuum* L.) Seedlings. *Horticulturae* 2021, 7, 339.

11. Liu, Y.; Liu, S.; Xiao, J.; Cheng, G.; Gong, Z. CaHSP18.1a, a Small Heat Shock Protein from Pepper (*Capsicum annuum* L.), Positively Responds to Heat, Drought, and Salt Tolerance. *Horticulturae* 2021, 7, 117.

12. Gong, C.; Pang, Q.; Li, Z.; Li, Z.; Chen, R.; Sun, G.; Sun, B. Genome-Wide Identification and Characterization of Hsf and Hsp Gene Families and Gene Expression Analysis under Heat Stress in Eggplant (*Solanum melongema* L.). *Horticulturae* 2021, 7, 149.

13. Wang, C.; Li, Y.; Bai, W.; Yang, X.; Wu, H.; Lei, K.; Huang, R.; Zhang, S.; Huang, Q.; Lin, Q. Comparative Transcriptome Analysis Reveals Different Low-Nitrogen-Responsive Genes in Pepper Cultivars. *Horticulturae* 2021, 7, 110.

14. Liu, C.; Yao, X.; Li, G.; Huang, L.; Wu, X.; Xie, Z. Identification of Major Loci and Candidate Genes for Anthocyanin Biosynthesis in Broccoli Using QTL-Seq. *Horticulturae* 2021, 7, 246.

15. Xie, Y.; Tian, L.; Han, X.; Yang, Y. Research Advances in Allelopathy of Volatile Organic Compounds (VOCs) of Plants. *Horticulturae* 2021, 7, 278.

2 Chapter

GENETICS AND MOLECULAR BREEDING OF BROCCOLI: RECENT ADVANCES

1. INTRODUCTION: AN OVERVIEW

Broccoli, scientifically known as *Brassica oleracea* L. var. *italica*, belongs to the Brassicaceae family and is extensively grown as a significant vegetable crop globally [1,2]. The plant produces edible enlarged reproductive structures (floral head and stalk) that have significant health advantages and are rich in nutrients such as vitamin A, vitamin K, calcium, magnesium, and bioactive chemicals with anticancer effects, including glucosinolates, sulforaphane, selenium, and flavonoids [3,4,5].

The italica group originates from the cultivation and domestication of Brassica oleracea (CC genome; 2n = 18) in the Mediterranean region. The cultivation of B. oleracea mustard plants can be dated back to the Hellenic culture, which began in the 6th century BC [6]. 'Broccoli' is likely a vernacular Latin term for the protruding shoots of the cabbage family, namely the *B. oleracea* varieties [6]. Over the course of the previous 2000 years, broccoli-like variations were created by choosing desirable forms of *B. oleracea*. These selections led to the development of different broccoli landraces, primarily in Italy [6,7,8,9]. The broccoli variety '*Vrocculi o Sparaceddi*' is recognized as the initial cultivated type of wild brassica, from which broccoli developed [10]. Over the course of the last three centuries, the quality of broccoli has significantly enhanced, mostly due to the careful selection carried out by Danish and English horticulturists [6]. Historically, the consumption of broccoli as a food was limited to the Italian peninsula, where it was mostly cultivated as sprouting broccoli varieties [10]. The breeding and enhancement of calabrese broccoli cultivars, which yield larger and denser heads resembling cauliflower, led to the global dissemination and increased popularity of broccoli [10,11]. Multiple broccoli landraces were brought to the United Kingdom throughout the 1700s and to the United States during the 1800s. Their popularity increased significantly after World War II [11]. Broccoli was first introduced to many southern provinces of China during the 1980s and has since become a popular vegetable that is extensively cultivated throughout the country. China has emerged as the world's greatest producer of broccoli, with a growing area of 80,000 hectares [12]. The market demand for broccoli is steadily expanding in major broccoli-producing countries such as China, the US, and India [12,13,14], driven by scientific proof of its health benefits.

Broccoli has been extensively studied due to advancements in molecular biology technologies and functional genomics. Marker-assisted selection (MAS) and genetic transformation were integrated with traditional breeding methods to enhance the productivity, quality, and tolerance of broccoli to biotic and abiotic stressors, among other factors. This paper discusses the latest advancements in the genetics and molecular breeding of broccoli, with a specific focus on important agronomic characteristics, male sterility, tolerance to abiotic stress, disease resistance, secondary metabolites, and genetic transformation. The review does not contain the new research on postharvest yellowing of broccoli, which focuses on prolonging its shelf life [2]. The analysis conducted in 2016 [1] examined the enhancement of broccoli using genetic engineering. Therefore, this review encompasses the significant progress made in the years after 2016 up to 2021.

2. BROCCOLI: GENETICS AND MOLECULAR BREEDING

2.1. Abiotic Stress Resistance

2.1.1. Heat Stress

The production of broccoli is confronted with the issues of meeting increased demand while expanding cultivation areas and ensuring production stability in the face of extreme weather events caused by climate change [13,15]. Broccoli thrives in chilly climates and requires temperatures between 15 and 23 °C for optimal growth during the initial stages of flower formation [16]. Broccoli quality is significantly affected by high temperatures above 25 degrees Celsius. This is because most broccoli varieties need to be exposed to temperatures below 23 degrees Celsius in order to undergo vernalization, a process necessary for proper growth and development. If exposed to temperatures above the optimal range, broccoli may fail to form heads altogether. Additionally, certain broccoli varieties that do not require vernalization still experience negative effects when exposed to high temperatures, particularly above 30 degrees Celsius. This can lead to undesirable traits such as bracting, uneven head surface and sizes of buds, discoloration, and even the formation of brown beads, rendering the broccoli products unsuitable for sale. Furthermore, high temperatures during the head maturity stage also result in reduced broccoli yield. These findings are supported by previous studies [15,17,18]. Significant advancements have been achieved in developing heat-resistant breeding lines and genetically manipulating heat tolerance in broccoli in recent years. In the United States, scientists have been working to develop sustainable methods for growing broccoli in the main production area on the east coast, specifically focusing on heat tolerance. These efforts have been supported by two projects: the National Institute of Food and Agriculture (NIFA) Project No. 2010-51181-21062 and the USDA Vegetable Brassica Research Project (CRIS No. 6080-21000-019-00D). References to these projects can be found in sources 16, 19, and 20. Researchers in Asia are attempting to bring broccoli to subtropical and tropical countries, including Taiwan, China, and Indonesia [21,22].

The capacity to generate superior heads through various broccoli germplasms under heat stress is regarded as a quantitative characteristic that governs many advantageous loci [13,15]. Lin et al. discovered 31 Quantitative Trait Loci (QTLs) associated with the size and weight of broccoli heads when cultivated in high-temperature seasons with an average daily temperature of 36.4 °C and nightly temperature of 25.9 °C [23]. Branham et al. created a detailed genetic map by using genotyping-by-sequencing on a DH broccoli population that was segregating for heat tolerance. They discovered five QTLs (quantitative trait loci) and one positive epistatic interaction between QHT_C03 and QHT_C05. These findings accounted for 62.1% of the observed variation in the phenotype [15]. Branham et al. conducted whole-genome resequencing of bulked segregants from a new DH population of broccoli. They discovered two previously unknown heat tolerance QTLs, one of which, QHT_C09.2, may be responsible for the inverse relationship between maturity and heat tolerance [13].

The researchers used reversed genetic techniques to clone a broccoli catalase gene that is connected to heat stress. They found that when this gene is expressed in Arabidopsis, it can improve the plant's ability to tolerate high temperatures. However, it is still unclear whether this gene also helps in maintaining a high-quality head when exposed to high temperatures. This information is based on references 24 and 25. Furthermore, by utilizing enhanced sequencing techniques and the availability of reference genomes, several researchers conducted studies linked to omics and successfully found microRNAs/genes that were expressed differently and possible pathways that had a role in heat tolerance [26,27].

2.1.2. Other Abiotic Stresses

Several studies have focused on broccoli resistance to other stresses, such as proteomic analysis for waterlogging stresses [28], microRNA analysis for salt stress [29] and transcriptome and metabolomics for wounding stress [30]; differentially expressed proteins/microRNAs/genes were identified as possibly related to resistance to these stresses [28,29,30].

In addition, cuticular waxes on the plant surface contribute to resistance to many environmental stresses, such as drought, UV light, high radiat2.2.ion and both bacterial and fungal pathogens [31]. Some loci and linked markers for this trait have been obtained. Using a natural glossy (cuticular wax defective) mutant, Branham and Farnham identified three candidates, Bo3g001070, Bo3g122030 and Bo3g008780, for this trait on C03 [32]. In the broccoli × Chinese kale-derived BolTBDH population, leaf color was segregated, which resulted from the differences in cuticular waxes between broccoli and Chinese kale; a locus for this trait, LC_C09@15.1, was identified on C09, explaining 45.64% of the phenotypic variation [33].

2.2. Desirable Agronomic Traits

2.2.1. Heading

Broccoli produces edible reproductive organs characterized by proliferation and developmental arrest of floral buds [17]. Floral head quality is the most important agronomic trait selected by breeders. With forward and reversed genetic approaches, some genes/loci related to head formation have been identified, but the genetic basis remains elusive [34].

Some works tried to identify homologs of the Arabidopsis floral meristem identity genes LEAFY (LFY), APETALA1 (AP1) and CAULIFLOWER (CAL) and implied that BoCAL and BoAP1 are involved in curding in cauliflower, a subspecies similar to broccoli but different in the developmental stage of the reproductive meristem at harvest [35,36,37]. Subsequent studies suggest that heading is quite complex in both cauliflower and broccoli, which seems not to be controlled solely by these floral genes [34].

In the 1990s, researchers started to construct genetic maps by crossing broccoli cultivars/inbred lines with various materials, including broccoli cultivar/landrace, cabbage, cauliflower, kale and Chinese kale, to detect loci of important traits, such as disease resistance, head morphology, nutritional quality and flowering/maturation time [38,39,40,41,42]. Several quantitative trait loci influencing head traits, including head weight, head height/width and floret height/width, have been identified, but the early constructed genetic maps are hard to unify [23,42,43] due to the differences in plant germplasm, marker types and linkage group nomenclature and the lack of *B. oleracea* reference genomes before 2014. Using a double-haploid BolTBDH mapping population derived from Early Big (broccoli DH line) and TO1000DH3 (nonhead Chinese kale), Stansell et al. identified heading-quality QTLs, including BU_C04@51.5, BR_C09@49.5, HC_C09@ 48.8, HU_C09@48.8, HE_C09@ 47.7 and OQ_C49.5 @09 (Table S1), and found genomic regions of approximately 49 Mb on C09 harboring FLOWERING LOCUS C (FLC) homologs Bo9g173400 and Bo9g173370, as hotspots contributing largely to over 40% phenotypic variance of the heading phenotype [33]. In another study, three head quality QTLs, qCQ-2, qCQ-3 and qCQ6-, associated with subtropical adaptation were identified [21]; and specific haplotype combinations of candidates BoFLC3 in the interval containing qCQ3- and PERIANTHIA (PAN, a bZIP-transcription factor required for AGAMOUS activation) in the interval containing qCQ6-, were supposed to adapt broccoli to high ambient temperature and short daylength. Along with these key head-related traits, QTL mapping for bud morphology was also reported by Stansell et al. and Lin et al. [21,33]. These studies provide genetic information and breeding materials for improving broccoli varieties.

2.2.2. Flowering Time

Flowering is an important agronomic trait of broccoli, as it influences maturity, head quality, hybrid seed production and geographical region adaptation. Flowering time is considered to be controlled by multiple QTLs. To detect QTLs/genes associated with this trait, segregation populations were generated by crossing broccoli with different germplasms, such as broccoli × cabbage, broccoli × Chinese kale and broccoli × broccoli. Different QTLs were detected in these studies, even using similar populations, such as broccoli × cabbage and broccoli × Chinese kale [**21,33,44,45,46,47**]. Most studies have implied that flowering is largely controlled by one or a few major QTLs [**21,33,44,45,46,47**]. As early as the late 1990s, using populations of broccoli (nonvernalization type) × cabbage (vernalization type), broccoli (late flowering type) × Chinese kale (early flowering type), several QTLs for flowering time were mapped [**44,45,46,47**]. Two subsequent studies used a similar population derived from broccoli × cabbage but obtained different results [**48,49**], possibly due to the differences of the specific germplasms used and the planting environmental conditions. Okazaki et al. detected six QTLs controlling flowering time (from February to July, 2001, Niigata, Niigata Prefecture, Japan), among which the major QTL in the interval BRMS215–F2-R4b, accounting for 36.8% of the phenotypic variance and *BoFLC2* in the interval is thought to be the candidate control of flowering time [**48**]. Similarly, using a broccoli × cabbage population, Shu et al. combined QTL-seq and a traditional linkage map to detect flowering time loci (from the spring of 2013 to the winter of 2014, Beijing China). A major QTL *Ef2.1* is located on C2.68–2.65 02 Mb, responsible for early flowering and explaining %51.5 of the phenotypic variation, and a homolog of *GROWTH-REGULATING FACTOR 6* (*BolGRF6*) is a possible candidate [**49**]. Using DH populations of broccoli × Chinese kale, Stansell et al. (2019) identified two QTLs *DM_C03@6.4* and *DM_C50.0@09* for days to maturity, two QTLs *DF_C6.4@03* and *DF_C50.0@09* for days to flowering, and the major QTLs *DM_C50.0@09* and *DF_C50.0@09* on C09 at approximately 50 Mb, explaining approximately %50 of the phenotypic variation [**33**].

Broccoli is usually sensitive and not feasible in high-temperature areas/seasons, which are thought to impede vernalization, resulting in defects in floral meristem development. Using tropical accessions in Taiwan, China, Lin et al. 2018 identified nonvernalization-responsive QTLs that contribute to subtropical adaptation (high ambient temperature and short day length) [21]. The candidate gene BoFLC3 identified in the major QTL qDCI-3 may function as an alternative pathway for the control of flowering in temperate and tropical environments [21].

2.2.3. Plant Architecture

Plant architecture is a complex trait attributed to stem and leaf morphologies, including plant height, leaf size, leaf shape, leaf angle, petiole length and lateral shoot growth. It affects the planting density, yield and quality of broccoli [50]. Several QTLs for plant architecture-relevant stem and leaf traits have been reported [23,27,32,51,52]. Before the release of reference genomes, researchers mapped QTLs associated with leaf lamina width on linkage groups C01 and C07 [51], stem width on LG5 [23], leaf apex on linkage groups C06 and C07 [43], leaf shape on linkage group C3 [43], leaf length on linkage group C7 [43], wing petiole length on linkage group C7 [43] and lobe number, wing number, leaf shape and lamina petiole length on linkage group C3 [43]. In recent years, in addition to focusing on heading traits, Stansell et al. mapped several QTLs for leaf morphology and lateral shoots. Four QTLs for leaf apex, two QTLs for leaf margin and leaf-associated hotspot genomic regions, *Lea3* on C1.7–0.7 03 Mb and *Lea7* on C39.5–37.0 07 Mb were identified. A *GRF-1INTERACTING FACTOR 1* (*GIF1*) homolog (*Bo7g093130*) within major QTL *LA_C07@36.6* may be responsible for the narrow leaf phenotype, and a *LATE MERISTEM IDENTITY1* ortholog (*BoLMI1, Bo3g002560*) near the major *LM_C03@0.7*, explaining over 40% phenotype variation, may be responsible for leaf margin phenotype [**33**]. Three lateral shoot growth-associated QTLs, *LT_C03@5.9*, *LT_C04@15.0* and *LT_C9.0@09*, are located on C03, C04 and C09, although no likely candidates were predicted [**33**]. Huang et al. constructed a genetic linkage map using a broccoli DH population and identified QTLs for plant height (PH), maximum outer petiole length (PL) and leaf width (LW), including major QTLs *phc1* for PH on chromosome 1, *plc2-6* for PL on chromosome 6 and *lwc1-3* for LW on chromosome **52**] 3].

2.2.4. Stem Development

Broccoli hollow stem is an undesirable phenotypic disorder showing symptoms of cracks in the internal stem tissue [53,54]. It reduces the quality of broccoli products because hollow stems can result in (1) yield reduction, as harvested broccoli comprise partially edible stalks; (2) secondary pathogen infection and rotting of stems and florets [38,39]. The incidence of hollow stems increases when plants grow rapidly, triggered by, for example, high levels of nitrogenous and warm weather but also varies in different broccoli accessions, indicating that this trait is largely genetically determined and can be controlled by breeding resistant varieties [53,54]. However, relevant studies on this trait are very limited. Yu et al. constructed a genetic map using specific locus-amplified fragment (SLAF) sequencing in a double-haploid segregation population of broccoli and defined nine QTLs on C02, C03, C05, C06 and C09 for hollow stems, among which *QHS. C09-2* could explain %14.1 of the phenotypic variation [**55**].

2.2.5. Head Color

Broccoli is rich in anthocyanin, an important nutritional value with antioxidant activity, can improve health, increase life expectancy and prevent diseases [**56**]. Anthocyanin accumulation in broccoli inflorescences, especially in septals, makes the appearance range from green/blue to purple. Some cultivars, such as 'Purple Sprouting Early', are selected for rich anthocyanin contents, producing obvious purple heads [56]. Purple traits in *B. oleracea* are attributed to the independent activation of *Brassica oleracea MYB DOMAIN PROTEIN 2* (*BoMYB2*) in subspecies of cabbage, cauliflower, kohlrabi and possibly broccoli [**57**].

On the other hand, broccoli cultivars producing heads with green-purple color are considered not beautiful and would be less attractive to consumers than the completely green type, especially in the market of China [**58**]. This green-purple type is sensitive to temperature, and cool weather would induce and deepen the purple degree. Yu et al. mapped this purple sepal trait using a DH population and SLAF sequencing; three QTLs were detected, with a major locus, *qPH. C01-2*, located on linkage group (LG)1, and two loci, *qPH. C01-4* and *qPH. C5-01*, located near *qPH. C01-2* [**59**].

2.3. Male Sterility and Fertility Restoration

Broccoli displays obvious heterosis and most commercial broccoli varieties are F1 hybrids. The production of broccoli F1 hybrids depends on self-incompatibility before the early 21st century and now nearly completely depends on male sterility-based breeding systems [11,12,59]. Male sterility comprises cytoplasmic male sterility (CMS) and genic male sterility (GMS) [60]. Among them, Ogura CMS, with the advantages of complete male gamete abortion, maternal inheritance and easy transfer, is now the most widely studied and applied male sterility source in broccoli seed production [60,61]. Ogura CMS is a natural mutation found in radish populations [62], which is caused by a mitochondrial gene named *orf138*, and can be fully restored by the nuclear gene *RFO* (*PPR-B*) [63,64].

Researchers have made efforts to introduce the CMS source to *B. oleracea* by distant hybridization and/or protoplast fusion, but the initially created CMSR1 and CMSR2 contain too much radish cytoplasm, displaying undesirable characteristics, including yellowing at low temperature, deformed flower shape and poor seed setting, which cannot be used in seed production [65,66,67]. Until the late 1990s, the American Asgrow company applied the method of asymmetric protoplast fusion to reduce the proportion of radish mitochondria, creating CMSR3 with normal fertility and pistil structure; this CMS has been transferred to many elite parent lines, playing a dominant role in the seed production of *B. oleracea* crops [**67,68**]. During the creation and transfer processes of Ogura CMS, specific *orf138* PCR markers were developed for MAS [69]. Additional mitochondrial markers were developed to distinguish the CMS types; detected by these six *orf138*-related and two simple sequence repeat markers in 2016, Shu et al. divided 39 CMS broccoli accessions into five groups, and observed that CMSR3 constituted 79.49% of the CMS accessions from China [67].

In addition to the Ogura CMS, GMS resources and GMS-based seed production systems were reported as promising alternatives [60,70,71,72,73]. A special dominant genic male sterility (DGMS) resource, 79–399–3, which arose in cabbage populations in China, has been successfully and widely applied in cabbage hybrid seed production [61,72]. The DGMS-based breeding system has been established in *B. oleracea* crops, including cabbage, broccoli and kohlrabi [61,72]. Compared with the Ogura CMS, the DGMS-based breeding system displayed advantages of much higher seed quality and yield [61]. However, its utilization is limited in broccoli, largely because homozygous DGMS plants must be preserved and reproduced by tissue culture, which is not effective for large-scale hybrid seed production [61]. Despite these disadvantages, this DGMS-based system has been preserved as an alternative for broccoli hybrid seed production. In recent decades, dozens of broccoli DGMS lines have been created, and several markers have been developed for MAS for the rapid creation of DGMS lines [61,74]. Shu et al. developed generic SSR markers linked to the male-sterile gene, with the marker scaffold10312a showed the highest accuracy of ≥96.43% [74]. By distinguishing the amplified products polyacrylamide gel, these markers were successfully used for identification of male and sterile plants in broccoli breeding lines DGMs8554, DGMs93219 and DGMs94174; enabled DGMS plants selection in the seedling stage. Han et al. developed a high-throughput kompetitive allele specific PCR (KASP) marker K6 with high accuracy and no genetic background bias applicable to all *B. oleracea* crops, including broccoli [61]. This marker was based on allele specific fluorescence on an Applied Biosystems Viia 7 real-time PCR system for high-throughput detection. In the DGMS-based breeding system, this marker was used for identifying homozygous DGMS plants from selfing progenies of heterozygous plants as an alternative to test crossing, which requires at least two years and additional labor in tissue culture [61]. These DGMS-specific markers enable effective selection in breeding programs.

On the other hand, there is increasing demand for the reutilization of CMS resources in *B. oleracea* crops. The Ogura CMS restorer gene *RFO* (*PPR-B*) was introduced from radish to rapeseed and recently to *B. oleracea* crops [75,76]. Liu et al. applied strategies of interspecific hybridization and backcrossing and introduced the *RFO* gene from rapeseed to broccoli. The foreground Rfo-specific markers BnRFO-AS2F/BnRFO-AS2F and BnRFO-AS2F/BnRFO-NEW-R, were used for detecting *Rfo*-positive interspecific hybrids; and 28 background SSR markers were used for detecting true intergeneric hybrids and assessing the genetic backgrounds of *Rfo*-positive interspecific hybrids. By evaluating polymorphism loci of the 28 background markers, the BC2 *Rfo*-positive individuals were found closer to the broccoli's genetic background [**76**].

2.4. Disease Resistance

2.4.1. Downy Mildew

Downy mildew, caused by the obligate fungus *Hyaloperonospora parasitica* (Pers. Fr.), is a destructive disease that affects brassica crops, including broccoli [77,78]. Broccoli plants are often stunned or killed when infected with downy mildew at the young seedling stage or infection can result in quality reduction and yield loss at the adult stage [79,80]. The disease is prevalent in cool weather, with initial symptoms of light green-yellow lesions on the upper leaf surface and later on the undersurface; the spot enlarges and turns yellow; white fungi are visible on the undersurface of leaves under high humidity conditions [79,80]. High resistance to downy mildew both at the young and adult stages is present in some broccoli germplasms and is controlled by a single dominant locus [79,81,82,83,84]. Resistance loci were mapped and linkage markers were developed for MAS, but the gene has not been cloned [79,82,83]. Giovannelli et al. identified 8 RAPD (random amplification of polymorphic DNA) markers linked to downy mildew resistance in broccoli (cotyledon and true leaf stage), among which two, UBC3596620 and OPM16750, were converted to SCAR (sequence characterized amplified regions) markers linked to the locus with 6.7 and 3.3 cM [82]. Farinhó et al. mapped the locus *Pp523* for downy mildew resistance to adult plants of broccoli and developed flanking RAPD markers OPK980_17 and AFLP marker AT. CTA_134/133, with genetic distances of 3.1 cM and 3.6 cM, respectively [83]; in a later study, new AFLP markers were developed and some of them were more user-friendly SCAR and CAPS (cleaved amplified polymorphic sequence) markers; sequencing indicated that *Pp523* is syntenic to the top arm end of *Arabidopsis thaliana* chromosome 79] 1]. We aligned the marker sequences to the broccoli HDEM reference genome [85] and found that the target *Pp523* region is 50.68–49.29 Mb on C8.

2.4.2. Clubroot

Clubroot, caused by the soil-borne pathogen *Plasmodiophora brassicae*, is one of the most devastating diseases of Brassica crops, including broccoli [86,87,88]. Plants infected by the pathogen form galls on roots, which prevent plant uptake of nutrients and water and become stunted and wilt under warm weather [89]. *B. olearcea* lacks germplasm highly resistant to clubroot, although it has been identified and studied for mining resistance loci/genes in its close relatives, such as turnip, radish and rapeseed [90,91,92,93]. The resistance gene *CRa* has been introduced from *B. rapa* to *B. olearcea* by distant hybridization and MAS [94]; in this process *CRa*-specific markers SC2930-Q-FW/SC2930-RV were applied for detection of *CRa* gene in the F1 and each backcross plants, enabled successful introgression of the *CRa* gene into the cabbage inbred lines. In recent years, commercial broccoli varieties with the *CRa* resistance gene, bred by the Syngenta Corporation, are available on the market of China, but the MAS process is not available.

While highly clubroot-resistant germplasms are lacking, some moderate clubroot resistance has been identified in *B. oleracea* [**95,96,97**]. There are two studies on genetic mapping for resistance loci related to broccoli, although both of them used broccoli as susceptible parents. These studies are useful for the rapid introduction of clubroot resistance from other subspecies/related species to broccoli with MAS [**95,96**]. Rocherieux et al. generated F2:3 segregation populations by crossing clubroot-resistant kale and clubroot-susceptible broccoli and constructed a restriction fragment length polymorphism (RFLP) based genetic map. The populations were infected by five isolates and two to five QTLs were identified depending on the isolates; one of these QTLs, *Pb-Bo1*, showed broad-spectrum resistance detected in all isolates [**95**]. Using populations of crossing resistant double-haploid line (Anju) with a susceptible double-haploid line (GC), Nagaoka et al. identified five CR-QTLs, *pb-Bo(Anju)1*, *PbBo(Anju)2*, *PbBo(Anju)3* and *PbBo(Anju)4* derived from Anju and *pb-Bo(GC)1* from the susceptible parent GC; this study also provided specific primer sequences linked to CR loci and a comparison with known *B. rapa* CR genes [**96**].

2.4.3. Black Rot

Black rot, caused by the bacterium *Xanthomonas campestris* pv. *campestris* (Pam.) Dowson (*Xcc*), is also one of the most destructive diseases of brassica crops in the world [**98,99**]. The pathogen often invades plants through hydathodes and spreads through vascular tissue, forming V-shaped lesions at the leaf margins, causing systemic infection and great loss of quality and yield [**98,99,100**]. While some resistant plant resources have been reported in *B. oleracea*, few loci/genes have been identified [**101,102,103**]. Camargo et al. identified genomic regions associated with young and adult plant resistance to black rot in linkage groups 1, 2 and 9 using a population of black rot-resistant cabbage line BI-16 and susceptible inbred broccoli line OSU Cr-7 [**38**]. Doullah et al. identified two genomic regions on LG 2 and LG 9 significantly associated with resistance to black rot, with a disease rating of populations from susceptible broccoli green commet P09 and resistant Reiho P01 [**104**]. In a later study using the same plant materials, Tonu et al. improved the previous genetic map and identified three QTLs, *XccBo(Reiho)1, XccBo(Reiho)2 and XccBo(Reiho)1,* for resistance to black rot, and the major QTL, *XccBo(Reiho)2*, was from parent Reiho [105]; comparison using common markers of the previous study by Camargo et al. revealed that *XccBo(Reiho)1* and *XccBo(GC)1* may be identical to the previously reported QTLs [**104,105**]. Iglesias-Bernabé et al. performed QTL analysis of black rot resistance (*Xcc* race 1) in the BolTBDH mapping population and identified four QTLs, including *Xcc1.1* showing overlap with the previously reported cabbage resistance locus *BRQTL-C1_1, BRQTL-C1_2* [106], *Xcc6.1* showing overlap with *BRQTL-C6, Xcc8.1* showing overlap with *XccBo(Reiho)*[**105**] *2*] and a novel locus, *Xcc9.1* [**107**]; in addition, this study indicated that resistance might be related to the synthesis of secondary metabolites [**107**].

2.5. Secondary Metabolites

Broccoli contains a number of beneficial secondary metabolites, including glucosinolates/sulforaphane, carotenoids, phenolic acids and flavonoids. Several loci/genes regulating the accumulation of these compounds in broccoli have been identified. Genetic models of secondary metabolite biosynthesis in *Arabidopsis* provide a convenient tool for homologous studies in broccoli [**108,109**]. Via a homologous cloning strategy, some broccoli genes are isolated directly, including *cytochrome P450 79F1* (*CYP79F1*), *cytochrome P450 83A1* (*CYP83A1*), *UDP-glucosyltransferase 74B1* (*UGT74B1*), *sulfotransferase 18* (*ST5b*) and flavin-containing monooxygenase *GS-OX1* (*FMOGS-OX1*), *cytochrome P450 83B1* (*BoCYP83B1*), *BoMY B51*, *GSL-PRO*, *GSL-ELO NG*, *GSL-ALK*, *GSL-OH*, *Myb28* and *Bo MY B51* for glucosinolate biosynthesis [**109,110,111,112,113**], and *Bo PAL*, *Bo DFR*, *BoT T8* and *BoT TG1* for anthocyanin biosynthesis [**114**].

Genetic loci determining the variation in these secondary metabolites were also detected by genetic mapping. Sotelo et al. performed genetic analysis to identify the genome regions regulating glucosinolate biosynthesis in the DH mapping population BolTBDH and detected eighty-two significant QTLs for individual and total glucosinolate synthesis in leaves, seeds and flower buds, and *QTL9.2* (proposed candidate as *GSL-ALK*) plays a central role in determining glucosinolate variation, showing epistatic interactions with other loci [**115**]. Brown et al. constructed a genetic linkage map with a broccoli mapping population, identified 14 QTLs associated with the accumulation of aliphatic, indolic or aromatic glucosinolates in florets, and a locus *GSL12* on C09 explains approximately %40 of the phenotypic variability of progoitrin [**116**]. Li et al. performed genetic mapping for sulforaphane metabolism with a DH population; 18 QTLs for sulforaphane metabolism in broccoli florets were identified, and six QTLs among them were detected in more than one environment [**117**]. Using the same population previously reported [**116**], Brown et al. constructed a genetic linkage map with an SNP array and identified three QTLs for carotenoid variation in broccoli florets [**118**]. Gardner et al. performed QTL analysis saturated with SNP markers in an Illumina 60 K array for total phenolic concentration and its individual components in the population previously reported by Brown et al. [**118**] and obtained twenty-three loci identified in at least two analyses [**119**]. In the BolTBDH mapping population, 33 QTLs were identified controlling phenolic concentrations in leaves, flower buds and seeds [**120**]. In addition, transcriptome analyses were performed to identify differentially expressed genes related to glucosinolate metabolism in broccoli seeds, sprouts and byproducts [**121,122,123**].

2.6. Development of Omics Research

Advances in techniques and reduced costs of high-throughput next- and third-generation sequencing have brought high-throughput tools for genomic-related studies and the improvement of broccoli. In 2014, *B. oleracea* draft genome-based short reads of the next generation were released [**124,125**]; in 2018, the first broccoli (HDEM) reference genome, a high-quality daft genome based on third-generation nanopore long reads and optical maps, was accessible [**85**]. These studies provided information on genome duplication and gene divergence and the direct prediction of genes related to phytochemicals and morphological variations and, as mentioned above, provided a reference for high density marker development [**97,116,118**]. Bulked-segregant analysis combined with whole genome resequencing (BSA-seq) for rapid gene/QTL mapping and candidate searching [**13,32**] and omics-related studies exploring differentially expressed genes/miRNAs related to important traits [**26,27,28,29,30,121,122,123**]. In addition, high-throughput strategies promote KASP marker-based fingerprinting for the essential broccoli germplasm [**126**], genetic diversity and population structure analysis for broccoli cultivars [**11,127**], and the genomic and morphological domestication syndrome of broccoli calabrese landraces, hybrids and sprouting broccoli [**11**].

2.7. Genome Editing

Genome editing is a powerful tool for efficient and targeted genome manipulations in living organisms. Depending on the genome editing tools, four engineered nucleases were developed: Meganucleases [**128**], zinc finger nucleases (ZFNs) [**129**], transcription activator-like effector-based nucleases (TALENs) [**130**] and short palindromic repeat (CRISPR)-associated protein (Cas9) systems [**131,132**]. CRISPR/Cas9 has proven to be a cost-effective and versatile tool for precise and efficient genome editing and in recent years, it has been extensively studied and applied to manipulate desired genes in plants [**133**]. While it has been realized in some *B. oleracea* crops [**134,135**], genome editing by CRISPR/Cas9 has not succeeded in complete broccoli background plants. Only one study applied this tool to broccoli-related plant material DH1012, a doubled haploid genotype from the crossing of *B. oleracea alboglabra* (A12DHd) with *B. oleracea italica* (Green Duke GDDH33), targeting *BolC.GA4.a* (*Bol038154*), resulting in dwarf stature [**136,137**].

2.8. Genetic Transformation

Agrobacterium-mediated transformation in broccoli was first reported by Metz et al. [**138**]. In the last decade, this genetic engineering tool has been applied for improving broccoli regarding (1) insect resistance by the genes *cry1A(c)*, *cry1C* and *cryIA(b)*; (2) fungal resistance by the Trichoderma harzianum endochitinase gene, *PR-1* and *PR-3*) ;2) abiotic stress resistance by *AtHSP101*; (4) herbicide resistance by *Bar* gene; (5) prolonged shelf-life/delayed postharvest yellowing by *ipt* (isopentenyl transferase) gene, *ACC synthase 1*, *BoCLH1* and *ACC oxidase* gene; and (6) flowering control by *CYP86MF*, *SLG*, *FCA* and *CONSTANS*, which has been reviewed by Kumar and Srivastava in 2016 [**1**]. Thus, we review the advances of broccoli transgenic improvement in recent years (**Table 1**).

Table 1. Broccoli improvement by genetic transformation in recent years.

Gene Transferred	Orign	Recipient Plant	Performance	References
BoAPX	broccoli	Broccoli	Enhanced resistance to downy mildew enhanced tolerance to heat stress	[139]
BoWRKY6	broccoli	broccoli	enhanced resistance to downy mildew	[140]
BoiCesA (RNAi)	broccoli	broccoli	enhanced salt tolerance; dwarf and smaller leaves	[141]
BoC3H	broccoli	broccoli	enhanced salt stress tolerance	[142]
BoC3H4	broccoli	broccoli	enhanced salt stress tolerance; more susceptible to *S. sclerotiorum*	[143]
BoERF1	broccoli	broccoli	enhanced salt stress tolerance; enhanced resistance to Sclerotinia stem rot	[144]
cryIAa	*Bacillus thuringiensis*	broccoli	resistance to diamondback moth	[145]
BoMYB29	wild *B. oleracea*	DH line AG1012, (partial broccoli background)	increased glucosinolate content	[146]

Gene Transferred	Orign	Recipient Plant	Performance	References
BoTSB1, BoTSB2	broccoli	Arabidopsis	increased glucosinolate content	[147]
BroMYB28 (transient overexpression)	broccoli	broccoli	increased glucoraphanin content	[148]
MAM1	broccoli	broccoli	increased sulforaphane content	[149]
FMOGS-OX2	broccoli	broccoli	increased sulforaphane content	[149]
Myrosinase	broccoli	broccoli	increased sulforaphane content	[149]
BoiDAD1F (RNAi)	broccoli	broccoli	recoverable male sterility	[150]
bol-miR171b	broccoli	broccoli	nearly completely male sterile and increased the chlorophyll content	[151]

2.8.1. Transgenic Breeding for Fungal Resistance

In two independent studies, Jiang et al. generated transgenic broccoli plants overexpressing the cytosolic ascorbate peroxidase gene *BoAPX* and the WRKY transcription factor gene *BoWRKY6*; both of them obtained enhanced resistance to downy mildew [**139,140**]. *BoAPX*-overexpressing broccoli, with a lower level of electrical conductivity and a higher level of APX enzyme activity, exhibited significantly higher resistance to *Hyaloperonospora parasitica* infection, as well as to heat stress, than wild-type plants [**139**]. *BoWRKY6*-overexpressing broccoli exhibited significantly increased resistance to downy mildew but varied from low to very high [**140**]; two of them, lines BWK14 and BWK31, exhibited very high resistance to downy mildew [**140**].

2.8.2. Transgenic Breeding for Abiotic Stress Resistance

Li et al. generated RNAi transgenic broccoli lines targeting the cellulose synthase gene *BoiCesA*; the *BoiCesA* knockdown plants showed a loss of cellulose content and significantly enhanced salt tolerance, and the expression of related genes (*BoiProH*, *BoiPIP2;2*, *BoiPIP2;3*) was significantly changed but also displayed phenotypic defects characterized by dwarfs and smaller leaves [**141**].

In three independent studies, Jiang et al. reported that the overexpression of the C3H-type zinc finger genes *BoC3H* and *BoC3H4* and the ethylene response transcription factor gene *BoERF1* enhanced salt stress tolerance [**142,143,144**].

The BoC3H-overexpression lines exhibited higher germination rates, dry weight and chlorophyll content under salt stress and less cell death in the leaves due to the decreased hydrogen peroxide level, relative electrical conductivity and malondialdehyde contents but increased free proline content and catalase, peroxidase and superoxide dismutase enzyme activities [**142**]. The *BoC3H4*-overexpression lines exhibited increased salinity stress tolerance, with an increase in proline and H2O2 and a decrease in chlorophyll loss, MDA and REC compared with WT plants; however, the lines were more susceptible to *S. sclerotiorum*, possibly due to the inhibited expression of the *BoPDF1.2* gene [**143**]. The *BoERF1*-overexpression lines exhibited a higher seed germination rate and less chlorophyll loss under salt stress, with less cell death in the leaves similar to the *BoC3H*-overexpression lines; in addition, the transgenic lines showed enhanced resistance to *Sclerotinia* stem rot [**144**].

2.8.3. Transgenic Breeding for Insect Resistance

Transgenic broccoli for insect resistance was extensively studied in the late 1990s and the beginning of the 21st century [**1**], but in recent years there have been few related studies. Kumar et al. generated transgenic broccoli overexpressing *cryIAa*, which showed effective resistance to infestation by diamondback moth (*Plutella xylostella*) larvae [**145**].

2.8.4. Transgenic Breeding for Enriched Glucosinolate/Sulforaphane Content

In recent years, improving the anticancer metabolite glucosinolate/sulforaphane content in broccoli by the genetic engineering of biosynthesis-/regulation-related genes has increased [**146,147,148,149,152**]. Zuluaga et al. reported that the overexpression of *BoMYB29* in DH line AG1012 resulted in the upregulation of the aliphatic glucosinolate pathway and higher production of methylsulphinylalkyl glucosinolates, including glucoraphanin [**146**]. Li et al. isolated two tryptophan synthase beta subunit (TSB) genes from broccoli and generated overexpression lines of *BoTSB1* or *BoTSB2* in Arabidopsis, which showed accumulation of tryptophan, indole-3-acetic acid (IAA) and indole glucosinolates; this study provides a target for improving glucosinolates, but no broccoli transgenic plants were generated [**147**]. Studies on *BroMYB28* revealed its possible role in the biosynthesis of glucoraphanin [**152**], but its function was not proven in broccoli until 2019 [**148**]. *Agrobacterium*-mediated transient overexpression of *BroMYB28* in broccoli results in the accumulation of glucoraphanin [**148**]. Cao et al. generated transgenic broccoli by overexpressing *MAM1*, FMO_{GS-OX2} and *Myrosinase* independently or in triple [**149**]. Compared with wild-type plants, independent transgenes of *MAM1* FMO_{GS-OX2} and *Myrosinase* enhanced sulforaphane content by -2.7–1.6 ,-3.4–1.7 and

-3.7fold, while transgenic plants with the triple gene enhanced sulforaphane content by -5.5–1.86fold [**149**].

2.8.5. Transgenic Breeding for Manipulating Male Fertility

Creation of a new male-sterile type by genetic engineering strategies can rapidly provide alternative resources for hybrid seed production. Male-sterile transgenic broccoli was reported by Chen et al. via RNAi of the jasmonic acid pathway gene *BoiDAD1F* [**150**]. These transgenic plants showed male sterility under normal conditions but recovered to fertility when treated with exogenous JA and were thus suitable for utilization in a two-line seed production system [**150**]. Li et al. reported that the overexpression of a microRNA *bol-miR171b* in broccoli resulted in nearly complete male sterility and increased the chlorophyll content [**151**].

3. CONCISE SUMMARY

Advancements have been achieved in the molecular breeding of broccoli to enhance its agronomic properties, secondary metabolites, male sterility, resistance to abiotic stress, disease, and insects. MAS enables the development of heat-stress-resistant and clubroot-resistant cultivars through selective breeding. Nevertheless, the advancement of broccoli's molecular breeding is hindered by insufficient fundamental research and an unclear understanding of the genetic foundation of its most desirable characteristics. Subsequent investigations on the molecular breeding of broccoli should focus on the following areas.

3.1. Mining Functional Loci/Genes

Several markers that are associated to desirable qualities and genes/QTLs that have been mapped have been reported. Additionally, advancements in sequencing technology have allowed for the prediction of candidates for the mapped genes/QTLs. Omics technologies, including transcriptomics, proteomics, and metabolomics, have been utilized to comprehend the underlying process of favorable characteristics. Although attempts have been made, a considerable number of genes in broccoli have been successfully cloned and their functionality has been validated. Future research should prioritize the exploration and functional validation of additional genes/QTLs to enhance the guidance and advancement of broccoli breeding efforts. (1) To identify and isolate the genes responsible for the desirable traits in broccoli, it is necessary to develop secondary mapping populations such as near-isogenic lines, introgression lines, and chromosome segment substitution lines. These populations will allow for a more precise mapping of the genes in question. (2) Advanced sequencing technology and databases that have been made available can be utilized in conjunction with reverse genetics approaches to identify additional functional genes. (3) Once target genes are obtained, it is important to verify their functionality through transient expression or genetic transformation. This verification process should not rely solely on fine mapping or homology cloning methods.

3.2. Enhancing the quality of broccoli by the utilization of landraces or other subspecies of *B. oleracea*.

Contemporary broccoli exhibits a limited range of genetic variation, which might lead to unfavorable characteristics in terms of quality, productivity, and resistance. Broccoli landraces, particularly those from Italy, and other subspecies offer a wide range of genetic resources that possess promising characteristics. These include variations in heading type, flowering/maturation period, high glucosinolate content, and robust disease resistance. Landraces and distinct subspecies can be used in Marker-Assisted Selection (MAS) to incorporate desirable genes into breeding materials, thereby enhancing the quality and expanding the genetic variety of broccoli. One specific example is the density of the head of calabrese broccoli, which is the most widely consumed type of broccoli. Over the past few decades, efforts have been made to increase the density of the head to make it easier to transport and store. One potential method to achieve this is by introducing genomic fragments from cauliflower, although there is no published literature on this topic.

3.3. Introducing disease resistance genes from closely related species

Broccoli exhibits susceptibility to several destructive diseases, such as clubroot and black rot. To protect broccoli genotypes from these diseases, remote hybridization and marker-assisted selection (MAS) can be employed to incorporate pyramid resistance genes/loci from closely related species. The clubroot pathogen P. brassicae has undergone evolutionary changes resulting in the development of various physiological races that exhibit distinct infection responses on host plants. Broccoli production is not viable against other races of P. brassicae due to the introduction of only one resistant locus, CRa, from B. rapa. Additional resistant genes/loci should be incorporated into broccoli from closely related species, such as turnip, radish, and rapeseed, and then combined by pyramiding. Strong resistance sources for black rot disease have been documented in the A and B genomes of Brassica species, whereas modest clubroot resistance has been identified in the C genome of cabbage. The resistance genes/loci can be transferred from Brassica carinata and cabbage to broccoli.

3.4. Enhancing the CRISPR/Cas9 Genome Editing System

The efficacy of the CRISPR/Cas9 technology in altering plant genomes has been well established. Recently, this technique of modifying genetic material has been effectively utilized in various crops, such as rice, maize, soybean, and tomato, to investigate gene functions and enhance crop quality. This includes improving features like high yield, disease resistance, herbicide resistance, optimal plant structure, and other desirable characteristics. Nevertheless, the CRISPR/Cas9 system has not yet been used in broccoli. Therefore, future research should prioritize enhancing and utilizing the CRISPR/Cas9 system to enhance broccoli.

REFERENCES

1. Kumar, P.; Srivastava, D.K. Biotechnological advancement in genetic improvement of broccoli (*Brassica oleracea* L. var. *italica*), an important vegetable crop. *Biotechnol. Lett.* 2016, *38*, 1049–1063.
2. Luo, F.; Fang, H.-X.; Wei, B.-D.; Cheng, S.-C.; Zhou, Q.; Zhou, X.; Zhang, X.; Zhao, Y.-B.; Ji, S.-J. Advance in yellowing mechanism and the regulation technology of post-harvested broccoli. *Food Qual. Saf.* 2020, *4*, 107–113.
3. Vallejo, F.; Garcia-Viguera, C.; Tomas-Barberan, F. Changes in Broccoli (*Brassica oleracea* L. var. *italica*) Health-Promoting Compounds with Inflorescence Development. *J. Agric. Food Chem.* 2003, *51*, 3776–3782.
4. Vallejo, F.; Tomas-Barberan, F.; Ferreres, F. Characterisation of flavonols in broccoli (*Brassica oleracea* L. var. *italica*) by liquid chromatography–UV diode-array detection–electrospray ionisation mass spectrometry. *J. Chromatogr. A* 2004, *1054*, 181–193.
5. Shapiro, T.A.; Fahey, J.W.; Wade, K.L.; Stephenson, K.K.; Talalay, P. Chemoprotective Glucosinolates and Isothiocyanates of Broccoli Sprouts. *Cancer Epidemiol. Prev. Biomark.* 2001, *10*, 501–508.
6. Buck, P.A. Origin and taxonomy of broccoli. *Econ. Bot.* 1956, *10*, 250–253.
7. Massie, I.H.; Astley, D.; King, G.J. Patterns of genetic diversity and relationships between regional groups and populations of Italian landrace cauliflower and broccoli (*Brassica oleracea* L. var. *botrytis* L. and var. *italica* Plenck). *Acta Hort.* 1996, *407*, 45–54.
8. Ciancaleoni, S.; Chiarenza, G.L.; Raggi, L.; Branca, F.; Negri, V. Diversity characterisation of broccoli (*Brassica oleracea* L. var. *italica* Plenck) landraces for their on-farm (in situ) safeguard and use in breeding programs. *Genet. Resour. Crop. Evol.* 2013, *61*, 451–464.
9. Maggioni, L.; Von Bothmer, R.; Poulsen, G.; Lipman, E. Domestication, diversity and use of *Brassica oleracea* L., based on ancient Greek and Latin texts. *Genet. Resour. Crop. Evol.* 2017, *65*, 137–159.
10. Branca, F.; Chiarenza, G.L.; Cavallaro, C.; Gu, H.; Zhao, Z.; Tribulato, A. Diversity of Sicilian broccoli (*Brassica oleracea* var. *italica*) and cauliflower (*Brassica oleracea* var. *botrytis*) landraces and their distinctive bio-morphological, antioxidant, and genetic traits. *Genet. Resour. Crop. Evol.* 2017, *65*, 485–502.
11. Stansell, Z.; Björkman, T. From landrace to modern hybrid broccoli: The genomic and morphological domestication syndrome within a diverse *B. oleracea* collection. *Hortic. Res.* 2020, *7*, 1–17.
12. Li, Z.; Mei, Y.; Liu, Y.; Fang, Z.; Yang, L.; Zhuang, M.; Zhang, Y.; Lv, H. The evolution of genetic diversity of broccoli cultivars in China since 1980. *Sci. Hortic.* 2019, *250*, 69–80.
13. Branham, S.E.; Farnham, M.W. Identification of heat tolerance loci in broccoli through bulked segregant analysis using whole genome resequencing. *Euphytica* 2019, *215*, 34.
14. Kumar, R.; Kandpal, K. Influence of foliar fertilization of boron on broccoli (*Brassica oleracea* var. *italica*) in boron deficient soil of Doon Valley, India. *Progress. Hortic.* 2017, *49*, 65.
15. Branham, S.E.; Stansell, Z.J.; Couillard, D.M.; Farnham, M.W. Quantitative trait loci mapping of heat tolerance in broccoli (*Brassica oleracea* var. *italica*) using genotyping-by-sequencing. *Theor. Appl. Genet.* 2017, *130*, 529–538.

16. Farnham, M.W.; Bjorkman, T. Breeding Vegetables Adapted to High Temperatures: A Case Study with Broccoli. *HortScience* 2011, *46*, 1093–1097.

17. Björkman, T.; Pearson, K.J. High temperature arrest of inflorescence development in broccoli (*Brassica oleracea* var. *italica* L.). *J. Exp. Bot.* 1998, *49*, 101–106.

18. Heather, D.; Sieczka, J.; Dickson, M.; Wolfe, D. Heat Tolerance and Holding Ability in Broccoli. *J. Am. Soc. Hortic. Sci.* 1992, *117*, 887–892.

19. Atallah, S.S.; Gómez, M.I.; Björkman, T. Localization effects for a fresh vegetable product supply chain: Broccoli in the eastern United States. *Food Policy* 2014, *49*, 151–159.

20. Ward, B.; Smith, P.; James, S.; Stansell, Z.; Farnham, M. Increasing Plant Density in Eastern United States Broccoli Production Systems to Maximize Marketable Head Yields. *Horttechnology* 2015, *25*, 330–334.

21. Lin, Y.-R.; Lee, J.-Y.; Tseng, M.-C.; Lee, C.-Y.; Shen, C.-H.; Wang, C.-S.; Liou, C.-C.; Shuang, L.-S.; Paterson, A.H.; Hwu, K.-K. Subtropical adaptation of a temperate plant (*Brassica oleracea* var. *italica*) utilizes non-vernalization-responsive QTLs. *Sci. Rep.* 2018, *8*, 13609.

22. Astarini, I.A.; Defiani, M.R.; Suriani, N.L.; Griffiths, P.; Stefanova, K.; Siddique, K. Adaptation of broccoli (*Brassica oleracea* var. *italica* L.) to high and low altitudes in Bali, Indonesia. *Biodiversitas J. Biol. Divers.* 2020, *21*, 5263–5269.

23. Lin, K.H.; Chang, L.C.; Lai, C.D.; Lo, H.F. AFLP mapping of quantitative trait loci influencing seven head-related traits in broccoli (*Brassica oleracea* var. *italica*). *J. Hortic. Sci. Biotechnol.* 2013, *88*, 257–268.

24. Lin, K.-H.; Huang, H.-C.; Lin, C.-Y. Cloning, expression and physiological analysis of broccoli catalase gene and Chinese cabbage ascorbate peroxidase gene under heat stress. *Plant Cell Rep.* 2010, *29*, 575–593.

25. Chiang, C.-M.; Chen, S.-P.; Chen, L.-F.O.; Chiang, M.-C.; Chien, H.-L.; Lin, K.-H. Expression of the broccoli catalase gene (*BoCAT*) enhances heat tolerance in transgenic *Arabidopsis*. *J. Plant Biochem. Biotechnol.* 2013, *23*, 266–277.

26. Chen, C.-C.; Fu, S.-F.; Norikazu, M.; Yang, Y.-W.; Liu, Y.-J.; Ikeo, K.; Gojobori, T.; Huang, H.-J. Comparative miRNAs analysis of Two contrasting broccoli inbred lines with divergent head-forming capacity under temperature stress. *BMC Genom.* 2015, *16*, 1026.

27. Lin, C.-W.; Fu, S.-F.; Liu, Y.-J.; Chen, C.-C.; Chang, C.-H.; Yang, Y.-W.; Huang, H.-J. Analysis of ambient temperature-responsive transcriptome in shoot apical meristem of heat-tolerant and heat-sensitive broccoli inbred lines during floral head formation. *BMC Plant Biol.* 2019, *19*, 3.

28. Lin, H.-H.; Lin, K.-H.; Chen, S.-C.; Shen, Y.-H.; Lo, H.-F. Proteomic analysis of broccoli (*Brassica oleracea*) under high temperature and waterlogging stresses. *Bot. Stud.* 2015, *56*, 1–11.

29. Tian, Y.; Tian, Y.; Luo, X.; Zhou, T.; Huang, Z.; Liu, Y.; Qiu, Y.; Hou, B.; Sun, D.; Deng, H.; et al. Identification and characterization of microRNAs related to salt stress in broccoli, using high-throughput sequencing and bioinformatics analysis. *BMC Plant Biol.* 2014, *14*, 1–13.

30. Torres-Contreras, A.M.; Senés-Guerrero, C.; Pacheco, A.; González-Agüero, M.; Ramos-Parra, P.A.; Cisneros-Zevallos, L.; Jacobo-Velázquez, D.A. Genes differentially expressed in broccoli as an early and late response to wounding stress. *Postharvest Biol. Technol.* 2018, *145*, 172–182.

31. Lee, J.; Yang, K.; Lee, M.; Kim, S.; Kim, J.; Lim, S.; Kang, G.-H.; Min, S.R.; Kim, S.-J.; Park, S.U.; et al. Differentiated cuticular wax content and expression patterns of cuticular wax biosynthetic genes in bloomed and bloomless broccoli (*Brassica oleracea* var. *italica*). *Process. Biochem.* 2015, *50*, 456–462.

32. Branham, S.E.; Farnham, M.W. Genotyping-by-sequencing of waxy and glossy near-isogenic broccoli lines. *Euphytica* 2017, *213*, 84.

33. Stansell, Z.; Farnham, M.; Björkman, T. Complex Horticultural Quality Traits in Broccoli Are Illuminated by Evaluation of the Immortal BolTBDH Mapping Population. *Front. Plant Sci.* 2019, *10*, 1104.

34. Duclos, D.V.; Bjoerkman, T. Meristem identity gene expression during curd proliferation and flower initiation in *Brassica oleracea*. *J. Exp. Bot.* 2008, *59*, 421–433.

35. Fujime, Y.; Okuda, N. The physiology of flowering in brassicas, especially about cauliflower and broccoli. *Acta Hortic.* 1996, 247–254.

36. Carr, S.M.; Irish, V.F. Floral homeotic gene expression defines developmental arrest stages in *Brassica oleracea* L. vars. *botrytis* and *italica*. *Planta* 1997, *201*, 179–188.

37. Smith, L.B.; King, G.J. The distribution of BoCAL-a alleles in *Brassica oleracea* is consistent with a genetic model for curd development and domestication of the cauliflower. *Mol. Breed.* 2000, *6*, 603–613.

38. Camargo, L.; Williams, P.; Osborn, T. Mapping of quantitative trait loci controlling resistance of *Brassica oleracea* to *Xanthomonas campestris* pv. *campestris* in the field and greenhouse. *Phytopathology* 1995, *85*, 1296–1300.

39. Bohuon, E.J.R.; Keith, D.J.; Parkin, I.A.P.; Sharpe, A.G.; Lydiate, D.J. Alignment of the conserved C genomes of *Brassica oleracea* and Brassica napus. *Theor. Appl. Genet.* 1996, *93*, 833–839.

40. Kianian, S.F.; Quiros, C.F. Generation of a *Brassica oleracea* composite RFLP map: Linkage arrangements among various populations and evolutionary implications. *Theor. Appl. Genet.* 1992, *84*, 544–554.

41. Slocum, M.K.; Figdore, S.S.; Kennard, W.C.; Suzuki, J.Y.; Osborn, T.C. Linkage arrangement of restriction fragment length polymorphism loci in *Brassica oleracea*. *Theor. Appl. Genet.* 1990, *80*, 57–64.

42. Brown, A.F.; Jeffery, E.; Juvik, J.A. A Polymerase Chain Reaction-based Linkage Map of Broccoli and Identification of Quantitative Trait Loci Associated with Harvest Date and Head Weight. *J. Am. Soc. Hortic. Sci.* 2007, *132*, 507–513.

43. Walley, P.G.; Carder, J.; Skipper, E.; Mathas, E.; Lynn, J.; Pink, D.; Buchanan-Wollaston, V. A new broccoli x broccoli immortal mapping population and framework genetic map: Tools for breeders and complex trait analysis. *Theor. Appl. Genet.* 2012, *124*, 467–484.

44. Kennard, W.C.; Slocum, M.K.; Figdore, S.S.; Osborn, T.C. Genetic analysis of morphological variation in *Brassica oleracea* using molecular markers. *Theor. Appl. Genet.* 1994, *87*, 721–732.

45. Camargo, L.; Osborn, T. Mapping loci controlling flowering time in *Brassica oleracea*. *Theor. Appl. Genet.* 1996, *92*, 610–616.

46. Bohuon, E.J.R.; Ramsay, L.D.; Craft, J.A.; Arthur, A.E.; Marshall, D.; Lydiate, D.J.; Kearsey, M.J. The Association of Flowering Time Quantitative Trait Loci with Duplicated Regions and Candidate Loci in *Brassica oleracea*. *Genetics* 1998, *150*, 393–401.

47. Rae, A.M.; Howell, E.C.; Kearsey, M.J. More QTL for flowering time revealed by substitution lines in *Brassica oleracea*. *Heredity* 1999, *83*, 586–596.

48. Okazaki, K.; Sakamoto, K.; Kikuchi, R.; Saito, A.; Togashi, E.; Kuginuki, Y.; Matsumoto, S.; Hirai, M. Mapping and characterization of FLC homologs and QTL analysis of flowering time in *Brassica oleracea*. *Theor. Appl. Genet.* 2006, *114*, 595–608.

49. Shu, J.S.; Liu, Y.M.; Zhang, L.L.; Li, Z.S.; Fang, Z.Y.; Yang, L.M.; Zhuang, M.; Zhang, Y.Y.; Lv, H.H. QTL-seq for rapid identification of candidate genes for flowering time in broccoli x cabbage. *Theor. Appl. Genet.* 2018, *131*, 917–928.

50. Hale, A.L.; Farnham, M.W.; Nzaramba, M.N.; Kimbeng, C.A. Heterosis for horticultural traits in Broccoli. *Theor. Appl. Genet.* 2007, *115*, 351–360.

51. Lan, T.-H.; Paterson, A.H. Comparative Mapping of Quantitative Trait Loci Sculpting the Curd of *Brassica oleracea*. *Genetics* 2000, *155*, 1927–1954.

52. Huang, J.; Sun, J.; Liu, E.; Yuan, S.; Liu, Y.; Han, F.; Li, Z.; Fang, Z.; Yang, L.; Zhuang, M.; et al. Mapping of QTLs detected in a broccoli double diploid population for planting density traits. *Sci. Hortic.* 2021, *277*, 109835.

53. Boersma, M.; Gracie, A.; Brown, P. Evidence of mechanical tissue strain in the development of hollow stem in broccoli. *Sci. Hortic.* 2013, *164*, 353–358.

54. Boersma, M.; Gracie, A.J.; Brown, P.H. Relationship between growth rate and the development of hollow stem in broccoli. *Crop. Pasture Sci.* 2009, *60*, 995–1001.

55. Yu, H.; Wang, J.; Zhao, Z.; Sheng, X.; Shen, Y.; Branca, F.; Gu, H. Construction of a High-Density Genetic Map and Identification of Loci Related to Hollow Stem Trait in Broccoli (*Brassic oleracea* L. *italica*). *Front. Plant Sci.* 2019, *10*, 10.

56. Chaudhary, A.; Choudhary, S.; Sharma, U.; Vig, A.P.; Singh, B.; Arora, S. Purple head broccoli (*Brassica oleracea* L. var. *italica* Plenck), a functional food crop for antioxidant and anticancer potential. *J. Food Sci. Technol.* 2018, *55*, 1806–1815.

57. Yan, C.; An, G.; Zhu, T.; Zhang, W.; Zhang, L.; Peng, L.; Chen, J.; Kuang, H. Independent activation of the BoMYB2 gene leading to purple traits in *Brassica oleracea*. *Theor. Appl. Genet.* 2018, *132*, 895–906.

58. Yu, H.; Wang, J.; Sheng, X.; Zhao, Z.; Shen, Y.; Branca, F.; Gu, H. Construction of a high-density genetic map and identification of loci controlling purple sepal trait of flower head in *Brassica oleracea* L. *italica*. *BMC Plant Biol.* 2019, *19*, 228.

59. Reshma, V.; Adarsh, M.; Dhanush, K. Development of hybrids in cole crops: A review. *Plant Arch.* 2018, *18*, 1–11.

60. Han, F.; Yuan, K.; Kong, C.; Zhang, X.; Yang, L.; Zhuang, M.; Zhang, Y.; Li, Z.; Wang, Y.; Fang, Z.; et al. Fine mapping and candidate gene identification of the genic male-sterile gene ms3 in cabbage 51S. *Theor. Appl. Genet.* 2018, *131*, 2651–2661.

61. Han, F.; Zhang, X.; Yuan, K.; Fang, Z.; Yang, L.; Zhuang, M.; Zhang, Y.; Wang, Y.; Liu, Y.; Li, Z.; et al. A user-friendly KASP molecular marker developed for the DGMS-based breeding system in *Brassica oleracea* species. *Mol. Breed.* 2019, *39*, 90.

62. Ogura, H. Studies on the new male sterility in Japanese radish, with special references on the utilization of this sterility towards the practical raising of hybrid seeds. *Mem. Fac. Agric. Kagoshima Univ.* 1968, *6*, 40–75.

63. Uyttewaal, M.; Arnal, N.; Quadrado, M.; Martin-Canadell, A.; Vrielynck, N.; Hiard, S.; Gherbi, H.; Bendahmane, A.; Budar, F.; Mireau, H. Characterization of *Raphanus sativus* Pentatricopeptide Repeat Proteins Encoded by the Fertility Restorer Locus for Ogura Cytoplasmic Male Sterility. *Plant Cell* 2009, *20*, 3331–3345.

64. Yamagishi, H.; Jikuya, M.; Okushiro, K.; Hashimoto, A.; Fukunaga, A.; Takenaka, M.; Terachi, T. A single nucleotide substitution in the coding region of Ogura male sterile gene, orf138, determines effectiveness of a fertility restorer gene, Rfo, in radish. *Mol. Genet. Genom.* 2021, *296*, 705–717.

65. Bannerot, H.; Boulidard, L.; Cauderon, Y.; Tempe, J. Transfer of cytoplasmic male sterility from *Raphanus sativus* to *Brassica oleracea*. *Proc. Eucarpia Meet. Crucif.* 1974, *25*, 52–54.

66. Sigareva, M.A.; Earle, E.D. Direct transfer of a cold-tolerant Ogura male-sterile cytoplasm into cabbage (*Brassica oleracea* ssp. *capitata*) via protoplast fusion. *Theor. Appl. Genet.* 1997, *94*, 213–220.

67. Shu, J.; Liu, Y.; Li, Z.; Zhang, L.; Fang, Z.; Yang, L.; Zhuang, M.; Zhang, Y.; Lv, H. Detection of the Diversity of Cytoplasmic Male Sterility Sources in Broccoli (*Brassica oleracea* var. *Italica*) Using Mitochondrial Markers. *Front. Plant Sci.* 2016, *7*, 927.

68. Wang, Q.; Zhang, Y.; Fang, Z.; Liu, Y.; Yang, L.; Zhuang, M. Chloroplast and mitochondrial SSR help to distinguish allo-cytoplasmic male sterile types in cabbage (*Brassica oleracea* L. var. *capitata*). *Mol. Breed.* 2011, *30*, 709–716.

69. Yao, X.; Li, Y.; Xie, Z.; Liu, L. Identification of specific SRAP marker associated with cytoplasmic male sterility gene of broccoli. *Mol. Plant Breed.* 2009, *7*, 941–947.

70. Sampson, D.R. Linkage of genetic male sterility with a seedling marker and its use in producing f1 hybrid seed of *brassica oleracea* (cabbage, broccoli, kale, etc.). *Can. J. Plant Sci.* 1966, *46*, 703.

71. Ruffio-Chable, V.; Bellis, H.; Herve, Y. A dominant gene for male sterility in cauliflower (*Brassica oleracea* var. *botrytis*): Phenotype expression, inheritance, and use in F1 hybrid production. *Euphytica* 1993, *67*, 9–17.

72. Fang, Z.; Sun, P.; Liu, Y.; Yang, L.; Wang, X.; Hou, A.; Bian, C. A male sterile line with dominant gene (Ms) in cabbage (*Brassica oleracea* var. *capitata*) and its utilization for hybrid seed production. *Euphytica* 1997, *97*, 265–268.

73. Ji, J.-L.; Yang, L.-M.; Fang, Z.-Y.; Zhuang, M.; Zhang, Y.-Y.; Lv, H.-H.; Liu, Y.-M.; Li, Z.-S. Recessive male sterility in cabbage (*Brassica oleracea* var. *capitata*) caused by loss of function of BoCYP704B1 due to the insertion of a LTR-retrotransposon. *Theor. Appl. Genet.* 2017, *130*, 1441–1451.

74. Shu, J.; Liu, Y.; Li, Z.; Zhang, L.; Fang, Z.; Yang, L.; Zhuang, M.; Lv, H. A generic SSR marker closely linked to a dominant genic male sterility gene (DGMs79-399-3) in broccoli (*Brassica oleracea* var. *italica*). *Mol. Breed.* 2016, *36*, 86.

75. Yu, H.-L.; Fang, Z.-Y.; Liu, Y.-M.; Yang, L.-M.; Zhuang, M.; Lv, H.; Li, Z.-S.; Han, F.-Q.; Liu, X.-P.; Zhang, Y.-Y. Development of a novel allele-specific Rfo marker and creation of Ogura CMS fertility-restored interspecific hybrids in *Brassica oleracea*. *Theor. Appl. Genet.* 2016, *129*, 1625–1637.

76. Liu, C.-Q.; Li, G.-Q.; Yao, X.-Q.; Huang, L.; Wu, X.-Y.; Xie, Z.-J. Characterization of Ogura CMS fertility-restored interspecific hybrids and backcross progenies from crosses between broccoli and rapeseed. *Euphytica* 2020, *216*, 1–12.

77. Monot, C.; Pajot, E.; Le Corre, D.; Silué, D. Induction of systemic resistance in broccoli (*Brassica oleracea* var. *botrytis*) against downy mildew (*Peronospora parasitica*) by avirulent isolates. *Biol. Control.* 2002, *24*, 75–81.

78. Monot, C.; Penguilly, D.; Silué, D. First confirmed report of downy mildew caused by *Hyaloperonospora parasitica* on broccoli, cauliflower and Romanesco-type cauliflower heads in France. *Plant Pathol.* 2010, *59*, 1165.

79. Farinhó, M.; Coelho, P.; Carlier, J.; Svetleva, D.; Monteiro, A.; Leitão, J.M. Mapping of a locus for adult plant resistance to downy mildew in broccoli (*Brassica oleracea* convar. *italica*). *Theor. Appl. Genet.* 2004, *109*, 1392–1398.

80. Wang, M.; Farnham, M.W.; Thomas, C.E. Phenotypic Variation for Downy Mildew Resistance among Inbred Broccoli. *HortScience* 2000, *35*, 925–929.

81. Wang, M.; Farnham, M.W.; Thomas, C.E. Inheritance of True Leaf Stage Downy Mildew Resistance in Broccoli. *J. Am. Soc. Hortic. Sci.* 2001, *126*, 727–729.

82. Giovannelli, J.L.; Farnham, M.W.; Wang, M.; Strand, A.E. Development of Sequence Characterized Amplified Region Markers Linked to Downy Mildew Resistance in Broccoli. *J. Am. Soc. Hortic. Sci.* 2002, *127*, 597–601.

83. Farnham, M.; Wang, M.; Thomas, C. A single dominant gene for downy mildew resistance in broccoli. *Euphytica* 2002, *128*, 405–407.

84. Coelho, P.S.; Monteiro, A.A. Inheritance of downy mildew resistance in mature broccoli plants. *Euphytica* 2003, *131*, 65–69.

85. Belser, C.; Istace, B.; Denis, E.; Dubarry, M.; Baurens, F.-C.; Falentin, C.; Genete, M.; Berrabah, W.; Chèvre, A.-M.; Delourme, R.; et al. Chromosome-scale assemblies of plant genomes using nanopore long reads and optical maps. *Nat. Plants* 2018, *4*, 879–887.

86. Lovelock, D.A.; Donald, C.E.; Conlan, X.; Cahill, D.M. Salicylic acid suppression of clubroot in broccoli (Brassicae oleracea var. *italica*) caused by the obligate biotroph *Plasmodiophora brassicae*. *Australas. Plant Pathol.* 2012, *42*, 141–153.

87. Crute, I.; Gray, A.; Crisp, P.; Buczacki, S. Variation in *Plasmodiophora brassicae* and resistance to clubroot disease in brassicas and allied crops-a critical review. *Plant Breed. Abstr.* 1980, *50*, 91–104.

88. Hirai, M. Genetic Analysis of Clubroot Resistance in Brassica Crops. *Breed. Sci.* 2006, *56*, 223–229.

89. Voorrips, R.E. *Plasmodiophora brassicae*: Aspects of pathogenesis and resistance in *Brassica oleracea*. *Euphytica* 1995, *83*, 139–146.

90. Diederichsen, E.; Frauen, M.; Linders, E.G.A.; Hatakeyama, K.; Hirai, M. Status and Perspectives of Clubroot Resistance Breeding in Crucifer Crops. *J. Plant Growth Regul.* 2009, *28*, 265–281.

91. Hirani, A.; Gao, F.; Liu, J.; Fu, G.; Wu, C.; McVetty, P.B.E.; Duncan, R.W.; Li, G. Combinations of Independent Dominant Loci Conferring Clubroot Resistance in All Four Turnip Accessions (*Brassica rapa*) from the European Clubroot Differential Set. *Front. Plant Sci.* 2018, *9*, 1628.

92. Gan, C.; Deng, X.; Cui, L.; Yu, X.; Yuan, W.; Dai, Z.; Yao, M.; Pang, W.; Ma, Y.; Yu, X.; et al. Construction of a high-density genetic linkage map and identification of quantitative trait loci associated with clubroot resistance in radish (*Raphanus sativus* L.). *Mol. Breed.* 2019, *39*, 1–12.

93. Li, L.; Luo, Y.; Chen, B.; Xu, K.; Zhang, F.; Li, H.; Huang, Q.; Xiao, X.; Zhang, T.; Hu, J.; et al. A Genome-Wide Association Study Reveals New Loci for Resistance to Clubroot Disease in Brassica napus. *Front. Plant Sci.* 2016, *7*, 1483.

94. Ren, W.; Li, Z.; Han, F.; Zhang, B.; Li, X.; Fang, Z.; Yang, L.; Zhuang, M.; Lv, H.; Liu, Y.; et al. Utilization of Ogura CMS germplasm with the clubroot resistance gene by fertility restoration and cytoplasm replacement in *Brassica oleracea* L. *Hortic. Res.* 2020, *7*, 1–10.

95. Rocherieux, J.; Glory, P.; Giboulot, A.; Boury, S.; Barbeyron, G.; Thomas, G.; Manzanares-Dauleux, M.J. Isolate-specific and broad-spectrum QTLs are involved in the control of clubroot in *Brassica oleracea*. *Theor. Appl. Genet.* 2004, *108*, 1555–1563.

96. Nagaoka, T.; Doullah, M.A.U.; Matsumoto, S.; Kawasaki, S.; Ishikawa, T.; Hori, H.; Okazaki, K. Identification of QTLs that control clubroot resistance in *Brassica oleracea* and comparative analysis of clubroot resistance genes between *B. rapa* and *B. oleracea*. *Theor. Appl. Genet.* 2010, *120*, 1335–1346.

97. Peng, L.; Zhou, L.; Li, Q.; Wei, D.; Ren, X.; Song, H.; Mei, J.; Si, J.; Qian, W. Identification of Quantitative Trait Loci for Clubroot Resistance in *Brassica oleracea* With the Use of Brassica SNP Microarray. *Front. Plant Sci.* 2018, *9*, 822.

98. Mirik, M.; Selcuk, F.; Aysan, Y.; Sahin, F. First Outbreak of Bacterial Black Rot on Cabbage, Broccoli, and Brussels Sprouts Caused by *Xanthomonas campestris* pv. *campestris* in the Mediterranean Region of Turkey. *Plant Dis.* 2008, *92*, 176.

99. Nagai, H.; Miyake, N.; Kato, S.; Maekawa, D.; Inoue, Y.; Takikawa, Y. Improved control of black rot of broccoli caused by *Xanthomonas campestris* pv. *campestris* using a bacteriophage and a nonpathogenic *Xanthomonas* sp. strain. *J. Gen. Plant Pathol.* 2017, *83*, 373–381.

100. Sakai, H.; Ikeda, K.; Urushibara, T.; Shiraishi, T. Black rot of broccoli (*Brassica oleracea* var. *italica*) caused by *Xanthomonas campestris* pv. *campestris*. *Jpn. J. Phytopathol.* 2006, *72*, 116–119.

101. Saha, P.; Kalia, P.; Sharma, M.; Singh, D. New source of black rot disease resistance in *Brassica oleracea* and genetic analysis of resistance. *Euphytica* 2016, *207*, 35–48.

102. Afrin, K.S.; Rahim, A.; Park, J.-I.; Natarajan, S.; Kim, H.-T.; Nou, I.-S. Identification of NBS-encoding genes linked to black rot resistance in cabbage (*Brassica oleracea* var. *capitata*). *Mol. Biol. Rep.* 2018, *45*, 773–785.

103. Afrin, K.S.; Rahim, A.; Park, J.-I.; Natarajan, S.; Rubel, M.H.; Kim, H.-T.; Nou, A.I.-S. Screening of Cabbage (*Brassica oleracea* L.) Germplasm for Resistance to Black Rot. *Plant Breed. Biotechnol.* 2018, *6*, 30–43.

104. Doullah, M.; Mohsin, G.; Ishikawa, K.; Hori, H.; Okazaki, K. Construction of a Linkage Map and QTL analysis for Black Rot Resistance in *Brassica oleracea* L. *Int. J. Nat. Sci.* 2011, *1*, 1–6.

105. Tonu, N.N.; Doullah, A.-U.; Shimizu, M.; Karim, M.; Kawanabe, T.; Fujimoto, R.; Okazaki, K. Comparison of Positions of QTLs Conferring Resistance to *Xanthomonas campestris* pv. *campestris* in *Brassica oleracea*. *Am. J. Plant Sci.* 2013, *4*, 11–20.

106. Lee, Y.H.; Hong, J.K. Differential defence responses of susceptible and resistant kimchi cabbage cultivars to anthracnose, black spot and black rot diseases. *Plant Pathol.* 2014, *64*, 406–415.

107. Iglesias-Bernabé, L.; Madloo, P.; Rodríguez, V.M.; Francisco, M.; Soengas, P. Dissecting quantitative resistance to *Xanthomonas campestris* pv. *campestris* in leaves of *Brassica oleracea* by QTL analysis. *Sci. Rep.* 2019, *9*, 2015.

108. Frerigmann, H.; Gigolashvili, T. MYB34, MYB51, and MYB122 Distinctly Regulate Indolic Glucosinolate Biosynthesis in *Arabidopsis thaliana*. *Mol. Plant* 2014, *7*, 814–828.

109. Guo, L.; Yang, R.; Gu, Z. Cloning of genes related to aliphatic glucosinolate metabolism and the mechanism of sulforaphane accumulation in broccoli sprouts under jasmonic acid treatment. *J. Sci. Food Agric.* 2016, *96*, 4329–4336.

110. Li, G.; Riaz, A.; Goyal, S.; Abel, S.; Quiros, C. Inheritance of Three Major Genes Involved in the Synthesis of Aliphatic Glucosinolates in *Brassica oleracea*. *J. Am. Soc. Hortic. Sci.* 2001, *126*, 427–431.

111. Traka, M.H.; Saha, S.; Huseby, S.; Kopriva, S.; Walley, P.G.; Barker, G.C.; Moore, J.; Mero, G.; van den Bosch, F.; Constant, H.; et al. Genetic regulation of glucoraphanin accumulation in Beneforte® broccoli. *New Phytol.* 2013, *198*, 1085–1095.

112. Xu, R.; Kong, W.W.; Peng, Y.F.; Zhang, K.X.; Li, R.; Li, J. Identification and expression pattern analysis of the glucosinolate biosynthetic gene BoCYP83B1 from broccoli. *Biol. Plant.* 2018, *62*, 521–533.

113. Yu, Q.; Hao, G.; Zhou, J.; Wang, J.; Evivie, E.R.; Li, J. Identification and expression pattern analysis of BoMYB51 involved in indolic glucosinolate biosynthesis from broccoli (*Brassica oleracea* var. *italica*). *Biochem. Biophys. Res. Commun.* 2018, *501*, 598–604.

114. Rahim, A.; Afrin, K.S.; Jung, H.-J.; Kim, H.-T.; Park, J.-I.; Hur, Y.; Nou, I.S. Molecular analysis of anthocyanin biosynthesis-related genes reveal BoTT8 associated with purple hypocotyl of broccoli (*Brassica oleracea* var. *italica* L.). *Genome* 2019, *62*, 253–266.

115. Sotelo, T.; Soengas, P.; Velasco, P.; Rodríguez, V.M.; Cartea, M.E. Identification of Metabolic QTLs and Candidate Genes for Glucosinolate Synthesis in *Brassica oleracea* Leaves, Seeds and Flower Buds. *PLoS ONE* 2014, *9*, e91428.

116. Brown, A.F.; Yousef, G.G.; Reid, R.W.; Chebrolu, K.K.; Thomas, A.; Krueger, C.; Jeffery, E.; Jackson, E.; Juvik, J.A. Genetic analysis of glucosinolate variability in broccoli florets using genome-anchored single nucleotide polymorphisms. *Theor. Appl. Genet.* 2015, *128*, 1431–1447.

117. Li, Z.; Liu, Y.; Yuan, S.; Han, F.; Fang, Z.; Yang, L.; Zhuang, M.; Zhang, Y.; Lv, H.; Wang, Y.; et al. Fine mapping of the major QTLs for biochemical variation of sulforaphane in broccoli florets using a DH population. *Sci. Rep.* 2021, *11*, 9004.

118. Brown, A.F.; Yousef, G.G.; Chebrolu, K.K.; Byrd, R.W.; Everhart, K.W.; Thomas, A.; Reid, R.; Parkin, I.A.P.; Sharpe, A.G.; Oliver, R.; et al. High-density single nucleotide polymorphism (SNP) array mapping in *Brassica oleracea*: Identification of QTL associated with carotenoid variation in broccoli florets. *Theor. Appl. Genet.* 2014, *127*, 2051–2064.

119. Gardner, A.M.; Brown, A.F.; Juvik, J.A. QTL analysis for the identification of candidate genes controlling phenolic compound accumulation in broccoli (*Brassica oleracea* L. var. *italica*). *Mol. Breed.* 2016, *36*, 81.

120. Francisco, M.; Ali, M.A.A.; Ferreres, F.; Moreno-Fernández, D.; Ángel; Velasco, P.; Soengas, P. Organ-Specific Quantitative Genetics and Candidate Genes of Phenylpropanoid Metabolism in *Brassica oleracea*. *Front. Plant Sci.* 2016, *6*, 1240.

121. Gao, J.; Yu, X.; Ma, F.; Li, J. RNA-Seq Analysis of Transcriptome and Glucosinolate Metabolism in Seeds and Sprouts of Broccoli (*Brassica oleracea* var. italic). *PLoS ONE* 2014, *9*, e88804.

122. Lee, Y.-S.; Ku, K.-M.; Becker, T.M.; Juvik, J.A. Chemopreventive glucosinolate accumulation in various broccoli and collard tissues: Microfluidic-based targeted transcriptomics for by-product valorization. *PLoS ONE* 2017, *12*, e0185112.

123. Li, Z.; Liu, Y.; Li, L.; Fang, Z.; Yang, L.; Zhuang, M.; Zhang, Y.; Lv, H. Transcriptome reveals the gene expression patterns of sulforaphane metabolism in broccoli florets. *PLoS ONE* 2019, *14*, e0213902.

124. Parkin, I.A.P.; Koh, C.; Tang, H.; Robinson, S.J.; Kagale, S.; Clarke, W.E.; Town, C.D.; Nixon, J.; Krishnakumar, V.; Bidwell, S.L.; et al. Transcriptome and methylome profiling reveals relics of genome dominance in the mesopolyploid *Brassica oleracea*. *Genome Biol.* 2014, *15*, R77.

125. Liu, S.; Liu, Y.; Yang, X.; Tong, C.; Edwards, D.; Parkin, I.A.P.; Zhao, M.; Perumal, S.; Yu, J.; Huang, S.; et al. The *Brassica oleracea* genome reveals the asymmetrical evolution of polyploid genomes. *Nat. Commun.* 2014, *5*, 3930.

126. Shen, Y.; Wang, J.; Shaw, R.K.; Yu, H.; Sheng, X.; Zhao, Z.; Li, S.; Gu, H. Development of GBTS and KASP Panels for Genetic Diversity, Population Structure, and Fingerprinting of a Large Collection of Broccoli (*Brassica oleracea* L. var. *italica*) in China. *Front. Plant Sci.* 2021, *12*, 655254.

127. Huang, J.; Liu, Y.; Han, F.; Fang, Z.; Yang, L.; Zhuang, M.; Zhang, Y.; Lv, H.; Wang, Y.; Jialei, J.; et al. Genetic Diversity and Population Structure Analysis of 161 Broccoli Cultivars Based on SNP Markers. *Hortic. Plant J.* 2021.

128. Silva, G.; Poirot, L.; Galetto, R.; Smith, J.; Montoya, G.; Duchateau, P.; Paques, F. Meganucleases and Other Tools for Targeted Genome Engineering: Perspectives and Challenges for Gene Therapy. *Curr. Gene Ther.* 2011, *11*, 11–27.

129. Kim, Y.G.; Cha, J.; Chandrasegaran, S. Hybrid restriction enzymes: Zinc finger fusions to Fok I cleavage domain. *Proc. Natl. Acad. Sci. USA* 1996, *93*, 1156.

130. Christian, M.; Cermak, T.; Doyle, E.L.; Schmidt, C.; Zhang, F.; Hummel, A.; Bogdanove, A.J.; Voytas, D.F. Targeting DNA Double-Strand Breaks with TAL Effector Nucleases. *Genetics* 2010, *186*, 757–761.

131. Mali, P.; Esvelt, K.M.; Church, G. Cas9 as a versatile tool for engineering biology. *Nat. Methods* 2013, *10*, 957–963.

132. Ma, X.; Zhang, X.; Liu, H.; Li, Z. Highly efficient DNA-free plant genome editing using virally delivered CRISPR–Cas9. *Nat. Plants* 2020, *6*, 773–779.

133. Liu, X.; Wu, S.; Xu, J.; Sui, C.; Wei, J. Application of CRISPR/Cas9 in plant biology. *Acta Pharm. Sin. B* 2017, *7*, 292–302.

134. Ma, C.; Zhu, C.; Zheng, M.; Liu, M.; Zhang, D.; Liu, B.; Li, Q.; Si, J.; Ren, X.; Song, H. CRISPR/Cas9-mediated multiple gene editing in *Brassica oleracea* var. *capitata* using the endogenous tRNA-processing system. *Hortic. Res.* 2019, *6*, 20.

135. Sun, B.; Jiang, M.; Zheng, H.; Jian, Y.; Huang, W.-L.; Yuan, Q.; Zheng, A.-H.; Chen, Q.; Zhang, Y.-T.; Lin, Y.-X.; et al. Color-related chlorophyll and carotenoid concentrations of Chinese kale can be altered through CRISPR/Cas9 targeted editing of the carotenoid isomerase gene BoaCRTISO. *Hortic. Res.* 2020, *7*, 161.

136. Lawrenson, T.; Shorinola, O.; Stacey, N.; Li, C.; Østergaard, L.; Patron, N.; Uauy, C.; Harwood, W. Induction of targeted, heritable mutations in barley and *Brassica oleracea* using RNA-guided Cas9 nuclease. *Genome Biol.* 2015, *16*, 258.

137. Sparrow, P.A.C.H.N.; Irwin, J.A. *Brassica oleracea* and *B. napus*. In *Methods in Molecular Biology*; Springer: Berlin/Heidelberg, Germany, 2014; Volume 1223, pp. 287–297.

138. Metz, T.D.; Dixit, R.; Earle, E.D. *Agrobacterium tumefaciens*-mediated transformation of broccoli (*Brassica oleracea* var. *italica*) and cabbage (*B. oleracea* var. *capitata*). *Plant Cell Rep.* 1995, *15*, 287–292.

139. Jiang, M.; Jiang, J.; He, C.; Guan, M. Broccoli plants over-expressing a cytosolic ascorbate peroxidase gene increase resistance to downy mildew and heat stress. *J. Plant Pathol.* 2016, *98*, 413–420.

140. Jiang, M.; Liu, Q.-E.; Liu, Z.-N.; Li, J.-Z.; He, C.-M. Over-expression of a WRKY transcription factor gene BoWRKY6 enhances resistance to downy mildew in transgenic broccoli plants. *Australas. Plant Pathol.* 2016, *45*, 327–334.

141. Li, S.; Zhang, L.; Wang, Y.; Xu, F.; Liu, M.; Lin, P.; Ren, S.; Ma, R.; Guo, Y.-D. Knockdown of a cellulose synthase gene BoiCesA affects the leaf anatomy, cellulose content and salt tolerance in broccoli. *Sci. Rep.* 2017, *7*, 41397.

142. Jiang, M.; Jiang, J.-J.; Miao, L.-X.; He, C.-M. Over-expression of a C3H-type zinc finger gene contributes to salt stress tolerance in transgenic broccoli plants. *Plant Cell Tissue Organ Cult. (PCTOC)* 2017, *130*, 239–254.

143. Jiang, M.; Miao, L.; Zhang, H.; Zhu, X. Over-Expression of a Transcription Factor Gene BoC3H4 Enhances Salt Stress Tolerance but Reduces Sclerotinia Stem Rot Disease Resistance in Broccoli. *J. Plant Growth Regul.* 2019, *39*, 1162–1176.

144. Jiang, M.; Ye, Z.-H.; Zhang, H.-J.; Miao, L.-X. Broccoli Plants Over-expressing an ERF Transcription Factor Gene BoERF1 Facilitates Both Salt Stress and Sclerotinia Stem Rot Resistance. *J. Plant Growth Regul.* 2018, *38*, 1–13.

145. Kumar, P.; Gambhir, G.; Gaur, A.; Sharma, K.C.; Thakur, A.K.; Srivastava, D.K. Development of transgenic broccoli with cryIAa gene for resistance against diamondback moth (*Plutella xylostella*). *3 Biotech* 2018, *8*, 299.

146. Zuluaga, D.L.; Graham, N.S.; Klinder, A.; Kloeke, A.E.E.V.O.; Marcotrigiano, A.R.; Wagstaff, C.; Verkerk, R.; Sonnante, G.; Aarts, M.G.M. Overexpression of the MYB29 transcription factor affects aliphatic glucosinolate synthesis in *Brassica oleracea*. *Plant Mol. Biol.* 2019, *101*, 65–79.

147. Li, R.; Jiang, J.; Jia, S.; Zhu, X.; Su, H.; Li, J. Overexpressing broccoli tryptophan biosynthetic genes BoTSB1 and BoTSB2 promotes biosynthesis of IAA and indole glucosinolates. *Physiol. Plant.* 2019, *168*, 174–187.

148. Kim, Y.-C.; Cha, A.; Hussain, M.; Lee, K.; Lee, S. Impact of *Agrobacterium*-infiltration and transient overexpression of BroMYB28 on glucoraphanin biosynthesis in broccoli leaves. *Plant Biotechnol. Rep.* 2020, *14*, 373–380.

149. Cao, H.; Liu, R.; Zhang, J.; Liu, Z.; Fan, S.; Yang, G.; Jin, Z.; Pei, Y. Improving sulforaphane content in transgenic broccoli plants by overexpressing MAM1, FMOGS-OX2, and Myrosinase. *Plant Cell Tissue Organ Cult. (PCTOC)* 2021, *146*, 461–471.

150. Chen, G.J.; Cao, B.H.; Xu, F.; Lei, J.J. Development of adjustable male sterile plant in broccoli by antisense DAD1 fragment transformation. *Afr. J. Biotechnol.* 2010, *9*, 4534–4541.

151. Li, H.; Zhang, Q.L.; Li, L.H.; Yuan, J.Y.; Wang, Y.; Wu, M.; Han, Z.P.; Liu, M.; Chen, C.B.; Song, W.Q.; et al. Ectopic overexpression of bol-mir171b increases chlorophyll content and results in sterility in broccoli (*Brassica oleracea* L. var. *italica*). *J. Agric. Food Chem.* 2018, *66*, 9588–9597.

152. Baskar, V.; Park, S.W. Molecular characterization of BrMYB28 and BrMYB29 paralogous transcription factors involved in the regulation of aliphatic glucosinolate profiles in *Brassica rapa* ssp. pekinensis. *Comptes Rendus Biol.* 2015, *338*, 434–442.

3 Chapter

GENE EDITING USING CRISPR-CAS9 FOR IMPROVING FRUIT AND VEGETABLE CROPS

1. INTRODUCTION: AN OVERVIEW

Fruit and vegetable crops include abundant amounts of cellulose, vitamins, trace elements, minerals, and other vital nutrients that are necessary for a healthy human diet [1]. Nevertheless, the production and availability of fruits and vegetables are under risk due to climate and environmental changes [2]. For a considerable period of time, humans have successfully tamed and nurtured untamed species. The utilization of cross-breeding technologies allows breeders to enhance variations by selectively crossing dominant strains [3]. Nevertheless, via extended artificial selection, the limitations of traditional breeding methods become more noticeable, particularly in the overwhelming reliance on naturally existing genetic variations. The availability of genetic germplasm resources for enhancing specific traits is limited, and traditional breeding methods can result in the expression of undesirable traits, therefore decreasing the effectiveness of breeding efforts [4]. While traditional breeding methods have proven effective in developing vegetable cultivars that exhibit high yield, good quality, and disease resistance, the growing global population and ongoing food-supply requirements necessitate the swift identification of novel variety to fulfill market demands [5]. Utilizing cutting-edge techniques in crop biotechnology can enhance the efficiency and accuracy of varietal breeding [6].

Genetic engineering has been employed to enhance the ability of plants to cope with both living and non-living stress factors, as well as to enhance the overall quality of fruits and vegetables. The Food and Drug Administration (FDA) approved a storage-resistant transgenic tomato in 1994 [7]. Approximately 80% of the papaya market was dominated by the supply of high-yield transgenic papaya that exhibited strong resistance to the cyclic spot virus [8]. Nevertheless, to guarantee the safety of planting procedures and the consumption of products, the development and implementation of genetically modified (GM) plants are subject to tight legislation and regulation. This meticulous oversight significantly prolongs the time it takes for transgenic cultivars to reach the market [9]. In 2013, researchers created CRISPR-Cas-mediated gene editing as a means to investigate the functionality of plant genes. During the following two years, numerous novel gene-edited agricultural germplasm resources were developed. In 2016, the US FDA granted approval for the CRISPR gene editing of a waxy corn null segregant line and an anti-browning mushroom (Agaricus bisporus) to be sold in the market. This approval was given without subjecting these products to the rigorous regulatory process that is typically necessary for genetically modified crops [10,11]. This suggests that CRISPR gene editing has already achieved success in facilitating the growth of crop varieties.

This paper provides a concise overview of the processes that underlie CRISPR technology, new applications in fruit and vegetable crops, and advancements in CRISPR-Cas systems. We provide a detailed overview of the CRISPR-associated regulatory frameworks that facilitate the commercialization of gene-edited crops in several nations. Lastly, we will address the future obstacles and possibilities for the introduction of favorable genetic variations and enhancement of numerous characteristics.

1.1. The Discovery and Development of CRISPR Technology

The discovery of CRISPR-associated (Cas) genes in the Escherichia coli genome dates back to 1987 and was officially designated by the Dutch scientist who made the identification [12]. In 2005, it was found that numerous CRISPR spacers include short sequences that closely resemble sequences derived from extrachromosomal DNA. The Cas-encoded protein can bind to both the CRISPR transcription products and the homologous foreign DNA sequences, resulting in the formation of a protein-RNA complex. This complex has the ability to cleave the foreign DNA pieces. The main role of the CRISPR complex in bacteria and archaea is to incorporate specific segments of foreign DNA (from invading phages or other sources) into their own genomes, resulting in the formation of interval sequences. Upon the invasion of foreign DNA, the particular recognition system is activated, thus initiating an acquired immune defense function [13,14,15].

The CRISPR-Cas technology has effectively been utilized for altering the genomes of humans, animals, and plants. Furthermore, it has been created for applications in drug screening, animal domestication, and food science research [16,17,18]. CRISPR-Cas systems can be classified into three primary categories. Types I and III employ a substantial multi-Cas protein complex for interference [19]. Type II utilizes a straightforward effector-module architecture to achieve interference through its two distinctive nuclease domains, RuvC and HNH [20]. SpCas9, a type II Cas9 nuclease derived from Streptococcus pyogenes, is the most often employed CRISPR nuclease in CRISPR-Cas technology [21]. The sgRNA-Cas complex identifies the protospacer adjacent motif (PAM) and Cas9 cuts the target DNA to create a double-strand break (DSB), which activates cellular DNA repair systems (Figure 1). Eukaryotes possess two primary healing processes for double-strand breaks (DSBs). One of the methods is nonhomologous end joining (NHEJ). When there is no available repair template that matches the damaged DNA, the non-homologous end joining (NHEJ) repair pathway is triggered at the site of the double-strand break (DSB), resulting in the disruption of gene activity. The second method is homology directed repair (HDR). If a donor DNA template that is similar in sequence to the area surrounding the double-strand break (DSB) site is present, the homology-directed repair (HDR) pathway is activated. This mechanism accurately introduces specified alterations, such as the insertion or substitution of desired sequences, into the break sites [22]. Gene targeting (GT) is a precise method of modifying a specific location in the DNA by using a donor DNA as a template. This technique is used to repair DNA double-strand breaks (DSBs).

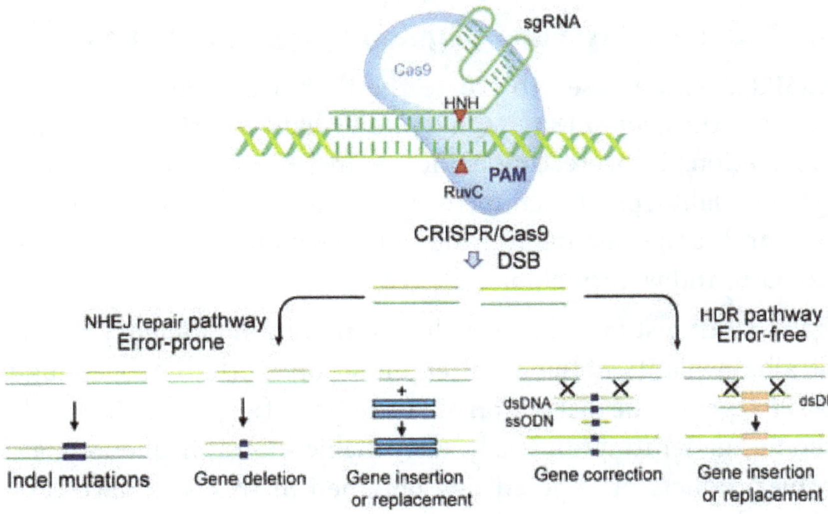

Figure 1. The potential applications of CRISPR-Cas systems in genome editing. CRISPR-Cas systems mediated genome modification depending on the two main double-strand break (DSB) repair pathways. Indel mutation and gene deletion are outcomes of the dominant nonhomologous end joining (NHEJ) repair pathway. Gene insertion, correction, and replacement, using a DNA donor template, are outcomes of the homology directed repair (HDR) pathway.

Various techniques are employed to enhance the frequency of homologous recombination between a genomic target and an external homologous template donor. The majority of the solutions aim to increase the quantity of donor repair templates by utilizing virus replicons [23], inhibiting the NHEJ pathway [24], and synchronizing the induction of double-strand breaks (DSBs) at target sites with the transport of the donor repair template in plant cells [25]. Ultimately, the recombination frequency can be increased by administering a Rad51-stimulatory compound1 (RS-1) [26].

1.2. Development of the CRISPR-Cas System in Plant Studies

Since the CRISPR-Cas system was first adopted for plant genetic engineering in 2013, numerous efforts have been made to develop it into a more powerful tool, for instance, to enable precisely targeted DNA mutations or genetic modifications [**27**]. CRISPR-Cas can now target the open reading frame, untranslated region, and promoter region of a target coding gene, as well as noncoding RNAs [**28,29,30**]. Single-base mutations at genomic targets have also been achieved by nickase Cas9 (nCas9) or catalytically inactive Cas9 (dead Cas9; dCas9) variants fused with cytosine or adenine deaminases, without inducing DSBs [**31**]. Cas9 proteins have been developed extensively to broaden PAM preferences. Cas9 orthologs which possess not only the canonical NGG PAM, but also NG and other PAMs, will expand the repertoire of CRISPR-Cas9 genome editing in plants [**32**].

1.3. CRISPR-Cas9 in Fruit and Vegetable Crop Improvement

In 2014, CRSPR-Cas9 was used to create the first needle-leaf mutant in tomato, by knocking out *Argonaute 7* [**33**]. Many studies have since been published on its possible applications in protecting plants against biotic and abiotic stresses, and improving fruit quality, plant architecture, and shelf life [**34**]. Currently, the system is in the research stage for many fruits and vegetables crops, such as cabbage, mustard, tomato, and watermelon.

Most gene-editing studies have evaluated mutation efficiency in terms of the number of albino plants obtained after mutation of the endogenous phytoene desaturase (*PDS)* gene. The disruption of *PDS* impairs the production of chlorophyll and carotenoid, generating an easily identifiable albinism phenotype in plants. However, the products of gene editing obtained in this way have no economic value [**35,36,37**]. Because of its high economic value and the availability of Agrobacterium-mediated transformation, tomato has become a model crop for testing CRISPR-Cas9 applications (**Figure 2**).

1.3.1. Improvement of Biotic Stress Resistance

Two strategies have been used to improve plant resistance to viruses: (1) designing sgRNAs and targeting the virus genome; or (2), modifying the fruit crop genes in the antiviral pathway. The binding of virus genome linked protein (VPg) to the plant protein 'eukaryotic translation initiation factor 4E' (eIF4E) is key in Y virus infection of plants. Mutation of a key site of eIF4E can affect the virus–plant interaction, and mediate plant resistance to this virus [**38**]. In cucumbers, using CRISPR-Cas to target the N′ and C′ ends of eIF4E-produced nontransgenic homozygous plants in the T3 generation; these showed immunity to cucumber vein yellow virus and pumpkin mosaic virus, and resistance to papaya ring spot mosaic virus (PRSV-W) [**39**].

CRISPR-Cas9 can generate mutations in the coding and noncoding regions of geminivirus, effectively reducing its pathogenicity. In *Nicotiana benthamiana*, sgRNA-Cas9 constructs target beet severe curly top virus (a geminivirus), inhibiting its accumulation in leaves [**40**]. Geminivirus noncoding-region mutations are believed to reduce or even inhibit its replication ability. Compared with coding-region mutations, noncoding-region mutations generate fewer viral variants [**41**].

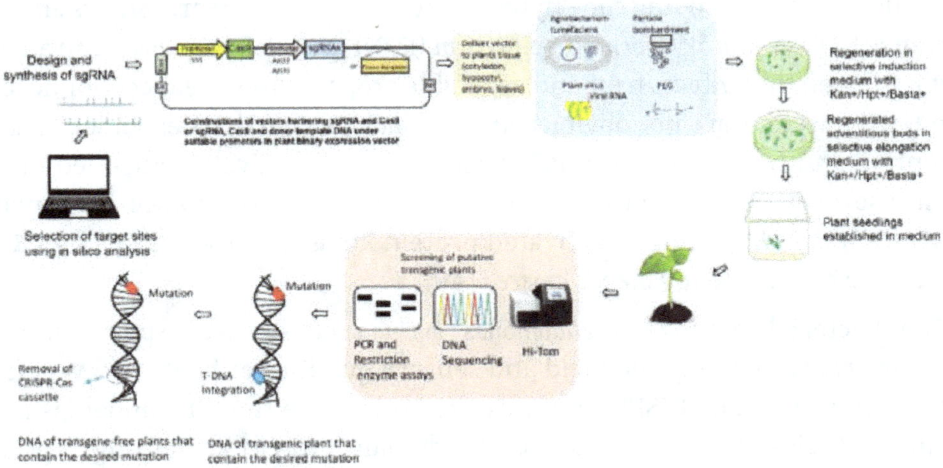

Figure 2. CRISPR-Cas9 mediated genome editing. (I) Selection of the desired genomic DNA target, and recognition of protospacer adjacent motif (PAM) sequences before 20 bp sequences. Design of the sgRNA using online bioinformatics tools. (II) Cloning of designed sgRNAs, and binary vector construction using promoters. (III) The delivery of CRISPR-Cas editing reagents into plant cells. The vector can be transferred into the plant via *Agrobacterium tumefaciens*, nanoparticles, biolistic bombardment, or polyethylene glycol (PEG). Alternatively, plant RNA viruses have been used to induce heritable genome editing. When the cassette harbouring the sgRNA, RNA mobile element, and tobacco rattle virus (TRV) is transformed into the Cas9 expressing plants, the systemic spread of sgRNA will introduce heritable genome editing. (IV) Plant transformation and development of transgenic plants. (V) Genotyping of transgenic plants. (VI) Transgene-free plants with the desired mutation are obtained.

Fungi cause many diseases, potentially causing severe losses in crop yield and quality. For instance, downy and powdery mildews cause serious economic losses in tomato [**42**]. *Arabidopsis thaliana DMR6* (down mildew resistant) is a member of the -2oxoglutarate oxygenase Fe(II)-dependent superfamily and is involved in salicylic acid homeostasis. Overexpression of *DMR6* in plants can reduce susceptibility to downy mildew [**43**]. The *DMR6* mutation obtained using CRISPR-Cas9 to knock out the homologous genes in tomato showed resistance to *Pseudomonas syringae*, *Phytophthora* and *Xanthomonas* spp. [**44**]. *Mlo1* (Mildew resistant locus 1) encodes a membrane-associated protein and is a powdery mildew disease-sensitivity gene. In tomato, *Mlo1* mutants obtained via gene editing exhibited resistance to the powdery mildew *Oidium neolycopersici*. Further, a mutant free of mlo1 T-DNA was obtained by selfing T0 generation plants [**45**].

The fungal pathogen *Fusarium oxysporum* can cause *Fusarium* wilt disease in fruit and vegetable crops [**46**]. In tomatoes, *Solyc08g075770*-knockout via CRISPR-Cas9 resulted in sensitivity to *Fusarium* wilt disease [**47**]. In watermelons, the knockout of *Clpsk1*, encoding the Phytosulfokine (PSK) precursor, confers enhanced resistance to *Fusarium oxysporum* f.sp.*niveum* (FON) [**48**]. *Botrytis cinerea*, an airborne plant pathogen that infects fruit and vegetable crops, causes great economic losses. Its initial symptoms are not obvious, and the lack of effective pesticides makes its prevention and control difficult. Pathogens can be effectively controlled in crops by the use of genetic resources that convey heritable resistance. In tomatoes, mutations in *MAPK3* (mitogen-activated protein kinase 3) produced using CRISPR-Cas9 induce resistance to *Botrytis cinerea* [**49**].

The bacterial pathogen *Pseudomonas syringae* causes leaf spot diseases in crops, severely impacting the yield and sensory qualities of fruits and vegetables. In *Arabidopsis thaliana*, CRISPR-Cas9 was used to mutate the C-terminal jasmonate domain (JAZ2Δjas) of *JAZ2* (jasmonate ZIM domain protein 2), causing expression of JAZ2 repressors; these repressors confer resistance to *Pseudomonas syringae* [**50**].

1.3.2. Abiotic Stress Resistance Improvement

With climate change, crop production is exposed to increased potential risks of abiotic stress. Although traditional breeding can to some extent ensure stable crop production, the application of new technologies to rapidly obtain new crop germplasm resources capable of responding to abiotic stress is essential for accelerating the cultivation of new varieties [**51**]. The emergence of CRISPR-Cas9 gene editing has shortened the time required to create new varieties. Brassinazole-resistant 1 gene (*BZR1*) participates in various brassinosteroid (BR) mediated development processes. The CRISPR mediated mutation in *BZR1* impaired the induction of *RESPIRATORY BURST OXIDASE HOMOLOG1*(*RBOH1*) and the production of H_2O_2. Exogenous H_2O_2 recovered the heat tolerance in tomato bzr1 mutant [**52**]. Further, new cold- and drought-tolerant germplasms can be created using gene-editing, for instance, of *CBF1* (C-repeat binding factor 1), which regulates cold tolerance in plants, and *MAPK3*, which participates in the drought stress response to protect plant cell membranes from peroxidative damage in tomatoes [**53,54**].

1.3.3. Herbicide Resistance Improvement

Weeds are an important cause of stress that affect vegetable yield and quality, and selective herbicides are often used to control weed growth during cultivation. To obtain herbicide-resistant fruits and vegetables for field production, CRISPR-

Cas9 gene editing was used for site-directed mutagenesis of the herbicide target gene acetolactate synthase (ALS) in watermelon, yielding a herbicide-resistant watermelon germplasm [**55**]. Cytidine base editing (CBE) was used for cytidine editing of key *ALS* sites in tomato and potato, resulting in amino acid mutations. Up to %71 of edited tomato plants exhibited resistance to the pesticide chlorsulfuron, and of the edited tomato and potato plants, %12 and %10, respectively, were free of GM components [**56**]. *Phelipanche aegyptiaca*, an obligate weedy plant parasite, requires the host roots to release the plant hormone strigolactone (SL) to promote seed germination; CRISPR-Cas9 was used to mutate carotenoid dioxygenase 8 (*CCD8*), a key enzyme in the carotenoid synthesis pathway that produces SLs in tomato, and More Axillary Growth1 (*MAX1*), which is involved in the synthesis of SLs, thereby significantly reducing SL content, and creating *P. aegyptiaca*-resistant tomato plants [**57,58**].

1.3.4. Fruit and Vegetable Quality Improvement

The primary goal in fruit and vegetable breeding is to improve quality and prolong shelf life after harvest. Quality refers to both external and internal factors. External quality refers to fruit size, colour, and texture, which can be discerned by the naked eye. Internal quality must be measured using equipment, and includes the levels of nutrients such as sugars, vitamins, and bioactive compounds including lycopene, anthocyanins, and malate. For example, in tomato, the ovary locule number, which determines 50% of the genetic variation in fruit size, is determined by multiple QTLs [**59**]. Researchers at Cold Spring Harbor Laboratory designed eight sgRNAs and used CRISPR-Cas9 to edit the promoter region of the tomato CLAVATA-WUSCHEL (CLV-WUS) stem cell gene *CLV3* to obtain fruits that are larger and more numerous than wild-type fruits [**60**]; editing of fruit-size determining QTLs, such as the QTLs for locule number (lc) and fasciated number (fas), generated germplasm resources with an increased number of locules [**61**].

Fruit and vegetable colour and texture are important traits for consumers. For example, European and American consumers prefer red tomatoes, whereas Asian consumers prefer pink tomatoes [**62,63**]. CRISPR-Cas was used to modify phytoene synthase 1 (*PSY1*), MYB transcription factor 12 (*MYB12*), and anthocyanin 2 (*ANT2*) to obtain yellow, pink, and purple tomatoes, respectively [**64,65,66**]. The carotenoid isomerase gene of Chinese kale (*BoaCRTISO*) is responsible for catalysis, then conversion of lycopene precursors to lycopene. When *BoaCRTISO* was targeted and edited, the colour of mutants changed from green to yellow [**67**]. The primary goal of improving the intrinsic quality of fruits and vegetables is to improve their nutrient and bioactive compound content. Carbohydrates and vitamins are essential nutrients. Many genes are involved in the synthesis and metabolism of sucrose and carotenoids. One of the carotenoids, provitamin A, can be absorbed by the human body and converted into vitamin A. For example, CRISPR-Cas was used to knock out *MPK20* (mitogen-activated protein kinase 20), blocking the transcription and protein products of multiple genes in the sucrose metabolism pathway [**68**]. Biofortification, the biotechnological improvement of the absorption, transport, and metabolism of minerals by plants, increases the levels of micronutrients that are beneficial to human health; long-term consumption of these micronutrients can effectively prevent cardiovascular disease and cancer [**69**].

Anthocyanins [**70**], malate [**71**], γ-aminobutyric acid (GABA) [**72**], and lycopene [**73**] are bioactive compounds. Adjusting key metabolic-pathway-related genes via CRISPR-Cas9 can enrich these nutrients in fruits. For example, in tomatoes, butylamine content was increased 19-fold through editing multiple genes in the GABA synthesis pathway, and malate content was improved by regulating aluminium-activated malate transporter (*ALMT9*) [**72**].

CRISPR-Cas9 can also be used to reduce the content of substances in vegetables that are not conducive to human health, by targeting mutations that inactivate genes in biosynthetic pathways. In potato tubers, for example, excessive content of steroidal glycoalkaloids (SGAs), such as α-solanine and α-chaconine, affects their taste and makes them less safe for human consumption, hence low content is an indicator of high quality. CRISPR-Cas9 was used to delete *St16DOX* (steroid 16α-hydroxylase) in the potato SGA biosynthetic pathway, resulting in SGA-free potato lines [**74**].

Prolonged shelf-life is an important breeding goal in fruit and vegetable production. CRISPR was used to knock out ripening inhibitor (*RIN*) or DNA demethylase (DNA demethylase 2, *DML2*) to slow fruit ripening, thereby prolonging their shelf life. However, regulating these two genes in fruit alters peel colour and reduces flavour and nutritional value, severely reducing the fruit's palatability and sensory qualities [**75,76**]. In tomatoes, inhibiting the expression of the pectate lyase (*PL*) and alcobaca (*ALC*) genes effectively extended shelf life, without affecting the sensory qualities or nutritional value [**77,78**].

1.3.5. Application of CRISPR-Cas9 to Crop Domestication

The domestication of wild species into commercial cultivated species requires changes in numerous crop traits, including seed setting, size, consistency of maturation, flowering, photoperiod sensitivity, and the nutritional value of the fruit [**79**]. Plant domestication mostly affects the genes controlling plant morphology, plant growth habit, floral induction, fruit size and number, dispersal, and architecture, as well as the nutritional composition. To achieve the ideotype, alleles controlling favourable nutritional attributes and stress resilience from wild relatives are introduced into cultivated species via traditional domestication technology, but this process is very time-consuming in bringing about changes to many loci. With its ability to precisely manipulate the genome, CRISPR-Cas9 can substantially accelerate de novo domestication.

The tomato is a model crop for artificial domestication using CRISPR-Cas9. In tomato plants, the joint is a weak region of the stem which allows the fruit to drop from the plant, making the fruit prone to fall after ripening, thus improving seed dispersal. Many years of artificial domestication based on harvesting habits has generated cultivars with jointless fruit stems, in which the fruit do not fall after maturation [**79,80**]. Roldan et al. [**81**] used CRISPR-Cas to mutate *MBP21* (MADS-box protein 21), obtaining a new jointless germplasm resource.

Parthenocarpy (fertilisation-independent seedless fruit development) is an important agronomic fruit and vegetable trait and can help ensure stable yield in fluctuating environments. It satisfies consumer preferences for seedless over seeded fruits and provides savings in energy consumption when separating the seeds for industrial production. In tomatoes, *SlAGL6* (SlAGAMOUS-like 6) is essential for parthenocarpy during high temperature stress. *SlAGL6*-mutant plants grow normally and have the same fruit weight and morphology as wild-type plants. Therefore, this gene is an important resource for creating new parthenocarpic germplasms. Homozygous or biallelic mutant plants obtained by modifying *SlAGL6* produced parthenocarpic fruits and fruits with a maximum of 10 seeds, respectively [**82**]. CRISPR-Cas has also been used to knock out *SlARF7* (auxin response factor 7) and *SlIAA9* (indole-3-acetic acid inducible 9) to obtain seedless tomatoes. Seedless tomatoes are obtained from the T0 generation of the biallelic and homozygous *SlIAA-9*mutant Micro-Tom cultivar and the commercial Ailsa Craig cultivar [**83,84**].

Plant yield depends primarily on the number of flowers, which in turn is determined by inflorescence structure. *BOP* (blade-on-petiole) is homologous to genes associated with leaf complexity and silique dehiscence in tomato and *Arabidopsis*. Knocking out *BOP* via CRISPR-Cas9 altered inflorescence morphology. CRISPR-Bop3/2/1 triple mutants flower faster and have simpler inflorescence structure than wild-type plants [**85**]. The site-directed editing of six key genes that determine yield in wild tomatoes (*Solanum pimpinellifolium*) has resulted in morphological changes in aspects such as size, fruit number, and nutritional composition [**86**]. In domesticated wild tomatoes, genes associated with morphology, number of flowers, fruit yield, and vitamin C synthesis have been improved by editing their coding sequences, cis-regulatory sequences, and upstream open reading frames [**81**]. APETALA2a (*AP2a*), NON-RIPENING (*NOR*), and FRUITFULL (*FUL1/TDR4* and *FUL2/MBP7*) have been modified to accelerate tomato maturation, producing plants that mature earlier in natural environments [**87**].

Table 1. List of target genes and traits modified via CRISPR-Cas9 technology in fruit and vegetable crops.

Crop Species	Target	Mutation	Transformation Method	Trait Modification	References
Tomato	*SlAGO7*	Loss of function	*Agrobacterium*-mediated transformation	Wiry phenotype	[33]
Cabbage	*BoPDS*	Loss of function	*Agrobacterium*-mediated hypocotyl transformation	Albino phenotype	[35]
Watermelon	*ClPDS*	Loss of function	*Agrobacterium*-mediated cotyledon transformation	Albino phenotype	[36]
Chinese kale	*BaPDS1BaPDS2*	Loss of function	*Agrobacterium*-mediated cotyledon transformation	Albino phenotype	[37]
Cucumber	*elF4E*	Loss of function	*Agrobacterium*-mediated cotyledon transformation	Resistance against cucumber vein yellowing virus, zucchini yellow mosaic virus, and papaya ring spot mosaic virus	[38]
Tomato	*ClDMR6*	Loss of function	*Agrobacterium*-mediated cotyledon transformation	Resistance against downy mildew	[44]
Tomato	*ClMlo1*	Loss of function	*Agrobacterium*-mediated cotyledon transformation	Resistance against downy mildew	[45]
Tomato	*Solyc08g075770*	Loss of function	*Agrobacterium*-mediated cotyledon transformation	Fusarium wilt susceptibility	[47]
Watermelon	*ClPSK1*	Loss of function	Agrobacterium-mediated cotyledon transformation	Resistance to *Fusarium oxysporum* f.sp. *niveum*	[48]
Tomato	*MAPK3*	Loss of function	*Agrobacterium*-mediated cotyledon transformation	Resistance to *Botrytis cinerea*	[49]
Tomato	*BZR1*	Loss of function	*Agrobacterium*-mediated cotyledon transformation	Decrease in heat stress tolerance	[52]
Tomato	*CBF1*	Loss of function	*Agrobacterium*-mediated cotyledon transformation	Decrease in chilling stress tolerance	[53]

Tomato	*SlMAPK3*	Loss of function	*Agrobacterium*-mediated cotyledon transformation	Decrease in drought stress tolerance	[54]
Watermelon	*ClALS*	Site-directed mutagenesis	*Agrobacterium*-mediated cotyledon transformation	Herbicide resistance	[55]
Tomato and Potato	*StALS2*	Site-directed mutagenesis	*Agrobacterium*-mediated transformation	Herbicide resistance	[56]
Tomato	Carotenoid cleavage dioxygenase8 (*CCD8*)	Loss of function	*Agrobacterium*-mediated transformation	Resistance against *Phelipanche aegytiaca*	[57]
Tomato	More Axillary Growth1 (*MAX1*)	Loss of function	*Agrobacterium*-mediated transformation	Resistance against *Phelipanche aegytiaca*	[58]
Tomato	*SP,SP5G,CLV3, WUS, GGP1*	Cis-regulatory variation and loss of function	*Agrobacterium*-mediated transformation	Introduction of desirable traits with morphology, flower number, fruit size and number, and ascorbic acid synthesis	[60]
Tomato	*S l W U S C a r G* element, *SlCLV3* promoter	Cis-regulatory variation	*Agrobacterium*-mediated transformation	Fruit size, inflorescence branching, and plant architecture	[61]
Tomato	*PSY1*	Different mutations in alleles	*Agrobacterium*-mediated cotyledon transformation	Yellow-coloured tomato	[64]
Tomato	*MYB12*	Different mutations in alleles	*Agrobacterium*-mediated cotyledon transformation	Pink-coloured tomato	[65]
Tomato	*ANT2*	Gene insertion	*Agrobacterium*-mediated cotyledon transformation	Purple-coloured tomato	[66]
Chinese kale	*BoaCRTISO*	Gene insertion and replacement	Agrobacterium-mediated transformation	Colour of mutants changed from green to yellow	[67]
Tomato	*MPK20*	Loss of function	*Agrobacterium*-mediated transformation	Repression of genes controlling sugar and auxin metabolism	[68]
Tomato	*G A D 2 , G A D 3 , S l y G A B A - TP1,SlyGABA-TP2,SlyGABA-TP3,SlyCAT9,SlySSADH*	Autoinhibitory domain deletion	*Agrobacterium*-mediated transformation	Increase in □-aminobutyric acid (GABA) content	[72]

Tomato	*SGR1,lycopene ε-cyclase (LCY-E), beta-lycopene cyclase(Blc), lycopene β-cyclase1(LCY-B1) and LCY-B2*	Loss of function	*Agrobacterium*-mediated apical segments of hypocotyls transformation	Lycopene content	[73]
Potato	*16α-hydroxylation* (St16DOX)	Loss of function	*Agrobacterium*-mediated shoot transformation	Steroidal glycoalkaloid (SGA) biosynthesis	[74]
Tomato	Ripening inhibitor *(RIN)*	Single base insertion or deletion of more than three bases	*Agrobacterium*-mediated transformation	MADS-box transcription factor regulating fruit ripening	[75]
Tomato	DNA demethylases *(SlDML2)*	Loss of function	*Agrobacterium*-mediated cotyledon transformation	Activation and inhibition of fruit ripening	[76]
Tomato	Enzymes pectate lyase (*PL*), Polygalacturonase 2a (*PG2a*), and β-galactanase (*TBG4*)	Generation of a range of CRISPR alleles	*Agrobacterium*-mediated transformation	Pectin degradation control	[77]
Tomato	Alcobaca *(SLALC)*	Loss of function	*Agrobacterium*-mediated hypocotyls transformation	Long shelf-life	[78]
Tomato	*SlMBP21*	Loss of function	*Agrobacterium*-mediated transformation	Jointless fruit stem	[81]
Tomato	*SlAGAMOUS-LIKE 6 (Sl AGL6)*	Loss of function	*Agrobacterium*-mediated transformation	Parthenocarpic	[82]
Tomato	*ARF7*	Loss of function	*Agrobacterium*-mediated transformation	Parthenocarpic	[83]
Tomato	*SlIAA9*	Loss of function	*Agrobacterium*-mediated leaf disk transformation	Parthenocarpic	[84]
Tomato	*Blade-on-petiole (SlBOP)*	Loss of function	*Agrobacterium*-mediated cotyledon segments transformation	Early flowering with simplified inflorescence architecture	[85]
Tomato	*Self-pruning 5G(SlSP5G)*	cis-regulatory variation	*Agrobacterium*-mediated transformation	Day-length-sensitive flowering	[88]

Crop sensitivity to photoperiod restricts their planting areas and regulating the photoperiod-associated genes can accelerate domestication. The disruption of self-pruning 5G (*SP5G*) generated a rapid surge in flowering that leads to an early fruit harvest [**88**].

Dwarf-crop breeding is an important direction in domestication research, as dwarf plants are resistant to lodging under high wind conditions. Compared with normal plants, it is more convenient to pick fruits from dwarfed plants [**89**]. Dwarf plants transport nutrients more readily over the shorter distances from the roots to the leaves [**89**]. The application of CRISPR-Cas9 technology in the genomes of several commercially important fruit and vegetables has been achieved as outlined in **Table 1**.

1.4. Improvements to CRISPR-Cas9 Gene-Editing Systems

1.4.1. Production of Non-GM Plants Using CRISPR-Cas9 Gene Editing

Most fruit crops have heterozygous genotypes, hybrid incompatibility, and long growth periods. Some have complex triploid or polyploid genomes [**90**]. During the last 20 years, innovative breeding technology employing genetic engineering has provided a favourable way to accelerate crop improvement for such species [**91**]. For example, the introduction of foreign DNA fragments when creating transgenic lines may block the function of endogenous genes and affect the expression of adjacent genes. In contrast to transgenic approaches, CRISPR/Cas9 technology is able to generate nontransgenic plants. Because CRISPR/Cas9 expression cassettes and their target sites are located at different positions of the genome, segregation and removal of the CRISPR/Cas9 cassettes is possible via subsequent selfing or crossing; however, this is not feasible in most fruit crops, because of their complex, highly heterozygous, and polyploid genomes, and because they are usually vegetatively propagated.

Fruit trees have a long juvenile stage and take several years to reach the reproductive stage. In such cases, the CRISPR/Cas9 components can be transiently expressed in the nucleus and function for a short time to induce precise mutations. This means that transgene-free edited plants can be generated, since the CRISPR/Cas9 expression cassette is not integrated into the genome. Currently, the overall efficiency of the transient system for the production of T-DNA-free edited apple lines is very low (0.4%) [**92**]; thus, the next step is to improve editing efficiency and to make this system suitable for other crops.

Preassembled CRISPR-Cas-sgRNA ribonucleoproteins (RNPs) can be delivered into plant cells and used for genome editing without the integration of foreign DNA because of degradation by endogenous proteases [**93**]. The protoplast transformation technique has been used to transform grape, apple, and lettuce with purified Cas9

RNPs. Sequencing analysis of transformed cells revealed mutagenesis efficiencies of 0.1 to 6.9% in grapevine and apple; however, due to the poor regeneration ability of protoplasts, no plants were regenerated [**93,94**].

Two methods have been used to obtain transgene-free plants with mutations via CRISPR/Cas9 gene editing. The first is based on the site-specific recombinase flippase (Flp) [**95**], which recognises 34 bp-long flippase recognition target site (FRT) sequences. The Flp/FRT system has been extensively used to remove undesired transgenic components in transgenic apple [**96,97**]. The second removal method relies on the Cas9 enzyme cleavage mechanism. Two additional synthetic target sites, referred to as cleavage target sites, were added next to the left border (LB) and right border (RB) sites of the CRISPR/Cas9 vector. When plants are transformed using CRISPR/Cas9, the Cas9 cleavage activity not only edits the endogenous target site, but also removes T-DNA by inserting two additional cleavage target sites, thereby resulting in T-DNA-free plants [**98**].

1.4.2. Novel Variants of Cas Protein and Applications

In commonly used CRISPR-Cas9 systems, the *Streptococcus pyogenes* (SpCas9)-gRNA complex generally recognises the region 20 nt upstream of the PAM sequence (5′-NGG-3′). To broaden the Cas9 protein recognition sequence and reduce the off-target editing rate, several approaches have been used to broaden PAM compatibility and enhance specificity. These approaches are based on the structural characteristics of SpCas9 binding to gRNA and target DNA. For example, the Cas9 variants SpCas9-VQR (NGA-PAM), SpCas9-EQR (NGAG-PAM), and SpCas9-VRER (NGCG-PAM) functioned, but their cleavage activity levels were lower compared to that of the wild-type SpCas9 in *Arabidopsis* and rice [**99,100,101,102**]. SpCas9-NG has a broader recognition sequence with enhanced compatibility, recognising NG-PAM, and has successfully generated targeted mutations in rice and *Arabidopsis* [**103,104,105**].

The variants SpCas9-HF1, eSpCas9, and HypaCas9 have been developed to enhance Cas9 protein-cleavage specificity. They show reduced off-target editing activities, indicating high specificity in plant cells [**106,107,108**]. Cas9 protein-directed evolution has been developed for Cas9 engineering, conferring high-specificity engineered SpCas9 proteins such as xCas9 [**109**], evoCas9 [**110**], and Sniper-Cas9 [**111**]. xCas9 recognises the NG, GAA, and GAT PAM sequences. Although the gRNA containing these PAM sequences can be mutated in plants, the mutation efficiencies and the preferences for different corresponding PAM sequences differ between cells. For example, the cleavage activity of xCas9 is lower in rice callus than in mammalian cells [**112**], and xCas9 does not recognise the NG-PAM sequence in tomatoes [**32**]. Cas9-NG has a stronger cleavage activity than xCas9, especially at CGG, AGC, TGA, and CGT sequence recognition sites [**98**], making Cas9-NG more suitable for genome editing at the NG-PAM site in plants. The single-base editing system developed based on Cas9 variants (SpCas9-NG and SpCas9-VQR) has been applied to precise base-editing of plant genomes [**103,113**].

At present, the most commonly used Cas9 protein comes from *Streptococcus pyogenes*, in order to broaden the Cas9 protein recognition sequence, orthologous Cas9 proteins have been isolated from other bacteria; for instance, NmCas9 has been isolated from *Neisseria meningitidis* [**114**], SaCas9 from *Staphylococcus aureus* [**115**], StCas9 from *Streptococcus thermophilus* [**116**], FnCas9 from *Francisella novicida* [**117**], and CjCas9 from *Campylobacter jejuni* [**118**]. These proteins are smaller than SpCas9, which is an advantage in cassette delivery. In *Arabidopsis*, the SaCas9 and SpCas9 systems do not interfere with each other [**119**], so they can fully utilise Cas9 orthologues that recognise different PAM sequences. Such simultaneous targeting by Cas9 orthologues with different PAM sequences would enable multiplex genome engineering by simultaneously targeting more than one site.

With the continued discovery and investigation of the functions of CRISPR protein family members, new types of Cas proteins have been discovered including the type VI CRISPR-Cas system Cas13 (C2c2) protein, which recognises RNA sequences and exhibits RNA editing activity without altering the genome sequence [**120,121**]. This system has been successfully applied to knockout gene function in rice and tobacco and promote resistance to RNA viruses in *Arabidopsis* [**121,122**]. Likewise, the CRISPR-Cas13 system created RNA-guided immunity against RNA viruses in plants. Type V CRISPR-Cas systems, such as Cas12c, Cas12g, Cas12h, Cas12i, and Cas14, are distinguished according to the type of their target template (ssRNA, ssDNA, dsDNA, or ssDNA) and cleavage activity strength. Their functions range from dsDNA nicking and cleavage, and can have collateral cleavage activity on ssRNA, ssDNA, dsDNA, and ssDNA [**123,124**]. In short, the functional differentiation of these Cas protein variants can be used to target mutations in different nucleic acid types, induce a small number of 100 kb sequence deletions, and expand the repertoire of plant genome-editing tools.

1.5. Regulatory Framework of CRISPR-Cas-Edited Crops

During CRISPR-Cas gene editing of plants, Cas cuts the target sequence to produce double-strand breaks, resulting in the loss of gene function. The CRISPR-Cas technique has been utilised to create modifications in the genome that are identical to natural genetic variation [**94**]. Similarly, in HDR (homologous DNA repair) of CRISPR-Cas-mediated dsDNA fragmentation, exogenously provided homologous DNA sequences are deemed transgenic; however, when the repair template is derived from the genes from the same species and related interbreeding species, the resultant crops are not regarded as transgenic crops [**125**]. Nevertheless, the regulatory framework regarding NHEJ and HDR-mediated gene editing contains differing definitions.

The United States Department of Agriculture (USDA), FDA, and Environmental Protection Agency state that the removal of transgenic elements in plants by CRISPR-mediated editing is equivalent to crop improvement by conventional breeding programs, and crops generated in this way are thus not considered GMOs for regulatory purposes [**126**]. In 2016, the USDA approved the marketing of gene-edited waxy corn without exogenous transgenic elements [**127**], and an *Agaricus bisporus* mushroom with an anti-browning trait obtained from CRISPR-Cas9 editing was exempted from GMO regulatory procedures [**128**]. This definition of gene-edited crops by the United States regulatory agencies promotes the genetic improvement of crops and accelerates the introduction of gene-edited crops to the market. Genome editing has emerged as a powerful and elegant technology to develop novel varieties or organisms with desirable traits in tomato, citrus, soybean, sugarcane, camelina, and rice. In May 2020, the USDA–APHIS issued the latest edition of biotechnology regulations that provide three exemptions for genetic modifications in any plant species: (i) resultant changes in DNA after DSB in the absence of an external repair donor template; (ii) single base pair substitution in targeted loci; and (iii) introduction of a known gene that exists in the plant's gene pool.

As a major producer of GM crops, Canada considers gene-edited products such as plants, animal feed, or human food as different from nonedited products, so they must undergo a premarket assessment. In Europe, there are stringent regulations regarding CRISPR/Cas9, and the European Court of Justice has included gene-edited crops in the scope of GM crop regulation [**129**]. Australia has taken a milder approach, allowing gene-editing without the introduction of any foreign genetic material [**130**]. In Asia, the attitude towards gene-edited crops has eased in China and Japan, and cases of gene-edited crops being planted in the field have been reported [**131**]. In addition, some countries have regulatory frameworks that are applied case-by-case, considering the breeding methodology used, new traits or characteristics introduced, and evidence of the genetic changes in the final product.

2. UPCOMING OBSTACLES IN THE IMPLEMENTATION OF CRISPR-CAS GENE EDITING

The use of CRISPR-Cas9 genome editing has facilitated the acquisition of ample germplasm resources containing genetic diversity, owing to the outcomes of comprehensive whole genome sequencing and functional genomics investigations in fruit and vegetable crops. However, there are two significant obstacles that the future implementation of this faces. The first step involves the precise identification of essential genes that are to be targeted for specific mutations, along with the associated types of mutations. Agronomic qualities of significant importance are typically intricate quantitative traits, therefore modifying a solitary gene does not result in observable changes in the phenotype. Hence, the utilization of efficient CRISPR-Cas-mediated target site-specific insertion and chromosome recombination techniques enables the accumulation of mutant alleles [132].

Utilizing gene-editing technology to suppress the expression of particular genes in plants diminishes their capacity to adapt. Thus, achieving accurate genome editing necessitates the effective and targeted control of gene activities. Genetic mutations occurring in the coding regions of genes can alter the functionality of proteins, whereas mutations at the junctions between exons and introns can lead to the production of alternative splicing variants. Cis-regulatory elements (CREs) are segments of DNA that do not code for proteins and are frequently involved in controlling the process of gene transcription. Single-nucleotide mutations, insertions, deletions, inversions, and epigenetic changes in gene regulatory areas are strongly linked to crop domestication in CREs [133]. The CRISPR-Cas system has been employed to induce mutations in the regulatory regions of promoters, resulting in the production of many alleles with diverse phenotypes. These alleles are valuable genetic assets that may be utilized in breeding programs. At now, the manipulation of CREs in fruit and vegetable crops through gene editing is in its early phases. Future research endeavors to establish the relationship between alterations in gene expression caused by various mutant CREs and the resulting phenotypic characteristics. Additionally, it aims to generate ample breeding resources by utilizing gene editing techniques on CREs. Targeting the induction of double-strand breaks (DSBs) using CRISPR-Cas not only leads to minor mutations like base substitutions, insertions, and deletions, but also causes significant changes in the genome, such as large deletions, chromosomal translocations, and inversions. This approach proves to be highly effective in completely eliminating unwanted genes, such as those responsible for allergens, in fruit and vegetable crops [134,135,136]. While chromosomal rearrangements by conventional targeted mutagenesis are more common, a lack of Ku70, a protein involved in the non-homologous end joining (NHEJ) pathway, leads to a higher occurrence of inversion and translocation (Figure 3) [135].

An additional obstacle in using gene-editing technology involves the conversion of the CRISPR-Cas gene-editing system for use in plant cells and the subsequent acquisition of regenerated plants. Conventional Cas9 or Cas variants, along with CBE, ABE, and prime editing, have been effectively used in model crops like rice, corn, tobacco, and tomatoes. This was achieved through methods such as Agrobacterium, gene gun, polyethylene glycol (PEG), and electroporation-based techniques. These successful applications serve as a basis for implementing these genetic mutation vector systems in other crops. Nevertheless, the task of creating a universally applicable and highly effective genetic transformation and regeneration system for fruit and vegetable crops is somewhat challenging. Specifically, the effectiveness of incorporating CRISPR-Cas cassettes into the genetic makeup and the capacity of altered cells to regenerate plants have emerged as constraining variables. In order to enhance the ability of explant regeneration and the activity of meristem induction following transformation, it is possible to simultaneously overexpress the plant morphogenesis regulating genes Bbm (baby boom) and Wus2 (Wuschel2) along with CRISPR-Cas expression cassette transformation [138]. The utilization of plant RNA and DNA viruses as vectors for plant cell transformation through CRISPR-Cas can offer an ample supply of sgRNAs and streamline the process of genetic transformation and regeneration. This approach is well-suited for in situ transformation and enables the production of gene-edited plants without the inclusion of transgenic elements [139].

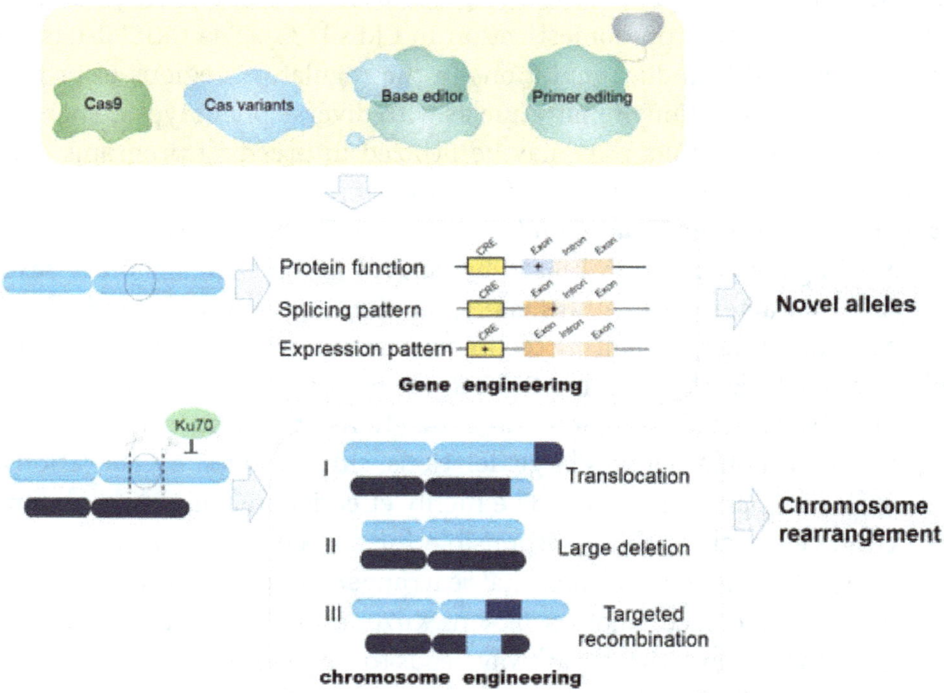

Figure 3. Genome-editing toolbox for trait improvement. CRISPR-Cas9, Cas9 variants, and base and primer editing enable precise gene modification and predictable rearrangement of genomes and chromosomes.

AT A GLANCE

Fruit and vegetable crops contain abundant dietary fiber, vitamins, and minerals that are essential for human health. Nevertheless, numerous biotic stressors, such as pests and diseases, as well as abiotic stressors, pose a significant danger to the growth, quality, and production of crops. Conventional breeding methods for enhancing agricultural characteristics involve a sequence of backcrosses and selection to incorporate advantageous features into high-quality genetic material. However, this process is time-consuming and requires significant resources. The CRISPR-Cas9 breeding technique has the capacity to enhance various traits in crops, including yield, quality, disease resistance, abiotic stress tolerance, and nutritional aspects, with speed and precision. This technique has been utilized to acquire novel germplasm resources by gene-directed mutation, because to its straightforward operation and great efficacy in inducing mutations. By utilizing whole-genome sequencing data and knowledge about gene function in relation to significant features, the use of CRISPR-Cas9 editing allows for the exact alteration of crucial genes. This process enables the quick creation of new germplasm resources, which may be utilized to enhance critical agronomic qualities. This evaluation examines the technology and its implementation in fruit and vegetable crops. We analyze the difficulties, current variations, and the corresponding regulatory structure, while also examining potential future uses.

REFERENCES

1. Giovannoni, J.; Nguyen, C.; Ampofo, B.; Zhong, S.; Fei, Z. The Epigenome and Transcriptional Dynamics of Fruit Ripening. *Ann. Rev. Plant Biol.* 2017, *68*, 61–84.
2. Karkute, S.G.; Singh, A.K.; Gupta, O.P.; Singh, P.M.; Singh, B. CRISPR/Cas9 Mediated Genome Engineering for Improvement of Horticultural Crops. *Front. Plant Sci.* 2017, *8*, 1635.
3. Meyer, R.S.; Purugganan, M.D. Evolution of crop species: Genetics of domestication and diversification. *Nat. Rev. Genet.* 2013, *14*, 840–852.
4. Tester, M.; Langridge, P. Breeding technologies to increase crop production in a changing world. *Science* 2010, *327*, 818–822.
5. Bigliardi, B.; Galati, F. Innovation trends in the food industry: The case of functional foods. *Trends Food Sci. Technol.* 2013, *31*, 118–129.
6. Parmar, N.; Singh, K.H.; Sharma, D.; Singh, L.; Kumar, P.; Nanjundan, J.; Khan, Y.J.; Chauhan, D.K.; Thakur, A.K. Genetic engineering strategies for biotic and abiotic stress tolerance and quality enhancement in horticultural crops: A comprehensive review. *3 Biotech.* 2017, *7*, 239.
7. Kramer, M.G.; Redenbaugh, K. Commercialization of a tomato with an antisense polygalacturonase gene: The FLAVR SAVR™ tomato story. *Euphytica* 1994, *79*, 293–297.
8. Millstone, E.; Stirling, A.; Glover, D. Regulating Genetic Engineering: The Limits and Politics of Knowledge. *Issues Sci. Technol.* 2015, *31*, 23–26.
9. Bawa, A.S.; Anilakumar, K.R. Genetically modified foods: Safety, risks and public concerns-a review. *J. Food Sci. Technol.* 2013, *50*, 1035–1046.
10. Waltz, E. Gene-edited CRISPR mushroom escapes US regulation. *Nature* 2016, *7599*, 293.
11. Waltz, E. With a free pass, CRISPR-edited plants reach market in record time. *Nat. Biotechnol.* 2018, *36*, 6–7.
12. Jansen, R.; Embden, J.D.V.; Gaastra, W.; Schouls, L.M. Identification of genes that are associated with DNA repeats in prokaryotes. *Mol. Microbiol.* 2002, *43*, 1565–1575.
13. Bolotin, A. Clustered regularly interspaced short palindrome repeats (CRISPRs) have spacers of extrachromosomal origin. *Microbiology* 2005, *151*, 2551–2561.

14. Mojica, F.J.; Díez-Villaseñor, C.; García-Martínez, J.; Soria, E. Intervening sequences of regularly spaced prokaryotic repeats derive from foreign genetic elements. *J. Mol. Evol.* 2005, *60*, 174–182.

15. Pourcel, C. CRISPR elements in Yersinia pestis acquire new repeats by preferential uptake of bacteriophage DNA, and provide additional tools for evolutionary studies. *Microbiology* 2005, *151*, 653–663.

16. Cong, L.; Ran, F.A.; Cox, D.; Lin, S.; Barretto, R. Multiplex Genome Engineering Using CRISPR/Cas Systems. *Science* 2013, *339*, 819–823.

17. Kaboli, S.; Babazada, H. CRISPR Mediated Genome Engineering and its Application in Industry. *Curr. Issues Mol. Biol.* 2018, *26*, 81–92.

18. Shan, Q.; Wang, Y.; Li, J.; Zhang, Y.; Chen, K.; Liang, Z.; Zhang, K.; Liu, J.; Xi, J.J.; Qiu, J.L.; et al. Targeted genome modification of crop plants using a CRISPR-Cas system. *Nat. Biotechnol.* 2013, *31*, 686–688.

19. Rouillon, C.; Zhou, M.; Zhang, J.; Politis, A.; Beilsten-Edmands, V.; Cannone, G.; Graham, S.; Robinson, C.V.; Spagnolo, L.; White, M.F. Structure of the CRISPR interference complex CSM reveals key similarities with cascade. *Mol. Cell* 2013, *52*, 124–134.

20. Gasiunas, G.; Barrangou, R.; Horvath, P.; Siksnys, V. Cas9-crRNA ribonucleoprotein complex mediates specific DNA cleavage for adaptive immunity in bacteria. *Proc. Natl. Acad. Sci. USA* 2012, *109*, 15539–15540.

21. Doudna, J.A.; Charpentier, E. Genome editing. The new frontier of genome engineering with CRISPR-Cas9. *Science* 2014, *346*, 1258096.

22. Yin, K.; Gao, C.; Qiu, J.L. Progress and prospects in plant genome editing. *Nat. Plants* 2017, 3, 17107.

23. Baltes, N.J.; Gil-Humanes, J.; Cermak, T.; Atkins, P.A.; Voytas, D.F. DNA replicons for plant genome engineering. *Plant cell* 2014, *26*, 151–163.

24. Endo, M.; Ishikawa, Y.; Osakabe, K.; Nakayama, S.; Kaya, H.; Araki, T.; Shibahara, K.-I.; Abe, K.; Ichikawa, H.; Valentine, L.; et al. Increased frequency of homologous recombination and T-DNA integration in Arabidopsis CAF-1 mutants. *EMBO J.* 2006, *25*, 5579–5590.

25. Gil-Humanes, J.; Wang, Y.; Liang, Z.; Shan, Q.; Ozuna, C.V.; Sánchez-León, S.; Baltes, N.J.; Starker, C.; Barro, F.; Gao, C.; et al. High-efficiency gene targeting in hexaploid wheat using DNA replicons and CRISPR/Cas9. *Plant J.* 2017, *89*, 1251–1262.

26. Ayathilaka, K.; Sheridan, S.D.; Bold, T.D.; Bochenska, K.; Logan, H.L.; Weichselbaum, R.R.; Bishop, D.K.; Connell, P.P. A chemical compound that stimulates the human homologous recombination protein RAD51. *Proc. Natl. Acad. Sci. USA* 2008, *105*, 15848–15853.

27. Donohoue, P.D.; Barrangou, R.; May, A.P. Advances in Industrial Biotechnology Using CRISPR-Cas Systems. *Trends Biotechnol.* 2018, *36*, 134–146.

28. Li, R.; Fu, D.; Zhu, B.; Luo, Y.; Zhu, H. CRISPR/Cas9-mediated mutagenesis of lncRNA1459 alters tomato fruit ripening. *Plant J.* 2019, *97*, 795.

29. Liang, W.; van Wersch, S.; Tong, M.; Li, X. TIR-NB-LRR immune receptor SOC3 pairs with truncated TIR-NB protein CHS1 or TN2 to monitor the homeostasis of E3 ligase SAUL1. *New Phytol.* 2019, *221*, 2054–2066.

30. Mao, Y.; Yang, X.; Zhou, Y.; Zhang, Z.; Botella, J.R.; Zhu, J.-K. Manipulating plant RNA-silencing pathways to improve the gene editing efficiency of CRISPR/Cas9 systems. *Genome Biol.* 2018, *19*, 149.

31. Hess, G.T.; Tycko, J.; Yao, D.; Bassik, M.C. Methods and Applications of CRISPR-Mediated Base Editing in Eukaryotic Genomes. *Mol. Cell* 2017, *68*, 26–43.

32. Niu, Q.; Wu, S.; Li, Y.; Yang, X.; Liu, P.; Xu, Y.; Lang, Z. Expanding the scope of CRISPR/Cas9-mediated genome editing in plants using an xCas9 and Cas9-NG hybrid. *J. Integr. Plant Biol.* 2020, *62*, 398–402.

33. Brooks, C.; Nekrasov, V.; Lippman, Z.B.; Van Eck, J. Efficient gene editing in tomato in the first generation using the clustered regularly interspaced short palindromic repeats/CRISPR-associated9 system. *Plant Physiol.* 2014, *166*, 1292–1297.

34. Kulus, D. Genetic resources and selected conservation methods of tomato. *J. Appl. Bot. Food Qual.* 2018, *91*, 135–144.

35. Ma, C.; Liu, M.; Li, Q.; Si, J.; Ren, X.; Song, H. Efficient BoPDS Gene Editing in Cabbage by the CRISPR/Cas9 System. *Hortic. Plant J.* 2019, *5*, 164–169.

36. Sun, B.; Zheng, A.; Jiang, M.; Xue, S.; Yuan, Q.; Jiang, L.; Chen, Q.; Li, M.; Wang, Y.; Zhang, Y.; et al. CRISPR/Cas9-mediated mutagenesis of homologous genes in Chinese kale. *Sci. Rep.* 2018, *8*, 16786.

37. Tian, S.; Jiang, L.; Gao, Q.; Zhang, J.; Zong, M.; Zhang, H.; Ren, Y.; Guo, S.; Gong, G.; Liu, F.; et al. Efficient CRISPR/Cas9-based gene knockout in watermelon. *Plant Cell Rep.* 2017, *36*, 399–406.

38. Bastet, A.; Zafirov, D.; Giovinazzo, N.; Guyon-Debast, A.; Nogué, F.; Robaglia, C.; Gallois, J.L. Mimicking natural polymorphism in eIF4E by CRISPR-Cas9 base editing is associated with resistance to potyviruses. *Plant Biotechnol. J.* 2019, *17*, 1736–1750.

39. Chandrasekaran, J.; Brumin, M.; Wolf, D.; Leibman, D.; Klap, C.; Pearlsman, M.; Sherman, A.; Arazi, T.; Gal-On, A. Development of broad virus resistance in non-transgenic cucumber using CRISPR/Cas9 technology. *Mol. Plant Pathol.* 2016, *17*, 1140–1153.

40. Ji, X.; Zhang, H.; Zhang, Y.; Wang, Y.; Gao, C. Establishing a CRISPR–Cas-like immune system conferring DNA virus resistance in plants. *Nat. Plants* 2015, *1*, 15144.

41. Ali, Z.; Ali, S.; Tashkandi, M.; Zaidi, S.S.E.A.; Mahfouz, M.M. CRISPR/Cas9-Mediated Immunity to Geminiviruses: Differential Interference and Evasion. *Sci. Rep.* 2016, *6*, 26912.

42. Borrelli, V.M.G.; Brambilla, V.; Rogowsky, P.; Marocco, A.; Lanubile, A. The Enhancement of Plant Disease Resistance Using CRISPR/Cas9 Technology. *Frontiers in Plant Science.* 2018, *9*, 1245.

43. Zeilmaker, T.; Ludwig, N.R.; Elberse, J.; Seidl, M.F.; Berke, L.; Van Doorn, A.; Schuurink, R.C.; Snel, B.; Van den Ackerveken, G. DOWNY MILDEW RESISTANT 6 and DMR6-LIKE OXYGENASE 1 are partially redundant but distinct suppressors of immunity in Arabidopsis. *Plant J.* 2015, *81*, 210–222.

44. Paula de Toledo Thomazella, D.; Brail, Q.; Dahlbeck, D.; Staskawicz, B. CRISPR-Cas9 mediated mutagenesis of a DMR6 ortholog in tomato confers broad-spectrum disease resistance. *BioRxiv* 2016, 064824.

45. Nekrasov, V.; Wang, C.; Win, J.; Lanz, C.; Weigel, D.; Kamoun, S. Rapid generation of a transgene-free powdery mildew resistant tomato by genome deletion. *Sci. Rep.* 2017, *7*, 1–6.

46. Chaudhary, R.; Hs, A. Resistance-Gene-Mediated Defense Responses against Biotic Stresses in the Crop Model Plant Tomato. *J. Plant Pathol. Microbiol.* 2017, *8*, 2.

47. Cahya, P.; Barbetti, M.J.; Barker, S.J. A Novel Tomato Fusarium Wilt Tolerance Gene. *Front. Microbiol.* 2018, *9*, 1226.

48. Zhang, M.; Liu, Q.; Yang, X.; Xu, J.; Liu, G.; Yao, X.; Ren, R.; Xu, J.; Lou, L. CRISPR/Cas9-mediated mutagenesis of Clpsk1 in watermelon to confer resistance to Fusarium oxysporum f.sp. niveum. *Plant Cell Rep.* 2020, *39*, 589–595.

49. Shujuan, Z.; Liu, W.; Ruirui, Z.; Wenqing, Y.; Rui, L.; Yujing, L.; Jiping, S.; Lin, S. Knockout of SlMAPK3 Reduced Disease Resistance to *Botrytis cinerea* in Tomato Plants. *J. Agric. Food Chem.* 2018, *66*, 8949–8956.

50. Andrés, O.; Selena, G.I.; Nathalie, L.; Roberto, S. Design of a bacterial speck resistant tomato by CRISPR/Cas9-mediated editing of SlJAZ2. *Plant Biotechnol. J.* 2018, *17*, 665–673.

51. Haque, E.; Taniguchi, H.; Hassan, M.M.; Bhowmik, P.; Karim, M.R.; Śmiech, M.; Zhao, K.; Rahman, M.; Islam, T. Application of CRISPR/Cas9 Genome Editing Technology for the Improvement of Crops Cultivated in Tropical Climates: Recent Progress, Prospects, and Challenges. *Front. Plant Sci.* 2018, *9*, 617.

52. Yin, Y.; Qin, K.; Song, X.; Zhang, Q.; Zhou, Y.; Xia, X.; Yu, J. BZR1 Transcription Factor Regulates Heat Stress Tolerance Through FERONIA Receptor-Like Kinase-Mediated Reactive Oxygen Species Signaling in Tomato. *Plant Cell Physiol.* 2018, *59*, 2239–2254.

53. Li, R.; Zhang, L.; Wang, L.; Chen, L.; Zhao, R.; Sheng, J.; Shen, L. Reduction of Tomato-Plant Chilling Tolerance by CRISPR-Cas9-Mediated SlCBF1 Mutagenesis. *J. Agric. Food Chem.* 2018, *66*, 9042–9051.

54. Wang, L.; Chen, L.; Li, R.; Zhao, R.; Yang, M.; Sheng, J.; Shen, L. Reduced drought tolerance by CRISPR/Cas9-mediated SlMAPK3 mutagenesis in tomato plants. *J. Agric. Food Chem.* 2017, *65*, 8674–8682.

55. Tian, S.; Jiang, L.; Cui, X.; Zhang, J.; Guo, S.; Li, M.; Zhang, H.; Ren, Y.; Gong, G.; Zong, M. Engineering herbicide-resistant watermelon variety through CRISPR/Cas9-mediated base-editing. *Plant Cell Rep.* 2018, *37*, 1353–1356.

56. Butler, N.M.; Atkins, P.A.; Voytas, D.F.; Douches, D.S. Generation and Inheritance of Targeted Mutations in Potato (Solanum tuberosum L.) Using the CRISPR/Cas System. *PLoS ONE* 2015, *10*, e0144591.

57. Bari, V.K.; Nassar, J.A.; Aly, R. CRISPR/Cas9 mediated mutagenesis of MORE AXILLARY GROWTH 1 in tomato confers resistance to root parasitic weed Phelipanche aegyptiaca. *Sci. Rep.* 2021, *11*, 3905.

58. Bari, V.K.; Nassar, J.A.; Kheredin, S.M.; Gal-On, A.; Ron, M.; Britt, A.; Steele, D.; Yoder, J.; Aly, R. CRISPR/Cas9-mediated mutagenesis of CAROTENOID CLEAVAGE DIOXYGENASE 8 in tomato provides resistance against the parasitic weed Phelipanche aegyptiaca. *Sci. Rep.* 2019, *9*, 11438.

59. Li, H.; Qi, M.; Sun, M.; Liu, Y.; Liu, Y.; Xu, T.; Li, Y.; Li, T. Tomato Transcription Factor SlWUS Plays an Important Role in Tomato Flower and Locule Development. *Front. Plant Sci.* 2017, *8*, 457.

60. Li, T.; Yang, X.; Yu, Y.; Si, X.; Zhai, X.; Zhang, H.; Dong, W.; Gao, C.; Xu, C. Domestication of wild tomato is accelerated by genome editing. *Nat. Biotechnol.* 2018, *36*, 1160–1163.

61. Rodríguez-Leal, D.; Lemmon, Z.H.; Man, J.; Bartlett, M.E.; Lippman, Z.B. Engineering Quantitative Trait Variation for Crop Improvement by Genome Editing. *Cell* 2017, *171*, 470–480.

62. Lin, T.; Zhu, G.; Zhang, J.; Xu, X.; Yu, Q.; Zheng, Z.; Zhang, Z.; Lun, Y.; Li, S.; Wang, X.; et al. Genomic analyses provide insights into the history of tomato breeding. *Nat. Genet.* 2014, *46*, 1220–1226.

63. Ballester, A.R.; Molthoff, J.; de Vos, R.; Hekkert, B.T.L.; Orzaez, D.; Fernandez-Moreno, J.P.; Tripodi, P.; Gran-dillo, S.; Martin, C.; Heldens, J.; et al. Biochemical and Molecular Analysis of Pink Tomatoes: Deregulated Expression of the Gene Encoding Transcription Factors SlMYB12 Leads to Pink Tomato Fruit Color. *Plant Physiol.* 2010, *152*, 71–84.

64. Čermák, T.; Baltes, N.J.; Čegan, R.; Zhang, Y.; Voytas, D.F. High-frequency, precise modification of the tomato genome. *Genome Biol.* 2015, *16*, 232.

65. Deng, L.; Wang, H.; Sun, C.; Li, Q.; Jiang, H.; Du, M.; Li, C.B.; Li, C. Efficient generation of pink-fruited tomatoes using CRISPR/Cas9 system. *J. Genet. Genom.* 2018, *45*, 51–54.

66. Filler Hayut, S.; Melamed Bessudo, C.; Levy, A.A. Targeted recombination between homologous chromosomes for precise breeding in tomato. *Nat. Commun.* 2017, *8*, 15605.

67. Sun, B.; Jiang, M.; Zheng, H.; Jian, Y.; Huang, W.L.; Yuan, Q.; Zheng, A.H.; Chen, Q.; Zhang, Y.T.; Lin, Y.X.; et al. Color-related chlorophyll and carotenoid concentrations of Chinese kale can be altered through CRISPR/Cas9 targeted editing of the carotenoid isomerase gene BoaCRTISO. *Hortic. Res.* 2020, *7*, 161.

68. Chen, L.; Yang, D.; Zhang, Y. Evidence for a specific and critical role of mitogen-activated protein kinase 20 in uni-to-binucleate transition of microgametogenesis in tomato. *Plant Biol.* 2018, *219*, 176–194.

69. Kris-Etherton, P.M.; Hecker, K.D.; Bonanome, A.; Coval, S.M.; Binkoski, A.E.; Hilpert, K.F.; Griel, A.E.; Etherton, T.D. Bioactive compounds in foods: Their role in the prevention of cardiovascular disease and cancer. *Am. J. Med.* 2002, *113*, 71–88.

70. Meng, X.; Yang, D.; Li, X.; Zhao, S.; Sui, N.; Meng, Q. Physiological changes in fruit ripening caused by overexpression of tomato SlAN2, an R2R3-MYB factor. *Plant Physiol. Biochem.* 2015, *89*, 24–30.

71. Ye, J.; Wang, X.; Hu, T.; Zhang, F.; Wang, B.; Li, C.; Yang, T.; Li, H.; Lu, Y. An InDel in the Promoter of Al-ACTIVATED MALATE TRANSPORTER9 Selected during Tomato Domestication Determines Fruit Malate Contents and Aluminum Tolerance. *Plant Cell* 2017, *29*, 2249.

72. Nonaka, S.; Arai, C.; Takayama, M.; Matsukura, C.; Ezura, H. Efficient increase of γ-aminobutyric acid (GABA) content in tomato fruits by targeted mutagenesis. *Sci. Rep.* 2017, *7*, 7057.

73. Li, X.; Wang, Y.; Chen, S.; Tian, H.; Fu, D.; Zhu, B.; Luo, Y.; Zhu, H. Lycopene Is Enriched in Tomato Fruit by CRISPR/Cas9-Mediated Multiplex Genome Editing. *Front. Plant Sci.* 2018, *9*, 559.

74. Nakayasu, M.; Akiyama, R.; Lee, H.J.; Osakabe, K.; Osakabe, Y.; Watanabe, B.; Sugimoto, Y.; Umemoto, N.; Saito, K.; Muranaka, T.; et al. Generation of α-solanine-free hairy roots of potato by CRISPR/Cas9 mediated genome editing of the St16DOX gene. *Plant Physiol. Biochem.* 2018, *131*, 70–77.

75. Ito, Y.; Nishizawa-Yokoi, A.; Endo, M.; Mikami, M.; Toki, S. CRISPR/Cas9-mediated mutagenesis of the RIN locus that regulates tomato fruit ripening. *Biochem. Biophys. Res. Commun.* 2015, *467*, 76–82.

76. Lang, Z.; Wang, Y.; Tang, K.; Tang, D.; Datsenka, T.; Cheng, J.; Zhang, Y.; Handa, A.K.; Zhu, J.K. Critical roles of DNA demethylation in the activation of ripening-induced genes and inhibition of ripening-repressed genes in tomato fruit. *Proc. Natl. Acad. Sci. USA* 2017, *114*, E4511–E4519.

77. Wang, D.; Samsulrizal, A.; Yan, B.C.; Allcock, C.; Craigon, D.J. Characterization of CRISPR Mutants Targeting Genes Modulating Pectin Degradation in Ripening Tomato. *Plant Physiol.* 2019, *179*, 544–557.

78. Yu, Q.H.; Wang, B.; Li, N.; Tang, Y.; Yang, S.; Yang, T.; Xu, J.; Guo, C.; Yan, P.; Wang, Q.; et al. CRISPR/Cas9-induced Targeted Mutagenesis and Gene Replacement to Generate Long-shelf Life Tomato Lines. *Sci. Rep.* 2017, *7*, 11874.

79. Ledford, H. Fixing the tomato: CRISPR edits correct plant-breeding snafu. *Nature* 2017, *545*, 394–395.

80. Soyk, S.; Lemmon, Z.H.; Oved, M.; Fisher, J.; Liberatore, K.L.; Park, S.J.; Goren, A.; Jiang, K.; Ramos, A.; van der Knaap, E.; et al. Bypassing Negative Epistasis on Yield in Tomato Imposed by a Domestication Gene. *Cell* 2017, *169*, 1142–1155.

81. Roldan, M.V.G.; Périlleux, C.; Morin, H.; Huerga-Fernandez, S.; Latrasse, D.; Benhamed, M.; Bendahmane, A. Natural and induced loss of function mutations in SlMBP21 MADS-box gene led to jointless-2 phenotype in tomato. *Sci. Rep.* 2017, *7*, 1–10.

82. Klap, C.; Yeshayahou, E.; Bolger, A.M.; Arazi, T.; Gupta, S.K.; Shabtai, S.; Usadel, B.R.; Salts, Y.; Barg, R. Tomato facultative parthenocarpy results from SlAGAMOUS-LIKE 6 loss of function. *Plant Biotechnol. J.* 2016, *15*, 634–647.

83. Hu, J.; Alon, I.; Naomi, O.; Sun, T.P. DELLA-ARF/IAA Interaction Mediates Crosstalk between Gibberellin and Auxin Signaling in Controlling Fruit Initiation in Solanum lycopersicum. *Plant Cell* 2018, *30*, 1710–1728.

84. Ueta, R.; Abe, C.; Watanabe, T.; Sugano, S.S.; Ishihara, R.; Ezura, H.; Osakabe, Y.; Osakabe, K. Rapid breeding of parthenocarpic tomato plants using CRISPR/Cas9. *Sci. Rep.* 2017, *7*, 507.

85. Xu, C.; Park, S.J.; Van Eck, J.; Lippman, Z.B. Control of inflorescence architecture in tomato by BTB/POZ transcriptional regulators. *Genes Dev.* 2016, *30*, 2048–2061.

86. Zsögön, A.; Čermák, T.; Naves, E.R.; Notini, M.M.; Edel, K.H.; Weinl, S.; Freschi, L.; Voytas, D.F.; Kudla, J.; Peres, L.E.P. De novo domestication of wild tomato using genome editing. *Nat. Biotechnol.* 2018, *36*, 1211–1216.

87. Wang, R.; Tavano, E.C.d.R.; Lammers, M.; Martinelli, A.P.; Angenent, G.C.; de Maagd, R.A. Re-evaluation of transcription factor function in tomato fruit development and ripening with CRISPR/Cas9-mutagenesis. *Sci. Rep.* 2019, *9*, 1696.

88. Soyk, S.; Müller, N.A.; Park, S.J.; Schmalenbach, I.; Jiang, K.; Hayama, R.; Zhang, L.; Van Eck, J.; Jiménez-Gómez, J.M.; Lippman, Z.B. Variation in the flowering gene SELF PRUNING 5G promotes day-neutrality and early yield in tomato. *Nat. Genet.* 2017, *49*, 162–168.

89. Zsögön, A.; Cermak, T.; Voytas, D.; Peres, L.E. Genome editing as a tool to achieve the crop ideotype and de novo domestication of wild relatives: Case study in tomato. *Plant Sci.* 2017, *256*, 120–130.
90. Bisognin, D.A. Breeding vegetatively propagated horticultural crops. *Crop Breed. Appl. Biotechnol.* 2011, *11*, 35–43.
91. Corte, E.D.; MMahmoud, L.; SMoraes, T.; Mou, Z.; WGrosser, J.; Dutt, M. Development of Improved Fruit, Vegetable, and Ornamental Crops Using the CRISPR/Cas9 Genome Editing Technique. *Plants* 2019, *8*, 601.
92. Charrier, A.; Vergne, E.; Dousset, N.; Richer, A.; Petiteau, A.; Chevreau, E. Efficient Targeted Mutagenesis in Apple and First Time Edition of Pear Using the CRISPR-Cas9 System. *Front. Plant Sci.* 2019, *10*, 40.
93. Malnoy, M.; Viola, R.; Jung, M.H.; Koo, O.J.; Kim, S.; Kim, J.S.; Velasco, R.; Nagamangala Kanchiswamy, C. DNA-Free Genetically Edited Grapevine and Apple Protoplast Using CRISPR/Cas9 Ribonucleoproteins. *Front. Plant Sci.* 2016, *7*, 1904.
94. Woo, J.W.; Kim, J.; Kwon, S.I.; Corvalán, C.; Cho, S.W.; Kim, H.; Kim, S.G.; Kim, S.T.; Choe, S.; Kim, J.S. DNA-free genome editing in plants with preassembled CRISPR-Cas9 ribonucleoproteins. *Nat. Biotechnol.* 2015, *33*, 1162–1164.
95. Pompili, V.; Dalla Costa, L.; Piazza, S.; Pindo, M.; Malnoy, M. Reduced fire blight susceptibility in apple cultivars using a high-efficiency CRISPR/Cas9-FLP/FRT-based gene editing system. *Plant Biotechnol. J.* 2020, *18*, 845–858.
96. Dalla Costa, L.; Piazza, S.; Campa, M.; Flachowsky, H.; Hanke, M.-V.; Malnoy, M. Efficient heat-shock removal of the selectable marker gene in genetically modified grapevine. *Plant Cell Tissue Organ Cult.* 2016, *124*, 471–481.
97. Herzog, K.; Flachowsky, H.; Deising, H.B.; Hanke, M.V. Heat-shock-mediated elimination of the nptII marker gene in transgenic apple (Malus×domestica Borkh.). *Gene* 2012, *498*, 41–49.
98. Dalla Costa, L.; Piazza, S.; Pompili, V.; Salvagnin, U.; Cestaro, A.; Moffa, L.; Vittani, L.; Moser, C.; Malnoy, M. Strategies to produce T-DNA free CRISPRed fruit trees via Agrobacterium tumefaciens stable gene transfer. *Sci. Rep.* 2020, *10*, 1–14.
99. Akihiro, Y.; Takashi, I.; Mika, Y.; Yuri, K.; Shinichiro, S. Developing Heritable Mutations in Arabidopsis thaliana Using a Modified CRISPR/Cas9 Toolkit Comprising PAM-Altered Cas9 Variants and gRNAs. *Plant Cell Physiol.* 2019, *60*, 2255–2262.
100. Hu, X.; Meng, X.; Liu, Q.; Li, J.; Wang, K. Increasing the efficiency of CRISPR-Cas9-VQR precise genome editing in rice. *Plant Biotechnol. J.* 2018, *16*, 292–297.
101. Hu, X.; Wang, C.; Fu, Y.; Liu, Q.; Jiao, X.; Wang, K. Expanding the Range of CRISPR/Cas9 Genome Editing in Rice. *Mol. Plant* 2016, *9*, 943–945.
102. Kleinstiver, B.P.; Prew, M.S.; Tsai, S.Q.; Topkar, V.V.; Nguyen, N.T.; Zheng, Z.; Gonzales, A.P.W.; Li, Z.; Peterson, R.T.; Yeh, J.-R.J.; et al. Engineered CRISPR-Cas9 nucleases with altered PAM specificities. *Nature* 2015, *523*, 481–485.
103. Hua, K.; Tao, X.; Han, P.; Wang, R.; Zhu, J.K. Genome Engineering in Rice Using Cas9 Variants that Recognize NG PAM Sequences. *Mol. Plant* 2019, *12*, 1003–1014.
104. Endo, M.; Mikami, M.; Endo, A.; Kaya, H.; Itoh, T.; Nishimasu, H.; Nureki, O.; Toki, S. Genome editing in plants by engineered CRISPR-Cas9 recognizing NG PAM. *Nat. Plants* 2019, *5*, 14–17.
105. Zhong, Z.; Sretenovic, S.; Ren, Q.; Yang, L.; Bao, Y.; Qi, C.; Yuan, M.; He, Y.; Liu, S.; Liu, X.; et al. Improving Plant Genome Editing with High-Fidelity xCas9 and Non-canonical PAM-Targeting Cas9-NG. *Mol. Plant* 2019, *12*, 1027–1036.
106. Kleinstiver, B.P.; Pattanayak, V.; Prew, M.S.; Tsai, S.Q.; Nguyen, N.T.; Zheng, Z.; Joung, J.K. High-fidelity CRISPR-Cas9 nucleases with no detectable genome-wide off-target effects. *Nature* 2016, *529*, 490–495.
107. Slaymaker, I.M.; Gao, L.; Zetsche, B.; Scott, D.A.; Yan, W.X.; Zhang, F. Rationally engineered Cas9 nucleases with improved specificity. *Science* 2016, *351*, 84–88.

108. Zhang, D.; Zhang, H.; Li, T.; Chen, K.; Qiu, J.L.; Gao, C. Perfectly matched 20-nucleotide guide RNA sequences enable robust genome editing using high-fidelity SpCas9 nucleases. *Genome Biol.* 2017, *18*, 191.

109. Hu, J.H.; Miller, S.M.; Geurts, M.H.; Tang, W.; Chen, L.; Sun, N.; Zeina, C.M.; Gao, X.; Rees, H.A.; Lin, Z.; et al. Evolved Cas9 variants with broad PAM compatibility and high DNA specificity. *Nature* 2018, *556*, 57–63.

110. Casini, A.; Olivieri, M.; Petris, G.; Bianchi, A.; Montagna, C.; Reginato, G.; Maule, G.; Lorenzin, F.; Prandi, D.; Romanel, A.; et al. Evocas9, a highly specific SpCas9 variant from a yeast in vivo screening. *Nat. Biotechnol.* 2018, *36*, 265–271.

111. Lee, J.K.; Jeong, E.; Lee, J.; Jung, M.; Shin, E.; Kim, Y.-H.; Lee, K.; Jung, I.; Kim, D.; Kim, S.; et al. Directed evolution of CRISPR-Cas9 to increase its specificity. *Nat. Commun.* 2018, *9*, 3048.

112. Wang, J.; Meng, X.; Hu, X.; Sun, T.; Li, J.; Wang, K.; Yu, H. xCas9 expands the scope of genome editing with reduced efficiency in rice. *Plant Biotechnol. J.* 2019, *17*, 709–711.

113. Wu, Y.; Xu, W.; Wang, F.; Zhao, S.; Yang, J. Increasing Cytosine Base Editing Scope and Efficiency With Engineered Cas9-PmCDA1 Fusions and the Modified sgRNA in RiceData_Sheet_1.docx. *Front. Genet.* 2019, *10*, 379.

114. Hou, Z.; Zhang, Y.; Propson, N.E.; Howden, S.E.; Chu, L.F.; Sontheimer, E.J.; Thomson, J.A. Efficient genome engineering in human pluripotent stem cells using Cas9 from Neisseria meningitidis. *Proc. Natl. Acad. Sci. USA* 2013, *110*, 15644–15649.

115. Ran, F.A.; Cong, L.; Yan, W.X.; Scott, D.A.; Gootenberg, J.S.; Kriz, A.J.; Zetsche, B.; Shalem, O.; Wu, X.; Makarova, K.S.; et al. In vivo genome editing using Staphylococcus aureus Cas9. *Nature* 2015, *520*, 186–191.

116. Müller, M.; Lee, C.M.; Gasiunas, G.; Davis, T.H.; Cradick, T.J.; Siksnys, V.; Bao, G.; Cathomen, T.; Mussolino, C. Streptococcus thermophilus CRISPR-Cas9 Systems Enable Specific Editing of the Human Genome. *Mol. Ther.* 2016, *24*, 636–644.

117. Hirano, H.; Gootenberg, J.S.; Horii, T.; Abudayyeh, O.O.; Kimura, M.; Hsu, P.D.; Nakane, T.; Ishitani, R.; Hatada, I.; Zhang, F. Structure and Engineering of Francisella novicida Cas9. *Cell* 2016, *164*, 950–961.

118. Kim, E.; Koo, T.; Park, S.W.; Kim, D.; Kim, K.; Cho, H.Y.; Song, D.W.; Lee, K.J.; Jung, M.H.; Kim, S.; et al. In vivo genome editing with a small Cas9 orthologue derived from Campylobacter jejuni. *Nat. Commun.* 2017, *8*, 14500.

119. Steinert, J.; Schiml, S.; Fauser, F.; Puchta, H. Highly efficient heritable plant genome engineering using Cas9 orthologues from Streptococcus thermophilus and Staphylococcus aureus. *Plant J.* 2015, *84*, 1295–1305.

120. Abudayyeh, O.O.; Gootenberg, J.S.; Konermann, S.; Joung, J.; Slaymaker, I.M.; Cox, D.B.; Shmakov, S.; Makarova, K.S.; Semenova, E.; Minakhin, L.; et al. C2c2 is a single-component programmable RNA-guided RNA-targeting CRISPR effector. *Science* 2016, *353*, aaf5573.

121. Cox, D.B.T.; Gootenberg, J.S.; Abudayyeh, O.O.; Franklin, B.; Kellner, M.J.; Joung, J.; Zhang, F. RNA editing with CRISPR-Cas13. *Science* 2017, *358*, 1019–1027.

122. Aman, R.; Ali, Z.; Butt, H.; Mahas, A.; Aljedaani, F.; Khan, M.Z.; Ding, S.; Mahfouz, M. RNA virus interference via CRISPR/Cas13a system in plants. *Genome Biol.* 2018, *19*, 1–9.

123. Harrington, L.B.; Burstein, D.; Chen, J.S.; Paez-Espino, D.; Ma, E.; Witte, I.P.; Cofsky, J.C.; Kyrpides, N.C.; Banfield, J.F.; Doudna, J.A. Programmed DNA destruction by miniature CRISPR-Cas14 enzymes. *Science* 2018, *362*, 839–842.

124. Yan, W.X.; Hunnewell, P.; Alfonse, L.E.; Carte, J.M.; Keston-Smith, E.; Sothiselvam, S.; Garrity, A.J.; Chong, S.; Makarova, K.S.; Koonin, E.V.; et al. Functionally diverse type V CRISPR-Cas systems. *Science* 2019, *363*, 88.

125. Kim, H.; Kim, J.S. A guide to genome engineering with programmable nucleases. *Nat. Rev. Genet.* 2014, *15*, 321–334.

126. Globus, R.; Qimron, U. A technological and regulatory outlook on CRISPR crop editing. *J. Cell Biochem.* 2018, *119*, 1291–1298.

127. USDA Re: Confirmation of Regulatory Status of Waxy Com Developed by CRISPR-Cas Technology. Available online: **https://www.aphis.usda.gov/biotechnology/downloads/reg_loi/15-352-01_air_response_signed.pdf** (accessed on 2 October 2020).

128. USDA Re: Request for Confirmation that Transgene-Free, CRISPR-Edited Mushroom Is Not a Regulated Article. Available online: **https://www.aphis.usda.gov/biotechnology/downloads/reg_loi/15-321-01_air_response_signed.pdf** (accessed on 2 October 2020).

129. Callaway, E. CRISPR plants now subject to tough GM laws in European Union. *Nature* 2018, *560*, 16.

130. Mallapaty, S. Australian gene-editing rules adopt 'middle ground'. *Nature* 2019.

131. Metje-Sprink, J.; Sprink, T.; Hartung, F. Genome-edited plants in the field. *Curr. Opin. Biotechnol.* 2020, *61*, 1–6.

132. Zhu, H.; Li, C.; Gao, C. Applications of CRISPR-Cas in agriculture and plant biotechnology. *Nat. Rev. Mol. Cell Biol.* 2020, *21*, 661–677.

133. Li, Q.; Sapkota, M.; van der Knaap, E. Perspectives of CRISPR/Cas-mediated cis-engineering in horticulture: Unlocking the neglected potential for crop improvement. *Hortic. Res.* 2020, *7*, 36.

134. Schmitz, D.J.; Ali, Z.; Wang, C.; Aljedaani, F.; Hooykaas, P.J.J.; Mahfouz, M.; de Pater, S. CRISPR/Cas9 Mutagenesis by Translocation of Cas9 Protein Into Plant Cells via the Agrobacterium Type IV Secretion System. *Front. Genome Ed.* 2020, *2*, 6.

135. Schmidt, C.; Pacher, M.; Puchta, H. Efficient induction of heritable inversions in plant genomes using the CRISPR/Cas system. *Plant J.* 2019, *98*, 577–589.

136. Beying, N.; Schmidt, C.; Pacher, M.; Houben, A.; Puchta, H. CRISPR–Cas9-mediated induction of heritable chromosomal translocations in Arabidopsis. *Nat. Plants* 2020, *6*, 638–645.

137. Chuang, Y.F.; Phipps, A.J.; Lin, F.L.; Hecht, V.; Hewitt, A.W.; Wang, P.Y.; Liu, G.S. Approach for in vivo delivery of CRISPR/Cas system: A recent update and future prospect. *Cell Mol. Life Sci.* 2021, *78*, 2683–2708.

138. Lowe, K.; Wu, E.; Wang, N.; Hoerster, G.; Hastings, C.; Cho, M.J.; Scelonge, C.; Lenderts, B.; Chamberlin, M.; Cushatt, J.; et al. Morphogenic Regulators Baby boom and Wuschel Improve Monocot Transformation. *Plant Cell* 2016, *28*, 1998–2015.

139. Ali, Z.; Abul-faraj, A.; Li, L.; Ghosh, N.; Piatek, M.; Mahjoub, A.; Aouida, M.; Piatek, A.; Baltes, N.J.; Voytas, D.F.; et al. Efficient Virus-Mediated Genome Editing in Plants Using the CRISPR/Cas9 System. *Mol. Plant* 2015, *8*, 1288–1291.

4 Chapter

IDENTIFICATION OF KEY GENETIC REGIONS AND POTENTIAL GENES RESPONSIBLE FOR THE PRODUCTION OF ANTHOCYANIN IN BROCCOLI THROUGH THE USE OF QTL-SEQ

1. INTRODUCTION: AN OVERVIEW

Broccoli, scientifically known as Brassica oleracea var. italic, is a well-liked vegetable that is distinct from other Brassica species such as Chinese cabbage, turnip, cabbage, cauliflower, and oilseed rape. Broccoli, in its various forms, is mostly derived from wild plants that are cultivated in the Mediterranean Basin. These plants are often grown as annuals and yield a sizable edible head consisting of florets, buds, leaves, stalks, and stems. Both broccoli and cauliflower cultivar groupings belong to the CC genome B. oleracea (2n = 18) coenospecies. There have been reports of cauliflower having high-quality reference genomes, with the assembled genome being 584.60 Mb in size [1]. Broccoli is a highly nutritious food that contains important vitamins and minerals. It also contains antioxidant phytochemicals, like glucoraphanin, which have the potential to prevent cancer [2]. Purple broccoli is gaining popularity as a functional food due to its attractive look and significant health benefits [3]. Anthocyanin buildup has been established as the cause of the purple coloration [4].

Anthocyanins are a type of flavonoid molecules that give color to plants and have a significant role in protecting plants from various biotic and abiotic challenges [5,6,7]. The genes involved in the biosynthesis of anthocyanin have been thoroughly studied in several plant species including Arabidopsis (Arabidopsis thaliana), maize (Zea mays), tomato (Solanum lycopersicum), eggplant (Solanum melongena), and others [8,9,10,11]. An important strategy for regulating anthocyanin production in the Brassica species is the activation of structural genes and transcription factors [12,13,14]. The cauliflower had a particular pattern of excessive anthocyanin synthesis in certain tissues due to a purple mutation [15]. The Harbinger transposon insertion leads to the activation of BoMYB2, which particularly upregulates BobHLHs and some downstream anthocyanin structural genes. This results in the abnormal buildup of anthocyanin. The activation of TT8 and MYB2 genes in red cabbages leads to the accumulation of anthocyanin [14]. Similarly, in a purple head of Chinese cabbage cultivar 11S91, the activation of BrMYB2 gene controls the expression of downstream genes such as BrTT8, BrF3'H, BrDFR1, BrANS1, BrUGTs, BrATs, and BrGSTs [16]. The purple hue in cabbages is exclusively attributed to the deletion or substitution of nucleotides in the exon of BoMYBL2-1 [17]. Furthermore, temperature and light exert significant influence on the accumulation of anthocyanin, making them the primary environmental influences. BrMYB2 and BrTT8 stimulated the activation of anthocyanin structural genes in purple head Chinese cabbage following exposure to low temperatures [18]. Increased temperature can inhibit the accumulation of anthocyanin through the signaling pathway involving COP1-HY5, and the down-regulation of MYBL2 largely controls the suppression of anthocyanin synthesis associated with high temperatures [19]. Additionally, several endeavors have resulted in the discovery of potential genes that control the pigmentation of Brassica species. Three quantitative trait loci (QTLs) have been genetically related to the purple sepal characteristic of the flower head on chromosome C01 in broccoli [20]. The genes responsible for pink and purple leaf colorations in ornamental kale have been precisely located on chromosomes C3 [21] and C9 [22], respectively. The stalk color trait in Zicaitai is strongly influenced by a key locus on chromosome 7, which is tightly regulated. This locus shows a significant correlation with the expression of bHLH49 in the F2 population [23]. The purple inner leaf phenotype in Chinese cabbage is controlled exclusively by the dominant gene BrPur, which has been located on chromosome A07, specifically between SSR markers A710 and A714. The genetic distance between these markers and the gene is 3.1 and 3.5 cM, respectively [24].

QTLs, which play a crucial role in crop breeding, are responsible for controlling most agronomic traits. They are essential for improvement in crop breeding through the use of marker-assisted selection (MAS). The conventional approach of QTL mapping is linkage mapping, which is arduous and time-consuming. Next-generation sequencing (NGS) is now the preferred method for identifying connections between agronomic parameters and biomarkers or genes. The effectiveness of the QTL-seq approach, which integrates bulk segregant analysis (BSA) with next-generation sequencing (NGS), in mapping and isolating quantitative trait loci (QTLs) has been established [25]. QTL-seq is conducted using two distinct groups of individual plants that exhibit opposing phenotypes for a specific feature of interest. These plants are selected from a segregating population, which can include F2 recombinant inbred lines, double haploid lines, or backcross populations. Using high-throughput SNP (Hi-SNP) technology, the genotype analysis of two mixed pools allows for the identification of the genomic position of the polymorphic molecular markers. Additionally, it helps identify the key QTL region where there is a considerable segregation of genotypes. Hi-SNP is a method used for genotyping huge numbers of single nucleotide polymorphisms (SNPs). It involves the use of multiplex-PCR, next generation sequencing, and bioinformatics tools. Amplicon sequencing, which involves the use of multiplex-PCR and NGS at a greater depth and with a lower false discovery rate, has proven to be more accurate than whole-genome resequencing (WGS). This technique has been employed for genotyping known SNPs in diploid crops, but its application in allopolyploid crops has not been documented. The QTL-seq method, used for SNP genotyping, offers several advantages including simple primer design, short reads sequences, high-throughput and high-depth sequencing. Additionally, it is easy to automate and process. For instance, this method has been recently employed in mapping QTLs associated with resistance to rice blast disease and seedling vigor [26], leaf spot resistance in peanuts [27,28], heat tolerance in broccoli [29], resistance to Fusarium oxysporum f. sp. niveum race 1 in watermelon [30], cucumber early flowering [31], and tomato fruit weight and lobule amounts [32].

The phenomenon of mutant analyses, resulting in the development of purple or red organs, has been thoroughly investigated in flowers, fruits, and model plants [15]. While there is much research on the fundamental processes involved in anthocyanin production in cauliflower, cabbage, kohlrabi, kale, and Chinese kale, limited attention has been given to broccoli. The purple broccoli mutant is a fascinating variation characterized by the buildup of anthocyanins in its flower buds, resulting in the mutant heads displaying a purple coloring. The genetic regulation of the purple mutation in broccoli was determined to be governed by Quantitative Trait Loci (QTLs). The symbol "Pur" is used to represent the Purple allele. This study utilized QTL-seq to specifically identify quantitative trait loci (QTLs) associated with the production of anthocyanin in the F2 population. The F2 population was generated by crossing the purple broccoli line BT 126 with SN 60, a cultivar characterized by green heads.

2. MATERIALS AND METHODS

2.1. Plant Materials

This study utilized two broccoli inbred lines, namely BT 126 (characterized by purple heads) and SN 60 (characterized by green heads), as parental lines. In order to create populations that exhibit separation in anthocyanin production, a cross was made between the purple-headed plant BT 126 and the green-headed plant SN 60 (Figure 1). The F1 plants underwent self-pollination, resulting in the production of two F2 populations, one with 302 individuals and the other with 280 individuals.

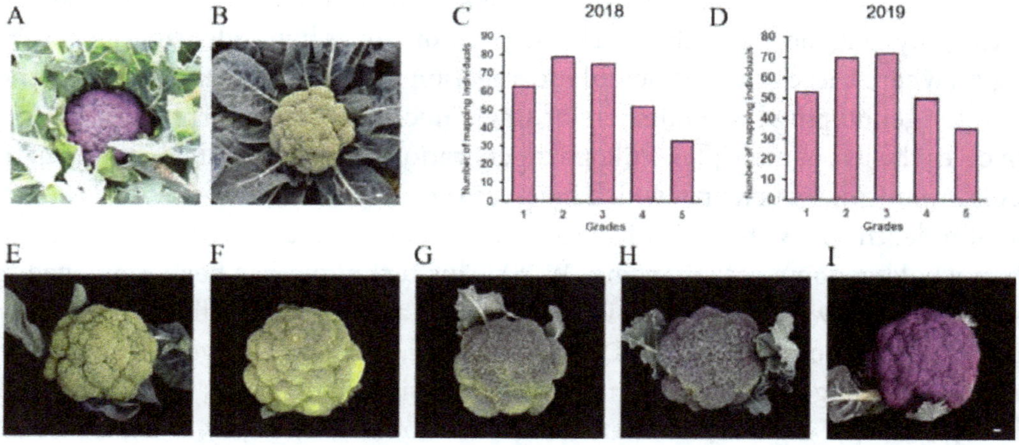

Figure 1. Phenotypes of the parents and F2 individuals and their frequency distributions. (A) Maternal line BT 126; (B) paternal line, SN 60; (C,D) the frequency distribution of the purple head trait in the F2 populations with 302 and 280 individuals planted in 2018 and 2019, respectively; (E–I) phenotype of two F2 populations with 1st to 5th grades of head color. The DNAs of 30 F2 individuals with extreme phenotypes (1st and 5th grade) were used to develop high anthocyanin and low anthocyanin bulks; scale bar = 1 cm.

Initially, a population of 302 F2 individuals, along with 10 plants from each parental line, were grown at Zhuanghang Experimental Base of Shanghai Academy of Agricultural Sciences, Shanghai in 2018. The phenotyping of each F2 individual was carried out on at least three separate days, after the flower head reached maturity in the field. Curd color, which showed a color distribution from green to purple, was visually scored from 1 to 5: 1–green, 2–slight purple, 3–light purple, 4–slightly darker purple, and 5–purple (**Figure 1**E–I). The total anthocyanin content has been determined by the high-performance liquid chromatography (HPLC) method at six biological replicates. The BT 126 is a purple broccoli variety, with an average content of anthocyanin 5.72 mg/g FW. It was crossed to SN 60, with very low anthocyanin content (0.64 mg/g FW), to generate F1 progeny. The average contents of anthocyanin, for the 5 phenotypic categories, were 5.38, 3.26, 2.29, 1.27, and 0.78 mg/g FW. All the above information was increased in the revised manuscript. Subsequently, two populations of 280 F2 and 580 F3 individuals were grown under routine management at Zhuanghang Experimental Base of Shanghai Academy of Agricultural Sciences, Shanghai in 2019 and 2020, respectively. We started the broccoli seeds in the plastic tunnel in August. Four weeks after germination, we transplanted them to field plots. The head color of each F2 and

F3 plants was visually phenotyped in 60 days, when grown from transplants, when the heads usually had a diameter larger than 2 ¼ inches. To avoid the sunlight and temperature effects, all the plants were grown in the same plastic tunnel. Leaf tissues were harvested and stored at −20 °C for DNA extraction. The head tissues of both parents (BT 126 and SN 60), at the full-size, mature stage, were collected from the same site of the top head, at three biological replicates, and used for RT-qPCR analysis. The anthocyanins in the broccoli heads were isolated and assessed, according to the method proposed by Liu C. et al. [**4**].

All materials were obtained from the Institute of Horticulture, Shanghai Academy of Agricultural Sciences.

2.2. DNA Purification, Library Generation, and Whole-Genome ReSequencing

Total DNA was extracted from fresh leaves using the CTAB method [**35**]. A total of 1.5 μg DNA, per specimen, was utilized for DNA sample preparations. The DNA quality was assessed by 1% agarose gel electrophoresis. Among the F2 segregating population, composed of 302 plants, 30 with purple heads (high anthocyanin biosynthesis, HAB) and 30 with green heads (low anthocyanin biosynthesis, LAB) were selected to extract equal amount of DNA, pooled to construct the DNA bulks. The DNA bulks, together with the two parental DNA, were used for whole-genome resequencing.

By ultrasonication, 350-bp fragments were obtained from the tested DNA sample. A sequencing library was generated using the Truseq Nano DNA HT Sample Preparation Kit (Illumina, San Diego, CA, USA), according to the manufacturer's instructions. The constructed library was sequenced on the Illumina HiSeq4000 platform (Illumina, CA, USA), and 150-bp paired-end reads were produced with approximately 350-bp inserts. Stringent quality control (QC) steps were applied to ensure the reliability and accuracy of the reads. Then, the filtered, high-quality sequences from the DNA bulks and parental genotypes were aligned and mapped to the public *B. oleracea* genome database (TO1000) (**http://plants.ensembl.org/Brassica_oleracea**) (accessed on 5 May 2021) with the Burrows-Wheeler alignment (BWA) tool [**36,37**]. Alignment files were converted into BAM files with SAMtools (Wellcome Trust Genome Campus, Cambridge, UK) [**38**]. SNP detection was carried out for each specimen, utilizing the UnifiedGenotyper function in the GATK3.8 software (The Broad Institute of Harvard and MIT, Cambridge, MA, USA) [**39**]. ANNOVAR (Children's Hospital of Philadelphia, Philadelphia, PA, USA) was employed for SNP annotation, based on the GFF3 file of the reference genome [**40**].

2.3. QTL-Seq Analysis

By applying an established QTL-seq method, based on SNP-index and Δ(SNP-index) estimates, candidate genomic regions harboring the main QTL(s) involved in anthocyanin biosynthesis in broccoli were identified. Only the SNPs homozygous in either parent and polymorphic between the parents were prepared for the further analysis. Then 1,117,709 SNPs, between both parents, were selected. The read depth for the above-mentioned homozygous SNPs, in LAB and HAB bulks, was obtained to evaluate the SNP-index. The SNP-index, calculated according to the reads order-checking depth information, utilized short reads covering the given nucleotide position, calculated the differs reads bar number, and accounted for the ratio of the total number [**26**]. The genotype of one parent was utilized as a reference to determine the number of reads for the parental genotype, or other genotypes, in the LAB and HAB libraries. Then, the number of the various reads was divided by the total read number, and the ratio constituted the SNP-index of the base sites. The SNP-index points below 0.3, in both libraries, were removed. The sliding window method was utilized to present the SNP-indexes for the entire genome. All SNP-indexes in a given window were averaged to obtain the SNP-index for that particular window. The window size of 1Mb and the step size of 1Kb were routinely utilized. The Δ (SNP-index) was then calculated using the following formula: [SNP index (HAB bulk)—SNP index (LAB bulk)].

2.4. High-Throughput Genotyping of SNPs in Intra-Specific Mapping Individuals

In order to validate the *BoPur* QTL, obtained by QTL-seq, classical QTL mapping was performed, via the selection of the SNP, with a polymorphism between the genotypes of the parents (BT 126 and SN 60). Through the Hi-SNP method, a total of 33 SNPs, differentiating BT 126 and SN60, were utilized for validation and high-throughput genotyping in the F2 segregating population with 280 individuals and parents. The HI-SNP technology was developed by Shanghai BioWing Applied Biotechnology Company (**http://www.biowing.com.cn/**) (accessed on 9 June 2021) the genotyping of 33 SNPs was carried out by multiplex PCR with NGS on Illumina X-10 (Illumina, CA, USA) [**41**]. The genotyping primers were shown in the **Supplementary Table S1**. Based on the phenotypic and genotypic data, linkage maps were constructed by QTL IciMapMaker (Chinese Academy of Agricultural Sciences, Beijing, China) with a logarithm of the odds (LOD) value [**42**]. The LOD score is a measure of the strength of the evidence for the presence of a QTL at a particular location (the LOD score = log10 likelihood ratio, comparing single-QTL model to the "no QTL anywhere" mode).

2.5. Analysis of the Candidate Gene

To infer the gene regulation patterns of *BoF3'H* during anthocyanin biosynthesis, a qRT-PCR assay was performed. The RNA was isolated from BT 126 and SN60 at three head developmental stages, with both low and high anthocyanin contents. Reverse transcription was conducted using oligo-dT primers and PrimeScript™ RT Master Mix (Takara BIO, Inc., Shiga, Japan). The primer pair sequences for *BoF3'H*, *BoANS2*, and *BoTTG1* used in this study were previously designed for the RT-qPCR analysis. *BoLBD-38.3*specific primers were designed based on the sequences in NCBI (accession number: XM_013739916). The qPCR primer pair sequences for *BoLBD38.3* were -'5GCCCAAACGGAGACGATTAG-3' (forward) and 5'-AACCGTTCACTGGCGATGTG-3' (reverse), gene-specific primers that were confirmed to produce specific gene products by sequencing [**15**]. Real-time PCR was performed using TB Green® Premix Ex Taq™ (Tli RNaseH Plus) (Takara, Dalian, China) on an ABI QuantStudio 5 real-time PCR system(Thermo Fisher Scientific, MA, USA) according to the manufacturer's instructions. Actin was used as a reference gene [**43**]. The data were processed using the 2-ΔΔCt method [**44**]. Sequence data of *BoF3'H* can be found in the GenBank data libraries, under accession number XM_013751545.

3. RESULTS

3.1. Inheritance of Head Color of BT 126 with SN 60

Significant differences in head color were observed among the BT 126 (parent 1, P1, purple head), SN 60 (parent 2, P2, green head), and F1 hybrids planted in 2018 and 2019 (**Figure 1**). The BT 126 purple line exhibited a high anthocyanin accumulation in the heads (5.72 mg/g FW) and developed normally when compared to green broccoli, under normal conditions, in the field and greenhouse [**4**]. It was crossed to SN 60, with very low anthocyanin content (0.241 mg/g FW), to generate the F1 progeny. F1 individuals displayed third grade color. As expected, the F2 population was divided into purple and green classes, with varying degrees of continuous distribution in 1st-5th color grades (**Figure 1**E–I). No transgressive segregation was observed in either direction of the parental genotype in the mapping population. We could infer that the purple trait was controlled by QTL.

3.2. QTL-Seq Analysis

The high-throughput, whole-genome resequencing of both parents, as well as the LAB and HAB libraries, yielded 158.49 to 261.57 million reads per sample. Based on the *B. oleracea* reference genome (estimated genome size~630 Mb), the read mapping ratio was >%93 (**Table 1**). The mean coverage of the reference genome, across all samples, was 33×. As depicted in **Table 1**, the 4× coverage of each specimen ranged from 86.85% to 92.43%. Meanwhile, the value of Q20 was above 93% in all specimens (data not shown). The comparative genome sequence analysis of the parental DNA and DNA bulks with the reference genome revealed 1,117,709 polymorphic SNPs.

Table 1. Whole-genome mapping statistics for parents, as well as LAB and HAB bulks.

Sample	Mapped Reads	Total Reads	Mapping Rate (%)	Average Depth (X)	Coverage 1× (%)	Coverage 4× (%)
SN 60	245,092,342	261,565,560	93.7	41.81	92.33	89.41
BT 126	148,780,556	158,487,152	93.88	26.2	91.51	86.85
LAB	218,978,598	232,601,048	94.14	34.9	95.03	92.43
HAB	192,459,282	203,800,728	94.44	31.13	94.77	91.78

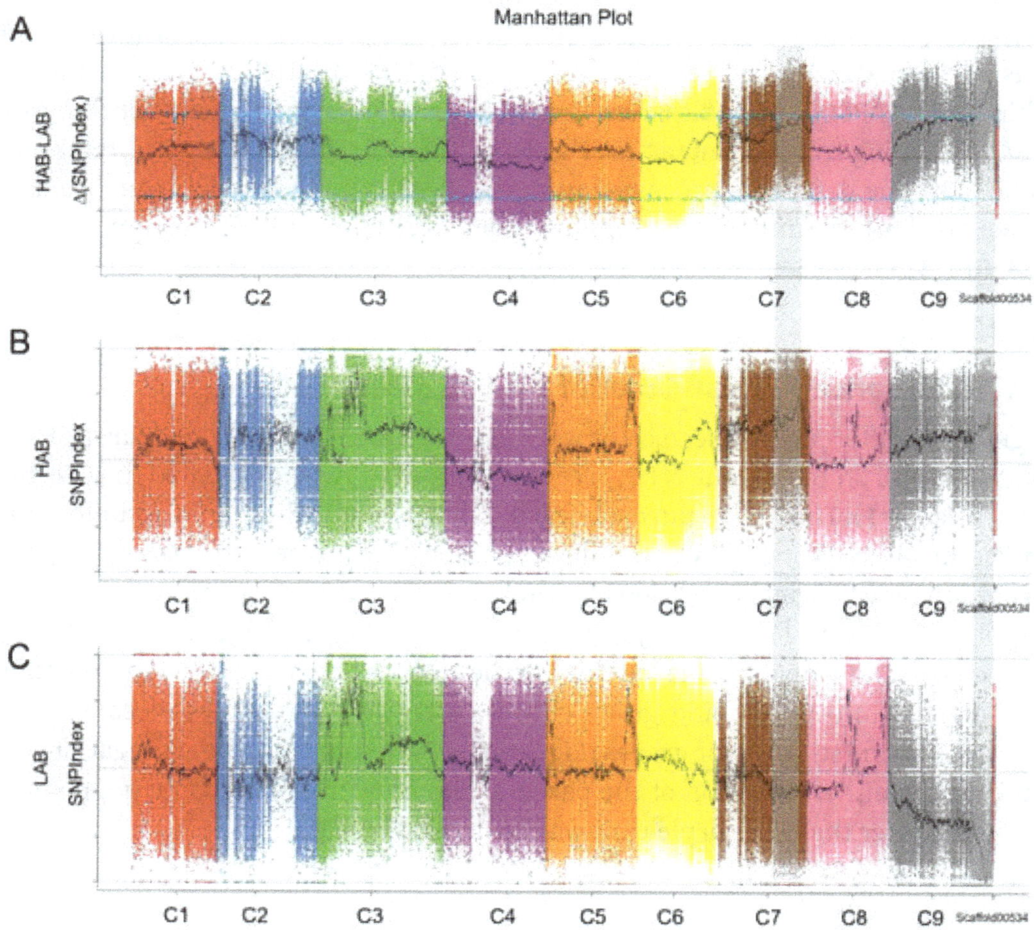

Figure 2. Distribution of two progeny SNP-index on the chromosome length, represented by the horizontal axis (Mb); vertical axis: SNP-index. (A) Δ (SNP-index) graphs generated from QTL-seq study; (B) SNP-index graphs depicting the HAB (high anthocyanin biosynthesis bulk); (C) SNP-index graphs depicting the LAB (low anthocyanin biosynthesis bulk). Gray shaded boxes indicate significant QTLs.

The SNP-index of each SNP, that distinguished LAB from HAB, was determined. The mean SNP-index, covering 1 Mb genomic sequence, was assessed separately in LAB and HAB, by the 1 kb sliding window method, and mapped against all *B. oleracea* reference chromosomes (9 in total). The Δ (SNP-index) was determined by combining the SNP-index data of LAB and HAB and was plotted against the positions (Mb) on the *B. oleracea* genome. The QTL-seq analysis detected two major genomic regions on chromosomes 7 (36,784,249–44,791,849) (*BoPur7.1*) and 9 (46,217,406–52,600,419) (*BoPur9.1*), with significant associations with anthocyanin biosynthesis in broccoli (**Figure 2**). Moreover, LAB and HAB mapping plants with reduced or elevated anthocyanin biosynthesis had the majority of SNP alleles from SNP 1 to SNP 26 (**Table S1**). An important genomic region [BoSNP5 (46,217,406 bp) to BoSNP 26 (52,600,419 bp)] harboring the QTL *Pur* on chromosome 9 showed a Δ (SNP-index) that was markedly different from 0 ($p < 0.05$) (**Figure 2**A and **Figure 3**B). The QTL-seq data confirmed an important QTL (*BoPur9.1*) located at the 6.38 Mb genomic interval [46,217,406 (BoSNP 5) to 52,600,419 (BoSNP 26) bp] on chromosome 9, which was associated with the regulation of anthocyanin biosynthesis in broccoli.

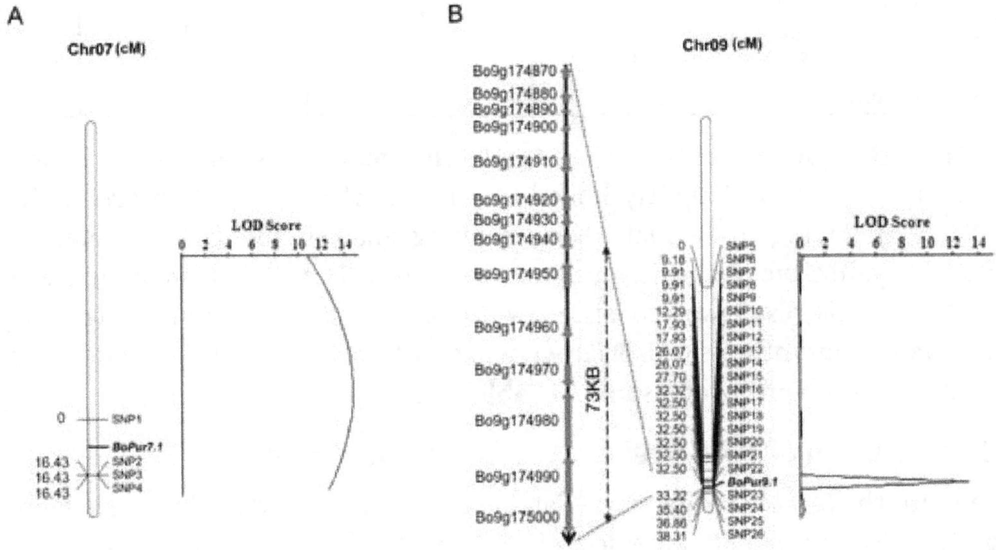

Figure 3. Maps of the Pur loci on C07 (A) and C09 (B) constructed using 26 SNP markers.

3.3. Validation of QTL-Seq-Derived Anthocyanin Biosynthesis QTL through High-Throughput SNP

To assess *BoPur7.1* and *BoPur9.1*, detected by QTL-seq, a high-throughput SNP was carried out. The genotyping data of 26 SNP markers were mapped to a highly dense intraspecific genetic region of chromosomes 7 and 9, depicting polymorphisms between parental genotypes and between LAB and HAB, and were combined with phenotypic features of a second F2 mapping population with 280 individuals. Classical QTL analysis, based on interval mapping and composite interval mapping, revealed two highly important genomic regions: [BoSNP37 (32.5 cM) to BoSNP38 (33.22 cM)], harboring a potent (LOD: 13.1) *Pur* QTL (*BoPur9.1*) on broccoli chromosome 9, and [BoSNP1 (0 cM) to BoSNP2 (16.43 cM)], harboring a potent (LOD: 14.6) *Pur* QTL (*BoPur7.1*) on broccoli chromosome 7. The detected QTLs had an interval of 0.72 cM with 72,683 bp [BoSNP22 (51,716,244 bp) to BoSNP23 (51,788,927 bp)] on chromosome 9 and an interval of 16.43 cM with 6,920,903 bp [BoSNP1 (36,784,249 bp) to BoSNP2 (43,705,152 bp)] on chromosome 7 (**Figure 3**). The proportions of phenotypic variants caused by the *BoPur9.1* and *BoPur7.1* QTLs were %28.19 and %38.12, respectively (**Table 2**).

Table 2. **Quantitative trait loci ($p < 0.05$) associated with anthocyanin biosynthesis in broccoli.**

Chromosome	Start	End	LOD	PVE (%)	Add	Dom
7	SNP1	SNP2	14.6193	38.1207	-0.8350	-0.4366
9	SNP22	SNP23	13.1231	28.1935	0.0150	1.0399
LOD-logarithm of odds; PVE-phenotypic variance explained; Add-additive effect; Dom-dominance effect.						

To further delineate the *BoPur9.1* QTL, the genotyping data of 17 selected SNP markers, the most tightly linked to *BoPur9.1* (SNP10–SNP26), were analyzed in 580 F3 mapping individuals. Loci mapping analysis identified a major QTL for anthocyanin biosynthesis, designed by two SNP markers, including SNP22 and SNP23. This result was consistent with the QTL analysis supporting a major anthocyanin biosynthesis QTL *BoPur9.1*, which was located in the genomic interval of 51.71–51.79 Mb on chromosome 9 (**Table S2**).

3.4. Prediction and Analysis of the Candidate Gene

Based on the *B. oleracea* genome databases (TO1000) (**http://plants.ensembl.org/Brassica_oleracea**) (accessed on 5 May 2021), a total of 3 anthocyanin-related and 14 predicted protein-coding genes were found in the interval of *BoPur7.1* and *BoPur9.1*, respectively (**Table 3**). According to domain annotations from InterPro and BLASTX (best hit) analyses, two of these fourteen genes have not been annotated (**Table 3**). The other fifteen candidate genes were as follows: *Bo7g096780* (homologous gene AT5G24520), encoding the transparent testa glabra 1; *Bo7g099150* (homologous gene AT3G49940), encoding the LOB domain-containing protein 38; *Bo7g108300* (homologous gene AT4G22880), encoding the anthocyanin synthase; *Bo9g174870* (homologous gene AT5G08000), encoding the X-8GPI family of proteins; *Bo9g174880*, *Bo9g174890*, and *Bo9g174900* (homologous gene AT5G07990), encoding flavonoid '3 hydroxylase (F3'H), which catalyzed the conversion of dihydrokaempferol to dihydroquercetin in anthocyanin biosynthesis [45]; *Bo9g174920* (homologous gene AT5G07960), encoding the Asterix-like protein; *Bo9g174940* (homologous gene AT5G07890), encoding the myosin heavy chain-like protein; *Bo9g174950* (homologous gene AT5G07830), encoding the glucuronidase 2; *Bo9g174960* (homologous gene AT5G07820), encoding the calmodulin-binding,protein-likeprotein; *Bo9g174970* (homologousgeneAT5G07800), encoding the flavin-containing monooxygenase; and *Bo9g174980*, *Bo9g174990*, and *Bo9g175000* (homologous gene AT5G07740), encoding the actin-binding protein. *BoF3'H* was annotated as a homologue of the anthocyanin biosynthesis gene.

Table 3. Annotation of *B. oleracea* genes in the candidate region.

Bo Genes	Chromosome	Gene Position (bp)	AT ID [a]	E-Value	AT GO [b] Annotation
Bo7g096780	C7	37466680–37466680	AT5G24520	0.0	TRANSPARENT TESTA GLABRA 1
Bo7g099150	C7	38873553–38874347	AT3G49940	0.0	LOB domain-containing protein 38
Bo7g108300	C7	42567442–42568605	AT4G22880	0.0	anthocyanin synthase
Bo9g174870	C9	51722173–51723584	AT5G08000	1×10^{-127}	X8-GPI family of proteins
Bo9g174880	C9	51725909–51727132	AT5G07990	0.0	flavonoid 3′ hydroxylase activity
Bo9g174890	C9	51728802–51729245	AT5G07990	0.0	flavonoid 3′ hydroxylase activity
Bo9g174900	C9	51731919–51732761	AT5G07990	0.0	flavonoid 3′ hydroxylase activity
Bo9g174910	C9	51736550–51738488	-	-	-
Bo9g174920	C9	51742405–51744195	AT5G07960	1×10^{-128}	Asterix-like protein
Bo9g174930	C9	51744445–51745784	-	-	-

Bo Genes	Chromosome	Gene Position (bp)	AT ID [a]	E-Value	A T GO [b] Annotation
Bo9g174940	C9	51747328–51749122	AT5G07890	0.0	myosin heavy, chain-like protein
Bo9g174950	C9	51751745–51754512	AT5G07830	0.0	glucuronidase 2
Bo9g174960	C9	51758770–51760263	AT5G07820	0.0	calmodulin-binding, protein-like protein
Bo9g174970	C9	51764406–51766923	AT5G07800	0.0	Flavin-containing monooxygenase
Bo9g174980	C9	51767719–51777672	AT5G07740	0.0	actin-binding protein
Bo9g174990	C9	51779465–51787686	AT5G07740	0.0	actin-binding protein
Bo9g175000	C9	51788944–51794354	AT5G07740	0.0	actin-binding protein
[a] The best hits of the seven *B. oleracea* genes compared to *A. thaliana* (AT). [b] GO annotations for seven Bo to AT best-hit genes obtained from TAIR.					

A primer pair, spanning the full-length CDS of *BoF3'H* was designed, and PCR was performed using the cDNA of BT 126 and SN 60 as a template. DNA sequencing revealed that the full-length of *BoF3'H* in purple-head SN 60 is 1669 bp, whereas it was 1600 bp in green-head SN 60. Compared with *BoF3'H* in BT 126, a deletion of 68 bp and 1bp was found at nucleotide 1080 and 1501, respectively, and nine SNPs were present in SN60 (sequences of BT 126 and SN60 were illustrated in **Table S3**). The polymorphism of the candidate gene BoF3'H was further confirmed in the segregating population. To delineate the potential candidate genes regulating the anthocyanin biosynthesis in broccoli, differ-ential expression profiling of *BoF3'H* was performed in two parental lines, at three head developmental stages. The expression of *BoF3'H*, *BoANS2*, and *BoLED38* in BT 126 were up-regulated at three head developmental stages of anthocyanin accumulation, and the expression levels were significantly higher than those in SN60 (**Figure 4**B–F).

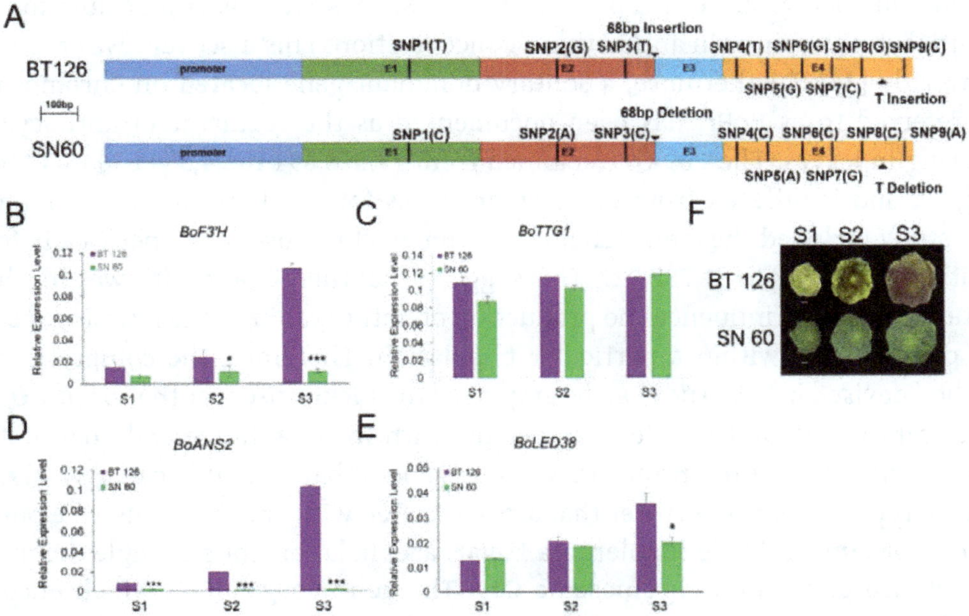

Figure 4. Structure identity and expression of broccoli gene *BoF3'H*. (A) Comparison of *BoF3'H* promoter and cDNA sequence between parental lines BT 126 and SN 60; (B–E) relative expression of *BoF3'H*, *BoTTG1*, *BoANS2*, and *BoLED38* at three head developmental stages. Significant differences between BT 126 and SN 60 are indicated by * based on Students *t* test (* $p < 0.05$, ** $p < 0.01$; *** $p < 0.001$); (F) the phenotype of the head during three developmental stages. The upper part depicts the head of BT 126, while the lower part depicts the head of SN 60. Scale bar = 1 cm.

4. DISCUSSION

A significant objective in nutriment breeding is to enhance the anthocyanin content in Brassica crops. This study identified two significant genomic areas, BoPur7.1 and Bopur9.1, located on chromosomes 7 and 9, respectively. These regions were shown to include QTLs (quantitative trait loci) associated with anthocyanin production. The identification was done using high-throughput QTL-seq, which is a whole-genome next-generation sequencing (NGS) approach, and intra-specific broccoli mapping individuals. The two Pur QTLs obtained using QTL-seq were then confirmed using Hi-SNP analysis. Prior research has indicated the presence of a dominant gene that regulates the production of anthocyanin coloring. The Pur gene, which encodes the transcription factor R2R3 MYB, was identified in purple cauliflower (B. oleracea var botrytis) [43]. The regulation of the purple sepal characteristic in the broccoli flower head is controlled by a primary locus and two secondary loci on chromosome C01 [20]. The purple gene in non-heading Chinese cabbage exhibits a single dominant inheritance pattern, although it does not adhere to Mendel's law [33]. A new genetic location, known as a locus, was identified in the R07 linkage group of purple turnip plants. This locus is responsible for the significant differences in anthocyanin concentration. This discovery was reported in a study [46]. Furthermore, a solitary dominant gene located on chromosome 09, referred to as BoPr, has been documented as the regulator of anthocyanin coloring in leaves. The C09 QTL spanned from 19,018,694 to 24,359,626 (5.3 Mb in length), and it differed from our primary QTLs for head anthocyanin contents. This study included the main genetic location on chromosome 9, specifically from position 51,716,244 to 51,788,927. This suggests that there is a specific way in which certain genes that influence the production of anthocyanin in the Brassica species are passed down within a particular population. Therefore, the comprehensive method devised in this study can be employed to quickly discover the specific QTLs, significant genes, and/or alleles associated with the qualitative and quantitative characteristics of different crops. Various genetic backgrounds can give rise to diverse types of purple features that are associated with anthocyanins. The purple gene in ornamental kale (B. oleracea L. var. acephala) exhibits a single dominant inheritance pattern on chromosome C09. The gene Bo9g058630, which encodes dihydroflavonol 4-reductase (DFR), has been identified as a potential candidate gene [22].

Combining QTL-seq with Hi-SNP analysis, the 0.73 Mb QTL *Bopur9.1* region [BoSNP51,716,244) 37 bp) to BoSNP51,788,927) 38 bp)], encompassing 14 genes, was detected on chromosome 9, which accounted for ~28.19% of all phenotypic variations in anthocyanin biosynthesis. Among these genes, according to their respective annotations (**Table 3**), three genes, *Bo9g174880*, *Bo9g174890*, and *Bo9g174900* were homologues of Arabidopsis *F3'H*, encoding flavanone 3'-hydroxylase. The *BoF3'H* gene has been suggested to have played a critical role in modifying plant coloration [**45**], which contained a coding sequence length of 1536 bp and encoded a protein of 511 amino acids with four exons. All of the 3 candidate genes became aligned to the different regions of *BoF3'H*. An increased expression of *F3'H* was found to be responsible for anthocyanin production in a number of anthocyanin-accumulating mutants. For example, the ectopic expression of apple *MdF3'H* can increase the production of flavonols and cyanidin-based pigments in the Arabidopsis *tt7* mutant, under nitrogen pressure [**47**]. The *VvF3'H* gene was identified from grapevine, and its ectopic expression results in high accumulation levels of anthocyanin and flavonols in the petunia *ht1* mutant line [**48**]. *OsF3'H* editing in the Heugseonchal or Sinmyungheugchal variety, using the CRISPR/Cas9 system, might be responsible for ocher seeds with lower anthocyanin contents than wild-type black rice plants [**49**]. Two copies of the *F3'H* gene were isolated in the barley genome and exhibited a tissue-specific expression pattern of anthocyanin synthesis [**50**]. Furthermore, nine SNPs and a 68-bp insertion, between the 3rd and 4rd exon of *BoF3'H* CDS sequences, were found, which may cause a gain-of-function mutation (**Figure 4**A). The 68 bp InDel was in the open reading frame (between E3 and E4), as well as the T InDel in the 3′ end. Therefore, a frame shift is expected, resulting in a putative truncated or extended protein. In addition, since most enzymes had conserved regions, anthocyanin discoloration might occur, as a result of *BoF3'H* enzyme inactivity, due to one of the two isoforms. In addition, the 6.92 Mb QTL *Bopur7.1* region [BoSNP36,784,249) 1 bp) to BoSNP43,705,152) 2 bp)], encompassing 3 anthocyanin-related genes, was detected on chromosome 7, which accounted for ~38.12% of all phenotypic variations in anthocyanin biosynthesis. Among these genes, *BoANS2* and *BoLED38* were activated at three head developmental stages of anthocyanin accumulation [**12**,**13**]. However, further experiments involving transformation are needed to verify whether the function of this gene is responsible for the purple head in broccoli.

Most of the agronomic traits are controlled by multiple QTLs. The traditional QTL fine-mapping method usually requires several generations of backcrossing, screening of a large population for recombinants and exhaustive field phenotyping, which is both time-consuming and labor-intensive [**51**]. Although the genotyping-by-sequencing (GBS) approach has been shown to be an efficient way to develop high-resolution genetic mapping [**52**], the genotyping of a large population with genome-wide markers is expensive. It is important to note that the QTL-seq approach has been successfully used to conduct a BSA analysis, as well as deployed for the mapping of a segregated population of homozygous parental lines with opposite phenotypes. Using QTL-seq, only two samples need to be sequenced. The sequence coverage is decided by genome size and complexity, but as an example for broccoli, sufficient data could be generated by sequencing each bulk to 30× coverage. Further, it's an effective method to identify markers most tightly linked to the trait.

Two specific quantitative trait loci (QTLs) were found to be located on chromosome 7 and 9, respectively, in this investigation. The hue and dimensions of the leaf veins exhibited significant differences among the F2 offspring. The temperature condition also had an impact on the anthocyanin content of purple head BT 126. Hence, environmental influences may exert a significant influence on the process of anthocyanin synthesis. The genetic process of anthocyanin production is intricate. The candidate genes responsible for regulating the purple characteristic can vary across different genetic backgrounds and exhibit distinct geographic and temporal expression patterns. To summarize, the aforementioned results suggest that the integration of QTL-seq, Hi-SNP analysis, and differential gene expression profiling can be used to discover potential genes that control anthocyanin production inside significant QTL regions in broccoli. Bo9g174880, Bo9g174890, and Bo9g174900 were first identified as the most promising candidate genes at the BoPur locus. An examination of the putative gene homologs, found within the confidence intervals of both QTLs, might offer new understanding of the intricate mechanisms involved in anthocyanin production in broccoli.

AT A GLANCE

Anthocyanins are responsible for the pigmentation of flowers, fruits, and purple vegetables. They also serve as crucial antioxidants that promote good health. BT 126 is a broccoli variety that has a significant amount of anthocyanins, specifically 5.72 mg/g FW. The objective of this work was to identify the specific genetic regions, known as quantitative trait loci (QTLs), that are responsible for the production of anthocyanin in the F2 population (n = 302). This population was created by crossing BT 126 with SN 60, which does not contain anthocyanin. The process of resequencing the entire genome of 30 purple and 30 green bulk segregates identified approximately 1,117,709 single nucleotide polymorphisms (SNPs) in the genome of B. oleracea. Two quantitative trait loci (QTLs), which are strongly associated with the production of anthocyanin ($p < 0.05$), were identified on chromosomes 7 (BoPur7.1) and 9 (BoPur9.1). Using high-throughput genotyping of SNPs technology, we conducted a high-resolution mapping of BoPur9.1 in the F2 population (n = 280) and F3 population (n = 580). This mapping narrowed down the key anthocyanin biosynthesis QTL region to a physical distance of 73 kb, which contains 14 genes. Out of these genes, Bo9g174880, Bo9g174890, and Bo9g174900 exhibited significant similarity to AT5G07990, which is a gene responsible for producing flavonoid 3′ hydroxylase. This gene, AT5G07990, has been identified as a potential candidate gene for BoPur9.1. The level of BoF3'H expression in BT 126 was markedly greater than in SN60. The QTLs identified in this study have multiple biomarkers associated with them, which could be suitable targets for marker-assisted selection (MAS) in the enhancement of anthocyanin production in broccoli. The current study offered genetic insights into the creation of new crop types with enhanced health-promoting characteristics and greater visual appeal.

REFERENCES

1. Sun, D.; Wang, C.; Zhang, X.; Zhang, W.; Jiang, H.; Yao, X.; Liu, L.; Wen, Z.; Niu, G.; Shan, X. Draft genome sequence of cauliflower (*Brassica oleracea* L. var. botrytis) provides new insights into the C genome in Brassica species. *Hortic. Res.* 2019, *6*, 1–11.

2. Fahey, J.W.; Holtzclaw, W.D.; Wehage, S.L.; Wade, K.; Stephenson, K.K.; Talalay, P. Sulforaphane Bioavailability from Glucoraphanin-Rich Broccoli: Control by Active Endogenous Myrosinase. *PLoS ONE* 2015, *10*, e0140963.

3. Rodríguez-Hernández, M.d.C.; Moreno, D.A.; Carvajal, M.; García-Viguera, C.; Martínez-Ballesta, M.d.C. Natural antioxidants in purple sprouting broccoli under Mediterranean climate. *J. Food Sci.* 2012, *77*, C1058–C1063.

4. Liu, C.; Yao, X.; Li, G.; Huang, L.; Xie, Z. Transcriptomic profiling of purple broccoli reveals light-induced anthocyanin biosynthetic signaling and structural genes. *PeerJ* 2020, *8*, e8870.

5. Steyn, W.J.; Wand, S.J.E.; Holcroft, D.M.; Jacobs, G. Anthocyanins in vegetative tissues: A proposed unified function in photoprotection. *New Phytol.* 2002, *155*, 349–361.

6. Pourcel, L.; Routaboul, J.-M.; Cheynier, V.; Lepiniec, L.; Debeaujon, I. Flavonoid oxidation in plants: From biochemical properties to physiological functions. *Trends Plant Sci.* 2007, *12*, 29–36.

7. Lev-Yadun, S.; Gould, K.S. Role of Anthocyanins in Plant Defence. In *Anthocyanins*; Springer: Berlin, Germany, 2008; pp. 22–28.

8. Shi, M.-Z.; Xie, D.-Y. Biosynthesis and metabolic engineering of anthocyanins in Arabidopsis thaliana. *Recent Pat. Biotechnol.* 2014, *8*, 47–60.

9. Sun, C.; Deng, L.; Du, M.; Zhao, J.; Chen, Q.; Huang, T.; Jiang, H.; Li, C.-B.; Li, C. A Transcriptional Network Promotes Anthocyanin Biosynthesis in Tomato Flesh. *Mol. Plant* 2020, *13*, 42–58.

10. Liu, X.; Li, S.; Yang, W.; Mu, B.; Jiao, Y.; Zhou, X.; Zhang, C.; Fan, Y.; Chen, R. Synthesis of Seed-Specific Bidirectional Promoters for Metabolic Engineering of Anthocyanin-Rich Maize. *Plant Cell Physiol.* 2018, *59*, 1942–1955.

11. Li, J.; Ren, L.; Gao, Z.; Jiang, M.; Liu, Y.; Zhou, L.; He, Y.; Chen, H. Combined transcriptomic and proteomic analysis constructs a new model for light-induced anthocyanin biosynthesis in eggplant (*Solanum melongena* L.). *Plant Cell Environ.* 2017, *40*, 3069–3087.

12. Guo, N.; Cheng, F.; Wu, J.; Liu, B.; Zheng, S.; Liang, J.; Wang, X. Anthocyanin biosynthetic genes in Brassica rapa. *BMC Genom.* 2014, *15*, 426.

13. Goswami, G.; Nath, U.K.; Park, J.-I.; Hossain, M.R.; Biswas, M.K.; Kim, H.-T.; Kim, H.R.; Nou, I.-S. Transcriptional regulation of anthocyanin biosynthesis in a high-anthocyanin resynthesized Brassica napus cultivar. *J. Biol. Res. Thessalon.* 2018, *25*, 19.

14. Yuan, Y.; Chiu, L.-W.; Li, L. Transcriptional regulation of anthocyanin biosynthesis in red cabbage. *Planta* 2009, *230*, 1141–1153.

15. Chiu, L.-W.; Zhou, X.; Burke, S.; Wu, X.; Prior, R.L.; Li, L. The Purple Cauliflower Arises from Activation of a MYB Transcription Factor. *Plant Physiol.* 2010, *154*, 1470–1480.

16. He, Q.; Wu, J.; Xue, Y.; Zhao, W.; Li, R.; Zhang, L. The novel gene BrMYB2, located on chromosome A07, with a short intron 1 controls the purple-head trait of Chinese cabbage (*Brassica rapa* L.). *Hortic. Res.* 2020, *7*, 1–19.

17. Song, H.; Yi, H.; Lee, M.; Han, C.-T.; Lee, J.; Kim, H.; Park, J.-I.; Nou, I.-S.; Kim, S.-J.; Hur, Y. Purple Brassica oleracea var. capitata F. rubra is due to the loss of BoMYBL2–1 expression. *BMC Plant Biol.* 2018, *18*, 82.

18. He, Q.; Ren, Y.; Zhao, W.; Li, R.; Zhang, L. Low Temperature Promotes Anthocyanin Biosynthesis and Related Gene Expression in the Seedlings of Purple Head Chinese Cabbage (*Brassica rapa* L.). *Genes* 2020, *11*, 81.

19. Kim, S.; Hwang, G.; Lee, S.; Zhu, J.-Y.; Paik, I.; Nguyen, T.T.; Kim, J.; Oh, E. High Ambient Temperature Represses Anthocyanin Biosynthesis through Degradation of HY5. *Front. Plant Sci.* 2017, *8*, 1787.

20. Yu, H.; Wang, J.; Sheng, X.; Zhao, Z.; Shen, Y.; Branca, F.; Gu, H. Construction of a high-density genetic map and identification of loci controlling purple sepal trait of flower head in *Brassica oleracea* L. italica. *BMC Plant Biol.* 2019, *19*, 228.

21. Zhu, P.; Cheng, M.; Feng, X.; Xiong, Y.; Liu, C.; Kang, Y. Mapping of Pi, a gene conferring pink leaf in ornamental kale (*Brassica oleracea* L. var. acephala DC). *Euphytica* 2015, *207*, 377–385.

22. Liu, X.-P.; Gao, B.-Z.; Han, F.-Q.; Fang, Z.-Y.; Yang, L.-M.; Zhuang, M.; Lv, H.-H.; Liu, Y.-M.; Li, Z.-S.; Cai, C.-C.; et al. Genetics and fine mapping of a purple leaf gene, BoPr, in ornamental kale (*Brassica oleracea* L. var. acephala). *BMC Genom.* 2017, *18*, 230.

23. Li, G.-H.; Chen, H.-C.; Liu, J.-L.; Luo, W.-L.; Xie, D.-S.; Luo, S.-B.; Wu, T.-Q.; Akram, W.; Zhong, Y.-J. A high-density genetic map developed by specific-locus amplified fragment (SLAF) sequencing and

identification of a locus controlling anthocyanin pigmentation in stalk of Zicaitai (*Brassica rapa* L. ssp. chinensis var. purpurea). *BMC Genom.* 2019, *20*, 343.

24. Wu, J.; Zhao, J.; Qin, M.; Ren, Y.; Zhang, H.; Dai, Z.; Hao, L.; Zhang, L. Genetic Analysis and Mapping of the Purple Gene in Purple Heading Chinese Cabbage. *Hortic. Plant J.* 2016, *2*, 351–356.
25. Michelmore, R.W.; Paran, I.; Kesseli, R.V. Identification of markers linked to disease-resistance genes by bulked segregant analysis: A rapid method to detect markers in specific genomic regions by using segregating populations. *Proc. Natl. Acad. Sci. USA* 1991, *88*, 9828–9832.
26. Takagi, H.; Abe, A.; Yoshida, K.; Kosugi, S.; Natsume, S.; Mitsuoka, C.; Uemura, A.; Utsushi, H.; Tamiru, M.; Takuno, S.; et al. QTL-seq: Rapid mapping of quantitative trait loci in rice by whole genome resequencing of DNA from two bulked populations. *Plant J.* 2013, *74*, 174–183.
27. Pandey, M.K.; Khan, A.W.; Singh, V.K.; Vishwakarma, M.K.; Shasidhar, Y.; Kumar, V.; Garg, V.; Bhat, R.S.; Chitikineni, A.; Janila, P. QTL-seq approach identified genomic regions and diagnostic markers for rust and late leaf spot resistance in groundnut (*A rachis hypogaea* L.). *Plant Biotechnol. J.* 2017, *15*, 927–941.
28. Clevenger, J.; Chu, Y.; Chavarro, C.; Botton, S.; Culbreath, A.; Isleib, T.G.; Holbrook, C.C.; Ozias-Akins, P. Mapping Late Leaf Spot Resistance in Peanut (*Arachis hypogaea*) Using QTL-seq Reveals Markers for Marker-Assisted Selection. *Front. Plant Sci.* 2018, *9*, 83.
29. Branham, S.E.; Farnham, M.W. Identification of heat tolerance loci in broccoli through bulked segregant analysis using whole genome resequencing. *Euphytica* 2019, *215*, 34.
30. Fall, L.A.; Clevenger, J.; McGregor, C. Assay development and marker validation for marker assisted selection of Fusarium oxysporum f. sp. niveum race 1 in watermelon. *Mol. Breed.* 2018, *38*, 130.
31. Lu, H.; Lin, T.; Klein, J.; Wang, S.; Qi, J.; Zhou, Q.; Sun, J.; Zhang, Z.; Weng, Y.; Huang, S. QTL-seq identifies an early flowering QTL located near Flowering Locus T in cucumber. *Theor. Appl. Genet.* 2014, *127*, 1491–1499.
32. Illa-Berenguer, E.; Van Houten, J.; Huang, Z.; Van Der Knaap, E. Rapid and reliable identification of tomato fruit weight and locule number loci by QTL-seq. *Theor. Appl. Genet.* 2015, *128*, 1329–1342.
33. Zhang, Y.; Hu, Z.; Zhu, M.; Zhu, Z.; Wang, Z.; Tian, S.; Chen, G. Anthocyanin accumulation and molecular analysis of correlated genes in purple kohlrabi (*Brassica oleracea* var. gongylodes L.). *J. Agric. Food Chem.* 2015, *63*, 4160–4169.
34. Tang, Q.; Tian, M.; An, G.; Zhang, W.; Chen, J.; Yan, C. Rapid identification of the purple stem (Ps) gene of Chinese kale (*Brassica oleracea* var. alboglabra) in a segregation distortion population by bulked segregant analysis and RNA sequencing. *Mol. Breed.* 2017, *37*, 153.
35. Porebski, S.; Bailey, L.G.; Baum, B.R. Modification of a CTAB DNA extraction protocol for plants containing high polysaccharide and polyphenol components. *Plant Mol. Biol. Rep.* 1997, *15*, 8–15.
36. Li, H.; Durbin, R. Fast and accurate short read alignment with Burrows-Wheeler transform. *Bioinformatics* 2009, *25*, 1754–1760.
37. Yu, J.; Zhao, M.; Wang, X.; Tong, C.; Huang, S.; Tehrim, S.; Liu, Y.; Hua, W.; Liu, S. Bolbase: A comprehensive genomics database for Brassica oleracea. *BMC Genom.* 2013, *14*, 664.
38. Li, H.; Handsaker, R.; Wysoker, A.; Fennell, T.; Ruan, J.; Homer, N.; Marth, G.; Abecasis, G.; Durbin, R. The Sequence Alignment/Map format and SAMtools. *Bioinformatics* 2009, *25*, 2078–2079.
39. McKenna, A.; Hanna, M.; Banks, E.; Sivachenko, A.; Cibulskis, K.; Kernytsky, A.; Garimella, K.; Altshuler, D.; Gabriel, S.B.; Daly, M.J.; et al. The Genome Analysis Toolkit: A MapReduce framework for analyzing next-generation DNA sequencing data. *Genome Res.* 2010, *20*, 1297–1303.
40. Wang, K.; Li, M.; Hakonarson, H. ANNOVAR: Functional annotation of genetic variants from high-throughput sequencing data. *Nucleic Acids Res.* 2010, *38*, e164.

41. Chen, K.; Zhou, Y.-X.; Li, K.; Qi, L.-X.; Zhang, Q.-F.; Wang, M.-C.; Xiao, J.-H. A novel three-round multiplex PCR for SNP genotyping with next generation sequencing. *Anal. Bioanal. Chem.* 2016, *408*, 4371–4377.

42. Meng, L.; Li, H.; Zhang, L.; Wang, J. QTL IciMapping: Integrated software for genetic linkage map construction and quantitative trait locus mapping in biparental populations. *Crop. J.* 2015, 3, 269–283.

43. Chiu, L.-W.; Li, L. Characterization of the regulatory network of BoMYB2 in controlling anthocyanin biosynthesis in purple cauliflower. *Planta* 2012, *236*, 1153–1164.

44. Livak, K.J.; Schmittgen, T.D. Analysis of relative gene expression data using real-time quantitative PCR and the 2- ΔΔCT method. *Methods* 2001, *25*, 402–408.

45. Schoenbohm, C.; Martens, S.; Eder, C.; Forkmann, G.; Weisshaar, B. Identification of the Arabidopsis thaliana Flavonoid 3′-Hydroxylase Gene and Functional Expression of the Encoded P450 Enzyme. *Biol. Chem.* 2000, *381*, 749–753.

46. Hayashi, K.; Matsumoto, S.; Tsukazaki, H.; Kondo, T.; Kubo, N.; Hirai, M. Mapping of a novel locus regulating anthocyanin pigmentation in *Brassica rapa*. *Breed. Sci.* 2010, *60*, 76–80.

47. Han, Y.; Vimolmangkang, S.; Soria-Guerra, R.E.; Rosales-Mendoza, S.; Zheng, D.; Lygin, A.V.; Korban, S.S. Ectopic Expression of Apple F3′H Genes Contributes to Anthocyanin Accumulation in the Arabidopsis tt7 Mutant Grown Under Nitrogen Stress. *Plant Physiol.* 2010, *153*, 806–820.

48. Bogs, J.; Ebadi, A.; McDavid, D.; Robinson, S.P. Identification of the Flavonoid Hydroxylases from Grapevine and Their Regulation during Fruit Development. *Plant Physiol.* 2005, *140*, 279–291.

49. Jung, Y.J.; Lee, H.J.; Kim, J.H.; Kim, D.H.; Kim, H.K.; Cho, Y.-G.; Bae, S.; Kang, K.K. CRISPR/Cas9-targeted mutagenesis of F3′H, DFR and LDOX, genes related to anthocyanin biosynthesis in black rice (*Oryza sativa* L.). *Plant Biotechnol. Rep.* 2019, *13*, 521–531.

50. Vikhorev, A.V.; Strygina, K.V.; Khlestkina, E.K. Duplicated flavonoid 3′-hydroxylase and flavonoid 3′, 5′-hydroxylase genes in barley genome. *PeerJ* 2019, *7*, e6266.

51. Zhang, H.; Wang, X.; Pan, Q.; Li, P.; Liu, Y.; Lu, X.; Zhong, W.; Li, M.; Han, L.; Li, J. QTG-Seq accelerates QTL fine mapping through QTL partitioning and whole-genome sequencing of bulked segregant samples. *Mol. Plant* 2019, *12*, 426–437.

52. Zhou, X.; Xia, Y.; Ren, X.; Chen, Y.; Huang, L.; Huang, S.; Liao, B.; Lei, Y.; Yan, L.; Jiang, H. Construction of a SNP-based genetic linkage map in cultivated peanut based on large scale marker development using next-generation double-digest restriction-site-associated DNA sequencing (ddRADseq). *BMC Genom.* 2014, *15*, 351.

5 Chapter

CREATION OF MOLECULAR MARKERS LINKED TO GRAY MOLD DISEASE RESISTANCE IN ONION (*ALLIUM CEPA* L.) USING RAPD-PCR AND TRANSCRIPTOME ANALYSIS

1. INTRODUCTION: AN OVERVIEW

Onions, scientifically known as *Allium cepa* L., hold significant economic and nutritional value on a global scale. Additionally, they are among the most ancient crops that have been produced and are employed as a constituent in many cuisines and sauces to augment taste and foster well-being, particularly in the context of reducing cholesterol levels [1,2,3]. Hence, it is crucial to cultivate and generate superior-grade onions in order to enhance their competitive edge in the market. Onion breeding is conducted for multiple objectives, including enhancing onion yield, improving characteristics such as size, taste, and color, achieving male sterility, and developing tolerance against biotic and abiotic challenges [3,4,5].

Onions are vulnerable to numerous infections and insects [3]. Consequently, there has been significant research on developing resistant onion varieties to minimize the harm caused by various illnesses, many of which are caused by the genus Botrytis. *Botrytis squamosa* is responsible for causing onion botrytis leaf blight, while onion neck rot is caused by *B. aclada, B. alli, B. squamosa*, and *B. porri* [6]. Specifically, *B. cinerea* is a necrotrophic disease that affects over 200 different crop species, resulting in significant harm to onions [7]. Gray mold, resulting from the presence of *B. cinerea* and *B. squamosa*, diminishes both the crop production and the storage capacity. The onion bulb is affected by gray mold disease, which is caused by *B. squamosa* during the creation and filling of the bulb, and by *B. cinerea* during later stages of culture and storage [8,9]. Prior research has made efforts to employ chemical and biological measures in order to avert onion damage caused by gray mold. In order to mitigate the spread of this disease, it is imperative to ensure that onions are stored in a dry environment, promptly remove any affected onions, and implement a crop rotation schedule every 3 to 4 years. Furthermore, a significant portion of the global fungicide market, around 10%, is dedicated to the management of gray mold caused by *B. cinerea* [10]. Nevertheless, despite these diligent endeavors, managing gray mold sickness proves to be challenging. Moreover, the use of artificial fungicides can give rise to issues such as worries about leftover traces and adverse effects on human well-being, the development and proliferation of pathogen populations that are resistant to treatment, and contamination of the environment [11,12,13,14]. Consequently, by selectively selecting onions that are resistant to illnesses like gray mold, we may minimize the harm caused by these diseases, boost onion production, and decrease the need for labor and the negative impact on the environment.

Onions, being a biennial plant, have a long breeding period for disease-resistant varieties. This is because a full generation of onions takes two years. Marker assistant selection (MAS) can expedite the breeding process of onions. Furthermore, doing a genome analysis of onions presents challenges due to their substantial genome size of 16Gb, which is 100 times greater than that of Arabidopsis genomes [3,4]. Various molecular markers, such as random amplified polymorphic DNA (RAPD) and sequence characteristic amplified region (SCAR), have been created for diverse applications. These markers are employed for Marker-Assisted Selection (MAS), which aids in the process of selecting desirable traits in breeding and reducing the duration of breeding cycles. RAPD, which stands for Random Amplified Polymorphic DNA, is a marker that uses short random primers in PCR. As a result, RAPD analysis may detect tiny genetic variations between big genomes, such the genome of onions [15,16]. RAPD markers suffer from limited repeatability. Hence, in this research, we created a SCAR marker by employing a primer that is more precise than RAPD, based on the findings of polymorphism investigations conducted on onion lines that are resistant and susceptible.

Furthermore, the utilization of NGS technologies, such as RNA sequencing, has facilitated the analysis of transcriptome data on a wide scale. This has resulted in enhanced gene identification efficiency, even in the absence of any prior knowledge regarding reference genome sequences [4,17,18]. This work utilized RNA-seq to establish molecular markers for the purpose of breeding onions that are resistant to gray mold. The DEG analysis and variation analysis, including SNPs and InDels, were performed using RNA-seq to compare the resistant and susceptible groups. The HRM markers validate the fluorescence of the melting curve of the PCR result, indicating the transition from double-stranded DNA to single-stranded DNA. The targeted DNA region was amplified using a fluorescent dye that binds to double-stranded DNA. The melting curve of the PCR result was then evaluated while it was gradually heated. The melting curve is subject to variation based on factors such as the existence of single nucleotide polymorphisms (SNPs), the GC content, the length, and the heterozygosity of the sequence [19,20]. This study involved the development of the HRM marker for breeding gray mold-resistant onions. The marker was created by analyzing selected transcripts from gene ontology (GO) and doing variation analysis utilizing RNA-seq data.

2. MATERIALS AND METHODS

2.1. Plant Materials and Genomic DNA Extraction

To develop molecular markers, four (S&P 7522, S&P 7521, S&P 7129, and S&P 7168) gray mold-resistant and three (S&P 7130, S&P 7175, and S&P 7483) susceptible onion (*Allium cepa* L.) lines grown in 'Seeds & People' Co., Ltd. (Yeonggwang, Korea) were used in this study. The characteristics of the seven onion lines of 'Seeds & People' Co. In addition, to confirm the versatility of the developed molecular markers, three (Asia-12, Asia-42, and Asia-53) resistant and four (Asia-30, Asia-35, Asia-45, and Asia-50) susceptible onion lines against gray mold were provided by 'Asia Seed' Co., Ltd. (Seoul, Korea) and analyzed. Genomic DNA was isolated from the leaf tissues of each line by using cetyltrimethylammonium bromide (CTAB) [**21**]. The concentration and purity of the gDNA were measured using a Nanodrop® ND1000-spectrophotometer (NanoDrop Technologies, Wilmington, NC, USA).

2.2. RAPD and Phylogenetic Analysis

The gDNA of each line was used for a RAPD analysis to identify the genetic relationship between the resistant and susceptible lines. For a RAPD analysis, OPERON random primers OPAN-1~OPAN-20 and OPL-1~OPL-20 were used. The sequences of the 40 OPERON random primers. The RAPD PCR was performed in 20 μL by using the Maxime PCR Premix (iNtRON Biotechnology, Seongnam, Korea) containing 2.5-U i-TaqTM *Taq* polymerase, -2.5mM dNTPs, 1X reaction buffer with 10 pmol primer, and 50 ng of gDNA. The amplification program was as follows: denaturation at 95 °C for 10 min, 40 cycles of 1 min at 95 °C, 1 min at 37 °C, 1 min at 72 °C, and a final extension at 72 °C for 10 min. The amplified RAPD PCR product was electrophoresed at 100 mA for 3 h in a 1% agarose gel. The polymorphic bands obtained from the RAPD-PCR of each line were converted into binary data, depending on the presence of a band; the presence of a polymorphic band was scored as 1, while its absence was scored as 0. The dendrogram was obtained using the unweighted pair group method (UPGMA) with an arithmetic mean [**22**] by using the Jaccard coefficient [**23**] through the XLSTAT program (Addinsoft, New York, NY, USA).

2.3. Development of the SCAR Marker

To develop the SCAR marker, the 2-Kb polymorphic bands obtained between four resistant and three susceptible lines by using the OPAN-1 primer (5′-ACT CCA CGT C-3′) were sequenced and identified. The amplified polymorphic products were eluted from a 1% agarose gel and purified using the NucleoSpin® Gel and PCR Clean-up Kit (Macherey-Nagel, Duren, Germany). The eluted products were ligated into the pGEM-T easy vector (Promega, Madison, WI, USA) for sequencing. The plasmid was purified using the Fast DNA-spinTM Plasmid DNA Purification Kit (iNtRON Biotechnology, Seongnam, Korea), and the fragments were sequenced by Macrogen® (Seoul, Korea). The SCAR primer set was designed from the sequence of the -2Kb polymorphic band between the resistant and susceptible lines. PCR using the SCAR primer set was performed under the following conditions: initial denaturation for 10 min at 95 °C, 35 cycles of 30 s at 95 °C, 30 s at 61 °C, 1 min at 72 °C, and a final extension for 10 min at 72 °C. Thereafter, the amplified products of the resistant onion lines were confirmed by electrophoresis in a %1 agarose gel. In addition, to confirm the versatility of the SCAR marker, three gray mold-resistant and four susceptible onion lines from the 'Asia Seed' Co. were also analyzed.

2.4. RNA Sequencing and Variant Analysis

For RNA sequencing, the leaves of four resistant and three susceptible lines were ground, and the total RNA was extracted using the RNease® kit (QIAGEN, Hilden, Germany). Before analyzing the RNA sequence, the OD values were measured using Dropsense96 (Trinean, Gentbrugge, Belgium), and the total RNA quality was checked using a Bioanalyzer RNA Chip (Agilent Technologies, Santa Clara, CA, USA). The library was constructed using the TruSeq Stranded mRNA kit (Illumina, San Diego, CA, USA), and RNA-seq was performed using HiseqX (Illumina, San Diego, CA, USA) by the DNAcare Company (Seoul, Korea). To remove the low-quality base and Illumina adapters, the Trimmomatic program (USADELLAB, Aachen, Germany) was used on the RNA-seq raw data of each line. After trimming, only paired reads containing at least 50 nucleotides were used for the analysis. In addition, quality trimming was performed by applying options such as a sliding window, average quality, and minimum read size. Thereafter, the generated trimmed data were used for the de novo assembly of resistant and susceptible lines by using the Trinity program. As a reference for mapping, the onion reference transcript (National Agricultural Biotechnology Information Center (NABIC), Rural Development Administration (RDA), JeonJu, Korea) and the data obtained through the de novo assembly were used. Read mapping was performed using the BWA-Mem algorithm. To remove the duplicated PCR reads from the produced BAM file, the Picard program (Broad Institute, Cambridge, MA, USA) was used. Thereafter, the variant information of seven lines was produced using haplotypeCaller of the Genome Analysis Tool Kit (GATK, Broad Institute, Cambridge, MA, USA), and the final raw variant call file (vcf) was generated by integrating the variant files of each line. Variant filtering was performed using Vcftools (VCFtools, 1000 Genomes Project Analysis Group, **http://vcftools.sourceforge.net/** accessed on 1 October

2021) to remove the low-quality genotypes, target missing levels of depth coverage (DP), genotype quality (GQ), and genotype data. The filtering conditions were set to min DP = 5, max DP = 100, and min GQ = 220, and the missing was set to %20. Various information was used to select the variants showing polymorphisms between the resistant and susceptible lines. For the selection condition for the variants, those showing the same genotype in each resistant and susceptible group and showing polymorphisms between these groups were selected.

2.5. Differentially Expressed Gene (DEG) and Gene Ontology (GO) Analysis for Selection of Transcript Related to Disease Resistance

Among the sequences generated from Trinity, sequences with a length of more than 100 amino acids were selected using TransDecoder. Based on these data and onion reference transcript data (NABIC, RDA, JeonJu, Korea, **https://nabic.rda.go.kr**, accessed on 7 February 2021), the sequences of each onion line were aligned with HISAT2. The read count of the transcript expression level was then calculated using the StringTie program. The transcripts obtained through StringTie were calculated at the transcript level, and a comparative analysis was performed between each onion line based on the read count of each transcript. After dividing into resistant and susceptible groups, a DEG analysis was performed using DEGseq (Bioconductor, **http://www.bioconductor.org/packages/release/bioc/html/DEGseq.html**, accessed on 2 October 2021) [**24**]. First, after normalizing the raw read count data, a correlation analysis was performed between each onion line based on the normalized data. The analysis was conducted using Pearson's correlation coefficient and the average linkage method. The DEGseq of the R package was used to confirm the statistical significance of the expression differences between resistant and susceptible groups. After comparing the average expression levels between the two groups, the conditions were set as follows to select genes using significantly different expressions. The test was conducted using the equation $\log_2\left(\frac{\text{Base Mean of R}}{\text{Base Mean of S}}\right)$. A negative value was set for the transcripts more expressed in the susceptible group than in the resistant group, and a positive value was set for the transcripts more expressed in the resistant group than in the susceptible groups. Thereafter, the transcripts with DEGs satisfying the conditions of $|\log2$ fold change$| \geq 2$ and PADJ < 0.05 were selected. From the results of variants analysis and DEG analysis, transcripts that commonly satisfied each condition were selected. A gene function analysis was performed to identify whether the selected variants were related to the disease-resistant mechanisms. To identify the functions related to disease resistance, the selected transcripts were analyzed using The Arabidopsis Information Resource (TAIR) ID derived from Arabidopsis, a model plant. Thereafter, GO annotation was performed using the transcripts' confirmed gene functions by TAIR ID, and the transcripts with functions related to disease resistance were selected.

2.6. HRM Primer Designs from Selected Transcripts and HRM Analysis

HRM primers were designed from 14 selected transcripts with SNPs that exist between the resistant and susceptible lines. The conditions of the designed primers were as follows: the amplified product size was between 80 and 200 bp, the variant region was inside the PCR product, and the annealing temperature was approximately 60 ± 1 °C. The sequences and annealing temperature information of the HRM primers are listed in **Table 1**. HRM was performed using a total 10-μL reaction mixture containing the BioFactTM 2X Real-Time PCR Master Mix (BIOFACT, Daejeon, Korea) with a -10pmol primer and 50 ng of gDNA. The reaction conditions were: pre-denaturation at 95 °C for 15 min, 40 cycles of denaturation at 95 °C for 20 s, annealing at 60 °C for 30 s, and extension at 72 °C for 20 s. Thereafter, the temperature was sequentially increased from 65 °C to 95 °C, and the melt curve and peak value were measured. At least six repetitions were performed for each line. The HRM results were statistically grouped using the ANOVA of the XLSTAT program. In addition, to confirm the versatility of the selected HRM marker, which was able to distinguish between resistant and susceptible onion lines, an additional analysis was performed using 'Asia Seed' Co. onion lines.

Table 1. HRM primers designed from 14 selected transcripts with SNPs that exist between the resistant and susceptible lines.

No.	SNP Name	Primer Name	Sequence (5'→3')	Tm	PCR Product Size
1	SNP-1	SNP 1 F	CTTTGAACTTCGGGCAATACCCG	60.5	199 bp
		SNP 1 R	CCTCATCAGGCGAGTGAGTGGAC	59.6	
2	SNP-2	SNP 2 F	AACGTCCGCCGAAGAAGCTGA	60.7	204 bp
		SNP 2 R	TTTGCTGGAGGAGGTGGTGGTG	60.1	
3	SNP-3	SNP 3 F	CGTTAGCTCAAGTGGGTTTGAGGTG	59.9	134 bp
		SNP 3 R	TTCTTCCAGCTCTTCTTCGCT	59.2	
4	SNP-4	SNP 4 F	AGGGTTCAGAACCAAAACAGCATCA	59.8	163 bp
		SNP 4 R	CGATGCTTTTTTGGTAACTGGGAAG	59.1	
5	SNP-5	SNP 5 F	TCGATGGCATTAAGGATGCTAAGGA	59.8	163 bp
		SNP 5 R	ATTGCCTTTGCTAGGGAGCCATAA	59.1	
6	SNP-6	SNP 6 F	TGGTGACAAGAAATTCTTCAACGGC	60.2	158 bp
		SNP 6 R	TCTCCATGCATCTCTTTCCCCACT	59.8	
7	SNP-7	SNP 7 F	TGAGCTCCTTTCAGACTCCTTTCCC	60.1	170 bp
		SNP 7 R	CGACCACCTTAACAGCTTGATCGTC	59.9	

No.	SNP Name	Primer Name	Sequence (5′→3′)	Tm	PCR Product Size
8	SNP-8	SNP 8 F	CTTTCTCAGGGTTAATAGAGGCGGG	60.1	171 bp
		SNP 8 R	GCCAAACTGGCTGAAAACCTTTTCT	59.9	
9	SNP-9	SNP 9 F	TTCATGGTCACAGAAACGCCAAGA	60.1	181 bp
SNP 9 R GGCAGAACTTCTTTGTTCATCCGCT 59.9					
10	SNP-10	SNP 10 F	AATCTCACAATCGAACCTCACTGCC	59.4	175 bp
		SNP 10 R	TGCGAGGTGAATTCCAGTCAAAGAG	60.4	
11	SNP-11	SNP 11 F	GCAACAAGGGCTGCAAATTACAGTT	60.1	169 bp
		SNP 11 R	GTTTGTGTGCATGAATCTGTGCAGG	59.9	
12	SNP-12	SNP 12 F	CGACTATGGCTGGGACACTGCA	59.6	164 bp
		SNP 12 R	TCCCCGAACTTGACCCCGTTAC	60.7	
13	SNP-13	SNP 13 F	AGCAATGTTGTCCGGTACTCCAAAG	59.4	185 bp
		SNP 13 R	CGCTCAAAAACCCAGCTCGTACA	60.1	
14	SNP-14	SNP 14 F	TATCGTACCTTCCTACCCTGAGCGA	60.1	182 bp
		SNP 14 R	TCCGAACATGGGCAGCTTCC	59.9	

2.7. Quantitative Real-Time PCR (qPCR) to Identify Expression Level of Transcripts

The total RNA was extracted from the leaves of the resistant and susceptible lines by using the Plant RNA Extraction Kit (Takara, Shiga, Japan). The concentration and purity of the RNA were measured using a Nanodrop® ND1000- spectrophotometer (NanoDrop Technologies, Wilmington, NC, USA). Thereafter, cDNA was synthesized from 500 ng of RNA extracted from each line by using the HiSenScript™ RH[-] RT PreMix Kit (iNtRON, Seongnam, Korea). PCR for cDNA synthesis was performed using the following cycle: reverse transcription step at 42 °C for 1 h and RTase inactivation extension step at 85 °C for 10 min. The synthesized cDNA was stored at -20 °C. Quantitative real-time PCR (qPCR) was performed to compare the expression levels of the SNP-3 transcript related to the aldo-keto reductase (AKR) gene with in silico data, which showed significant HRM results in 14 transcripts. The qPCR primer conditions were as follows: the amplified product size was less than 200 bp, and the annealing temperature was approximately 60 ± 1 °C. The forward primer was 5′-CGT TAG CTC AAG TGG GTT TGA GGT G-3′, and the reverse primer was 5′-CTC CAG CAC ACG CCC TCC A-3′. Before the qPCR analysis, it was confirmed that the 171-bp band targeted by the qPCR primer set was amplified by RT-PCR using the Maxime™ PCR PreMix Kit (iNtRON, Seongnam, Korea). The reaction cycle was as follows: 10 min of initial denaturation at 95 °C, followed by 40 cycles of 95 °C for 30 s, 60 °C for 30 s, 72 °C for 40 s, and a final extension at 72 °C for 10 min. The amplified product was electrophoresed using 1% agarose gel.

The qPCR analysis was performed with the TransStart Top Green qPCR Super Mix (TransGen, Beijing, China) containing 1 pmol of qPCR primer and 500 ng of cDNA by using Roter-GeneTM6000 (Corbett, Melbourne, Vic, Australia), and the qPCR program proceeded at 95 °C for 10 s, followed by 40 cycles consisting of 2 steps: 95 °C for 10 s and 60 °C for 30 s. After amplification, a fluorescence melting curve was obtained by heating the samples from 60 °C to 95 °C. AKR gene expression was identified using the Ct value. To compare the differences in the gene expression levels between the resistant and susceptible lines, the actin gene, the housekeeping gene, was used as a control. After obtaining the Ct value using Roter-Gene Q Series Software, the ΔCt value ((Ct value of target gene)-(Ct value of actin gene)) was calculated. To compare the expression levels based on the susceptible S&P 7483 lines, the ΔΔCt value ((ΔCt of each onion line)-(ΔCt of S&P 7483)) was obtained. The difference in the expression levels was confirmed by calculating the 2ΔΔCt. The expression level of each line was calculated from the results of six repetitions. Data were presented as the means with standard errors, and the means were compared using Duncan's multiple comparison test ($p \leq 0.05$).

3. RESULTS

3.1. RAPD and Phylogenetic Analysis

Four (S&P 7522, S&P 7521, S&P 7129, and S&P 7168) gray mold-resistant and three (S&P 7130, S&P 7175, and S&P 7483) susceptible onion lines were amplified using 40 OPERON random primers, and the polymorphic bands were confirmed by electrophoresis. Most of the amplified band sizes ranged from 200 bp to 2500 bp, and 124 bands (38%) of the 330 bands showed polymorphisms. The RAPD analysis was repeated at least three times to obtain reproducibility, and the presence or absence of amplified bands was transformed into binary data. The binary data were converted into dendrograms by using the UPGMA methods of the XLSTAT program using Jaccard coefficients (Figure 1). The genetic similarity within the group was 78.27%, and the genetic similarity between the groups was 21.73%. Based on the phylogenetic analysis, a genetic relationship between the seven lines was confirmed. The results of the phylogenetic analysis were compared with the pedigrees of the seven breeding lines to confirm the reliability of the RAPD (Figure S1). The phylogenic analysis results were consistent with the genetic relationship of the pedigree of the breeding lines.

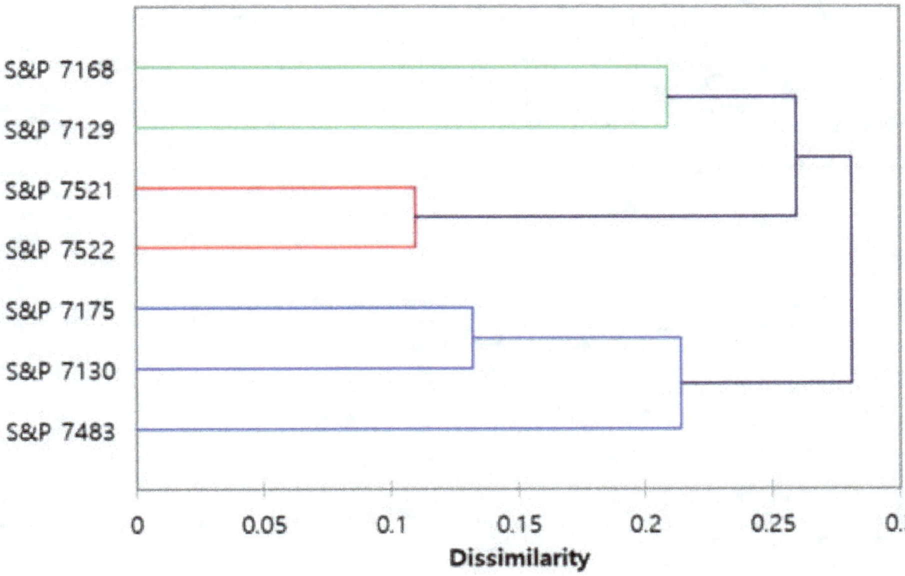

Figure 1. Phylogenetic tree of the resistant and susceptible lines provided by 'Seeds & People' Co. by a RAPD analysis. The resistant lines S&P 7168 and S&P 7129 showed a close genetic relationship, and also, resistant lines S&P 7522 and S&P 7521 showed a close genetic relationship. The susceptible onion lines S&P 7175, S&P 7130, and S&P 7483 appeared to have a closer genetic relationship.

3.2. Development of a SCAR Marker for the Selection of Resistant Lines

From the RAPD results, a SCAR marker was developed using a specific RAPD product showing polymorphisms between the resistant and susceptible lines (Figure S2). The 2-Kb bands were amplified in the resistant lines S&P 7522, S&P 7521, S&P 7129, and S&P 7168 by using the OPAN-1 primer. The amplified polymorphic bands were eluted, purified, and sequenced. From the sequencing results, 5'-ACT CCA CGT C-3', which is a sequence of random primer OPAN-1, was identified at both ends of the fragment sequence (Figure 2). In addition, the sequences were analyzed using BLAST at the National Center for Biotechnology Information (NCBI), and insignificant results were obtained. Therefore, further studies are required to analyze this unknown DNA sequence. The SCAR marker primer was designed to include the 5′-ACT CCA CGT C-3′ sequence of the OPAN-1 random primer. The forward primer was 5′-ACT CCA CGT CAT CGA TTC GAA-3′, and the reverse primer was 5′-ACT CCA CGT CCG AAC TAC AGA A-3′. The primer set for the SCAR marker was designed for considering the annealing temperature, GC content, the possibility of dimer formation, and hairpin loops. The developed SCAR marker was designated as SCAR-OPAN1.

ACTCCACGTC ATCGATTCGA AGGAGCATCG ATACGTATGT ATCGATATGC AGCCATCAGT CCACTCGCAA AATCAACTTT
GATCGATCAT AACTCGAAAA CAACAAGTTT AAATGCTATG AAACTTCCAT AGATGCTTGC TTATTCGCTT AAGAATGTCC
ATGATCTACC GCAGCTCAAA TGGAGCTAAA CCTGATATAT ACGAAGATGA TAGAGATCGA TCCTAAACGA CCCTAGATCC
ATCTGGTGGG CCCATCCACG TATGAGAAGA GGTATTTCGT GGTTTGAGAG AGATATTTTT AGATCTACAC CTATAAATAG
CACTTCTACG CCTGTAGAAA AAAAGACTTT TTGGCTTTTC TTCGTCGTTA CTTTGCTGAA ACTCCATTGT AACTTGGATG
TCATCGATTG TACTATTGTA AATTAATTGA GAATCATTAA TAAACATTGA ACTCTTCAAA GTGGACGTAG GCTACAACGC
CGAACCACTA TAAATCTCGT ATCTTTACTG CTTTCTTTTT AATGTTTGTC GTTCTTCATA TTTATATTTT CTTCTCATAT
TCTATCTGTG ATCTGTGACC GTGAATTGGT GTTTTCATTC GCAATTGGGT ATTTGTGATC AATTTTGACG CAAACAGTGT
GTATCGCTGA AGACATAATT AGGTTTATCT TCATCTTCAT CTATCATAGG TTTTTTGTCT AGACATTGAT ATGCAAAACT
CGTAAAACCT ACCCCACCCC AATATTTTAA CTGTCGCTAG AACTTTATTG GTATTTCAGG CGTTATGCAT ATGGTTACGC
ATATAAAGTG TGGCTTATCT TCACAAAATA AACTTCGCCA TCGACCATTT TATAGATTGA CTTGTGAGTT AGGTAACTAA
GCTTGTGGTT AAAGTTTTTT TTTTTGCCCG GGGGAAAAAA TTTTTTAATA AAAAAAAAAA TTCCCGGAAA CGGGGGGCCT
TGTTTTTTTT GAAATTAAAA ATTTCCAGGG AAAATTTTCA GTTTTCCGAC TTCGATGAGA TTTCAGAAAT TATTGGTTTG
TACTTTACAT CCAGCTGAGA TACATTTTTA TGTCGGTCCT TTTAGCCTGC TAGTAGATTC AATCAAATTA AAAAGGATTT
TCGTACATAG ATTTGGCATA GGTTAAAATT GTAGCTGAAT GTAACTACAG CTATTAGGCG AAATATACCT ACAGTTACAG
GGTGCATCAA ATGAGAAGAA AAAAAAATTG TCTTAACCAT TTTTTTCTCT ATAGCTTCTT TTTACATCAC TGTTTCTATC
AAATCTTAAA CTTTAGTTCT GTAATAGTGT AATTAACTTG CACAATTGGT ATCAAGATAA AATTATGTTT TAGCTTCCAT
CAGCTGCAGC GTCGCAAGTT CTGATGATCC ATTTAAATAA ATATTTACTT CCGGTGAATT AATAGTTTTC AATGTCGACC
CTTTGATTGT GCATTGCATG GATTGTATGC ACTGGGTTCC ACATGTTTAC CATACTAGTT GTCCGCTCCT TTGTACGTCA
GAAATTAATG GATTACTAAT TACAACATTT CACTTTACCG TTTTTTGATT TGTTACATTT TCACTTTCCT TAGAATTTTG
TTCTGCAAAA ATTAATGTTA ATGACATTAT TAAAATTCTT GCTTTTTTGC CTAGTGAATG AATCATTGGC GGGCAAACAT
AGTTTATCAT GTTCGATTTC ATTTAGAAAA TGGTGGAGGT ATCATCATAA TTTGGTGTAT ATATTTTGTA TGACATAAAA
GCCCATTTTA TCTTTCAAAG AATATGTGTA GTAGTTGTAA ATCGGTTATA ATGTAAAATC ATTAATCTAG GAAACCCATA
ATCTTTCGTT TAAATTTTTT TTACTGCACA CTATAAATGA GCAAAAATAC AGGCGCCAAT AAATAGAATT ATAATCCGTG
TTGTAAACTT ATGTATCTTT GTTAGTACAC AATATATTGA AGTTTTTGTG TGATATACAG ACCCGTTGCA GGCAAATAAT
TCTGTAGTTC GGACGTGGAG T

Figure 2. The SCAR-OPAN-1 primer and sequence of a 2-Kb product amplified in four resistant lines by an OPAN-1 random primer. The SCAR-OPAN-1 primer contained the OPAN-1 random primer sequence. The sequence of the designed SCAR-OPAN-1 primer is shown in bold. The underlined sequence is the sequence of OPAN-1 random primer, 5′-ACT CCA CGT C-3′.

3.3. Validation of the Developed SCAR Marker Using the Resistant and Susceptible Onion Lines

The SCAR-OPAN1 marker was used to screen the resistant and susceptible lines. The 2-Kb-sized polymorphic band was amplified in the resistant lines S&P 7522, S&P 7521, S&P 7129, and S&P 7168 and was not amplified in the susceptible lines S&P 7483, S&P 7130, and S&P 7175 (Figure 3a). The developed SCAR-OPAN1 marker amplified the specific band only from the resistant lines, demonstrating its potential as a molecular marker for the selection of resistant lines. To validate the SCAR-OPAN1 marker, three resistant lines and four susceptible lines of 'Asia Seed' Co. were screened using the SCAR-OPAN1 marker (Figure 3b). Similar to the screening results of the 'Seeds & People' Co. onion lines, a product size of 2 Kb was amplified only in the resistant lines (Asia-12, Asia-42, and Asia-53). These polymorphic bands between the resistant and susceptible lines allow the selection of gray mold-resistant onion lines for breeding programs.

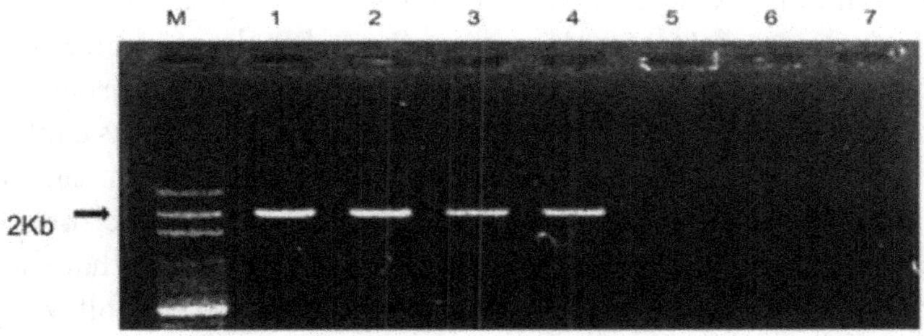

(A)

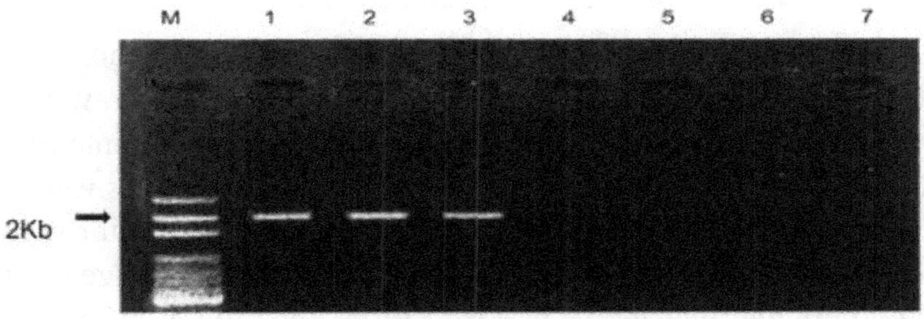

(B)

Figure 3. Identification of resistant onion lines of 'Seeds & People' Co. and 'Asia Seed' Co. using the SCAR-OPAN1 marker. (A) The 2-Kb-sized polymorphic products were amplified in the resistant lines (S&P 7522, S&P 7521, S&P 7129, and S&P 7168) and were not amplified in the susceptible lines (S&P 7483, S&P 7130, and S&P 7175). M: 100-bp DNA ladder marker, Lane 1: S&P 7522, Lane 2: S&P 7521, Lane 3: S&P 7129, Lane 4: S&P 7168, Lane 5: S&P 7483, Lane 6: S&P 7130, and Lane 7: S&P 7175. (B) The 2-Kb products were amplified only in the resistant lines (Asia-12, Asia-42, and Asia-53) and were not amplified in the susceptible lines (Asia-30, Asia-35, Asia-45, and Asia-50). M: 100-bp DNA ladder marker, Lane 1: Asia-12, Lane 2: Asia-42, Lane 3: Asia-53, Lane 4: Asia-30, Lane 5: Asia-35, Lane 6: Asia-45, and Lane 7: Asia-50.

3.4. Preprocessing of Raw Data of the RNA Sequence

After the quality check of the extracted RNA was completed, a library was constructed, and the quality of distribution by cDNA size and concentration of the constructed library were measured using a Bioanalyzer DNA Chip (Agilent Technologies, Santa Clara, CA, USA). Thereafter, RNA sequencing was performed using the constructed library information and HiseqX (Illumina, San Diego, CA, USA), and the raw RNA sequencing data of each onion line were obtained. The raw RNA-seq data are shown in Table S3. The raw reads of the seven onion lines ranged from 28,899,894 to 41,664,542, the total reads of the trimmed reads ranged from 20,744,300 to 27,932,656, and the mapped reads ranged from 16,852,911 to

21,958,205. Trimming was performed by removing low-quality bases and Illumina adapters from the raw data. Mapping was performed on the reference data of NABIC (RDA, JeonJu, Korea, https://nabic.rda.go.kr, accessed on 7 February 2021), and the number of paired reads containing at least 50 nucleotides was confirmed. As a result of mapping, the lengths of the mapping reads ranged from 267 nucleotides to 5478 nucleotides, the average length of the reads was 2236 nucleotides, and the median was 1988 nucleotides. The mapping results determined that the length of the mapped reads was sufficient for the next analysis, and the following analysis was performed.

3.5. Variant Analysis and Filtering for Selection of Transcripts

A variant analysis was conducted to confirm the genetic variations between the gray mold disease-resistant and -susceptible lines. The total raw variants was 359,288, which was generated using haplotypeCaller of the Genome Analysis Tool Kit (GATK, Broad Institute, Cambridge, MA, USA). These variants were confirmed to have occurred in 28,602 transcripts out of a total of 87,427 acquired transcripts. To remove the low-quality genotypes, the minimum depth coverage was set to five, and the maximum depth coverage was set to 100. The minimum genotype quality was filtered to 20%. As a result, the total filtered-out variants were 28,602, including 27,303 SNPs and 1299 indels. Among them, the variants that matched the following conditions were selected: four resistant lines showed the same genotype within a group, and three susceptible lines also showed the same genotype within a group but showed differences between the resistant and susceptible groups. A total of 233 variants were obtained that matched these conditions, and 118 corresponding transcripts were selected.

3.6. Selection of Transcripts Related to Disease Resistance through DEG Analysis and GO Annotation

A DEG analysis between the resistant and susceptible lines was performed to develop molecular markers related to disease resistance. TransDecoder was used to select 109,521 sequences with a length of 100 or more amino acids among the sequences generated using Trinity de novo assembly. As a result of calculating the read count of the transcript expression level using the StringTie program, we confirmed that the overall average mapping rate was 85%, and a total of 87,427 transcript read counts were obtained. Based on the read count of the transcript of each line, a comparative analysis between each onion line was performed using DESeq2. First, the raw read count data were normalized through size factor and dispersion. Thereafter, a correlation analysis between each line was performed based on the standardized value. In addition, from the RNA-seq results, the DEGs were analyzed for the resistant and susceptible groups. Using the DEGseq of the R package, it was confirmed that the difference in resistance versus susceptibility was statistically significant. Significantly expressed genes were verified using the MA plot results (**Figure S3**). In the MA plot, a false discovery rate (FDR) of less than 0.05 was indicated in red. Through the identification of the significantly expressed genes shown in red, it was confirmed that it was sufficient for use in the subsequent resistance-related gene analysis. In addition, a heatmap was constructed using z-scores to analyze the differences in expression for each line by using a group of significantly expressed genes (**Figure 4**). The heatmap was analyzed by comparing and analyzing the Pearson correlation coefficients for each line and gene after hierarchical clustering using the average linkage method. In addition, a volcano plot of the DEGs was obtained (**Figure S4**). The results of the volcano plot are shown in different colors according to the following conditions: FDR < 0.05 and |log2 fold change| ≥ 2. In the volcano plot, it was expressed in different colors according to the following conditions: red: FDR < 0.01 and |log2 fold change| ≤ 2, green: FDR < 0.01 and |log2 fold change| > 2, and orange: FDR ≥ 0.01 and |log2 fold change| > 2. Through this visualized DEG result, it was confirmed that there was a transcript showing a significant DEG difference between the gray mold-resistant and -susceptible groups.

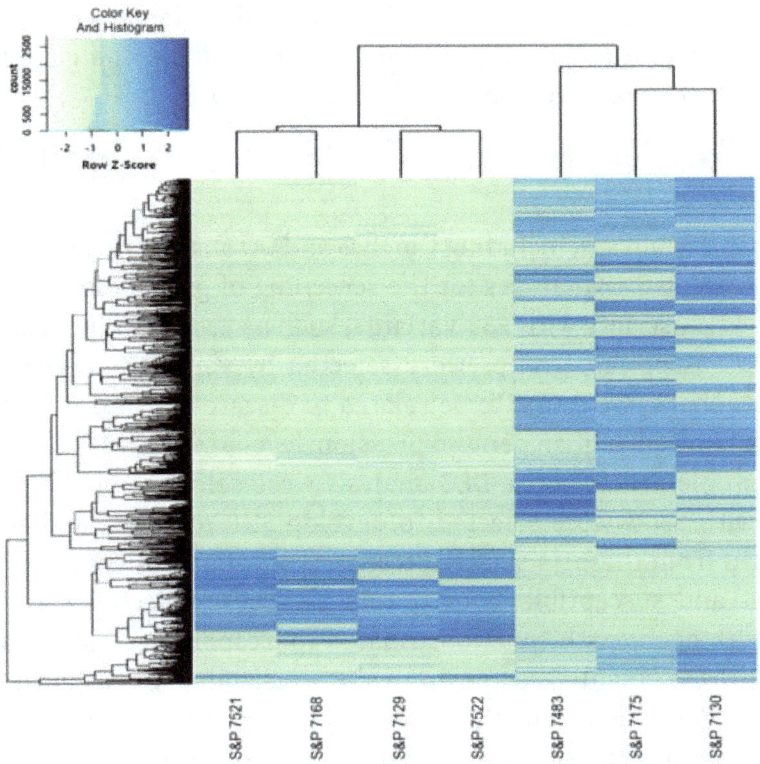

Figure 4. The heatmap constructed using z-scores to analyze the differences of expression in each line by using a group of significantly expressed genes.

The differences between resistant and susceptible lines of 'Seeds & People' Co. were significant, and it could be statistically confirmed that a resistance-related mechanism analysis was possible. Among the obtained 87,427 transcripts, the transcripts with significantly different expression levels between the resistant and susceptible lines were selected to satisfy the following conditions: PADJ < 0.05 and |log2 fold change| ≥ 2. A total of 1636 transcripts were selected. Among the 1636 transcripts, 320 transcripts showed higher expression levels in the resistant group, and 1316 showed higher expression in the susceptible group. Among the 320 transcripts that showed a higher level in the resistant group, only 182 transcripts matched to TAIR ID, while 138 did not, suggesting an unknown onion gene. In addition, among the 1316 transcripts that showed higher expression levels in the susceptible group, only 897 transcripts matched, while 419 did not. The matched transcripts were analyzed by GO annotations of the cellular components, molecular functions, and biological processes using TAIR ID.

To select genes with increased expression in relation to resistance, among the 182 transcripts with increased expression levels, 22 transcripts related to 'response to stress' and seven transcripts related to 'response to biotic stimuli' were analyzed by GO annotation (Figure S5). Finally, 29 resistance-related transcripts were confirmed to be related to disease resistance, and variants were also observed between the resistant and susceptible groups. In addition, the gene functions of the identified transcripts were analyzed using TAIR ID

3.7. Transcripts Selection to Develop SNP Markers and a HRM Analysis

Based on variant analyses, including the DEG and GO analyses, the transcripts were selected to develop SNP markers for the screening of gray mold-resistant onions. A total of 118 transcripts with 233 variants, such as SNPs and indels, between the resistant and susceptible groups were selected by variant analyses. Thereafter, 29 transcripts that were found to be related to disease resistance through the GO analysis and showed higher gene expression levels in the resistant group than in the susceptible group in the DEG analysis were selected. Finally, a total of 14 transcripts with SNPs were selected to develop SNP markers, and HRM primers were designed from these 14 selected transcripts with SNPs that existed between the resistant and susceptible lines (**Table 1**). These selected transcripts were associated with genes with functions related to 'response to biotic stimulus' and 'response to stress'. Plants are known to show resistance to pathogens in various ways. Although it prevents infection by pathogens structurally, such as through cell walls, it has several defense systems that show resistance to invading pathogens. The gene functions of the selected transcripts were 'lipoxygenase 3', 'Glutathione

S-transferase', and 'systemic acquired resistance', which were genes related to plant resistance (**Table 2**).

Table 2. Gene functions of 14 selected transcripts with SNPs that exist between the resistant and susceptible lines.

No	Transcripts No	TAIR ID [z]	Gene Function
1	SNP-1 transcript	AT1G17420	Lipoxygenase 3
2	SNP-2 transcript	AT1G42970	Glyceraldehyde-3-phosphate dehydrogenase B subunit
3	SNP-3 transcript	AT1G59960	Aldo/keto reductase
4	SNP-4 transcript	AT1G67090	Ribulose bisphosphate carboxylase small chain 1A
5	SNP-5 transcript	AT1G68090	Annexin Arabidopsis thaliana 5
6	SNP-6 transcript	AT1G78380	Glutathione S-transferase
7	SNP-7 transcript	AT3G16640	Translationally controlled tumor protein
8	SNP-8 transcript	AT4G39260	Glycine-rich protein
9	SNP-9 transcript	AT5G13750	Zinc induced facilitator-like 1
10	SNP-10 transcript	AT5G52810	Systemic acquired resistance
11	SNP-11 transcript	AT1G14290	Sphingoid base hydroxylase2
12	SNP-12 transcript	AT1G29930	Chlorophyll A/B protein
13	SNP-13 transcript	AT5G25220	KNAT3
14	SNP-14 transcript	AT5G38430	Rubisco small subunit 1B

[z] The Arabidopsis Information Resource ID.

Each HRM amplicon obtained from the 14 HRM primers was confirmed to match the expected size by electrophoresis. In addition, the sequence analysis confirmed that SNPs existed between the resistant and susceptible lines and revealed the amino acid sequences that were altered by the SNP (Figure 5). The SNPs targeted by the HRM primers are shown in green, and the nonsynonymous mutations are shown in red. Synonymous mutations are shown in blue and were present in SNP-1, SNP-2, SNP-4, SNP-12, and SNP 14. Nonsynonymous mutations were identified in SNP-3, SNP-5, SNP-6, SNP-7, SNP-8, SNP-9, SNP-10, SNP-11, and SNP-13. The melting peak values obtained from the HRM analysis were confirmed by ANOVA grouping. As a result, SNP-3 primers targeting the AKR gene transcript capable of distinguishing resistance and susceptibility were selected. The expected amplicon size of the SNP-3 primer was 134 bp. The HRM amplicon sizes of all seven onion lines were identified as the same. In addition, the sequences of the four resistant and three susceptible lines targeted by the SNP-3 primer were analyzed (Figure 6). Two SNPs were identified within the sequence. One SNP showed the 'T' allele in the resistant lines and the 'C' allele in the susceptible lines, while the other SNP showed the 'C' allele in the resistant lines and the 'G' allele in the susceptible ones.

Figure 5. SNPs and amino acid sequences targeted by the HRM primers in the resistant and susceptible lines. The SNPs targeted by HRM primers are shown in green. The synonymous mutations are shown in blue, and the nonsynonymous mutations between the resistant and susceptible groups are shown in red.

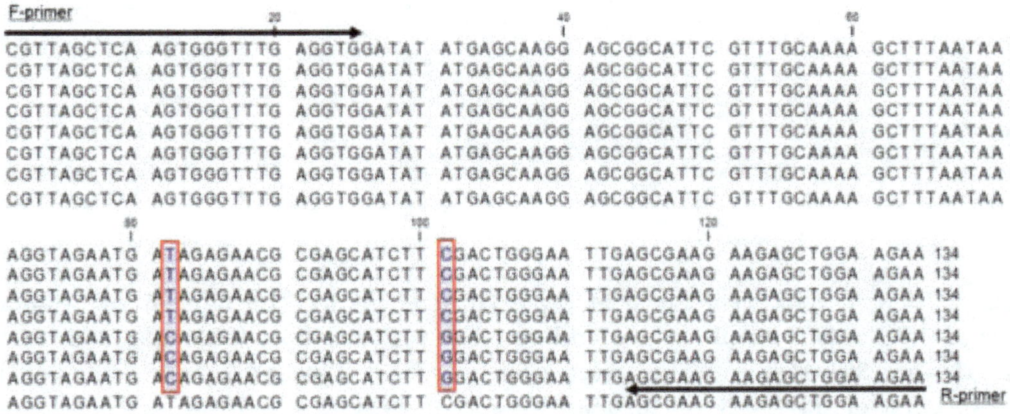

Figure 6. Sequences of the products amplified with the SNP-3 HRM primer in four resistant and three susceptible onion lines. Two SNPs were identified within the sequence. One SNP showed the 'T' allele in the resistant lines and the 'C' allele in the susceptible lines, while the other SNP showed the 'C' allele in the resistant lines and the 'G' allele in the susceptible ones. The forward (F) and reverse (R) primers for the SNP-3 HRM marker are underlined with the arrow.

In the results of the HRM analysis, which was repeated six times using the SNP-3 primer, the melting values of the resistant strains were 79.642–79.838 and those of the susceptible strains were 80.353–80.472. Upon statistically grouping these values by the Duncan method using ANOVA, significant results were obtained by grouping four resistant lines into one group (A) and three susceptibility lines into another group (B) (Table 3a). In the HRM result graph, the resistant lines (red) show a distinct curve from the susceptible lines (blue) (Figure 7a). Compared with the sequence results, it was confirmed that the resistant group that showed 'T' and 'C' at the SNP position had a lower melting temperature than the susceptible group that had 'C' and 'G'. In addition, for validation, the 'Asia Seed' Co. onion lines were analyzed using the SNP-3 HRM marker. As a result of ANOVA grouping, the resistant lines (Asia-12, Asia-48, and Asia-53) were grouped into one group (A), and the susceptible lines (Asia-30, Asia-35, Asia-45, and Asia-50) were also grouped together (B) (Table 3b). Similar to the HRM results of the 'Seeds & People' Co. onion lines, the HRM curve graph of the 'Asia Seed' Co. onion lines showed that the resistant lines were melted at a lower temperature than the susceptible lines (Figure 7b). Based on this result, it was confirmed that the SNP-3 HRM marker is versatile and can be used to select resistant onions.

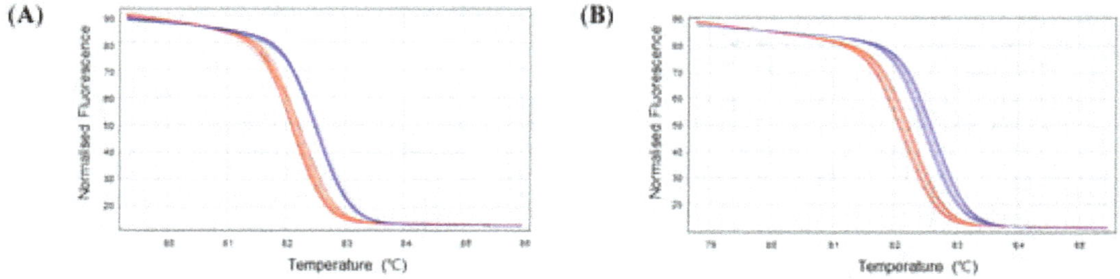

Figure 7. HRM analysis of the 'Seeds & People' Co. and 'Asia Seed' Co. onion lines. (A) HRM curve of the 'Seeds & People' Co. onion lines. Red: resistant lines (S&P 7522, S&P 7521, S&P 7129, and S&P 7168) and blue: susceptible lines (S&P 7130, S&P 7175, and S&P 7483). (B) HRM curve of the 'Asia Seed' Co. onion lines. Red: resistant lines (Asia-12, Asia-42, and Asia-53) and, blue: susceptible lines (Asia-30, Asia-35, Asia-45, and Asia-50).

Table 3. HRM analysis with the SNP-3 primers and statistically grouped by using ANOVA.

(A) HRM analysis results of the 'Seeds & People' Co. onion lines.				
HRM Marker	Line	LS Means	Group *	
SNP-3	S&P 7522	79.642	A	
	S&P 7521	79.768	A	
	S&P 7129	79.782	A	
	S&P 7168	79.838	A	
	S&P 7483	80.388		B
	S&P 7130	80.353		B
	S&P 7175	80.472		B
(B) HRM analysis results of the 'Asia seed' Co. onion lines.				
HRM Marker	Line	LS Means	Group *	
SNP-3	Asia-12	82.183	A	
	Asia-53	82.217	A	
	Asia-48	82.253	A	
	Asia-50	82.473		B
Asia-35 82.500				B
	Asia-45	82.527		B
	Asia-30	82.550		B
* In each column, means with the same letter are not significantly different.				

3.8. Confirmation of AKR Gene Expression Level in the Onion Lines through qPCR Analysis

The SNP-3 HRM marker was selected from the AKR transcript and showed higher expression levels in the resistant line group in the in silico data. qPCR was performed to confirm the gene expression levels of AKR. Before performing qPCR, RT-PCR was performed to confirm the amplification using the primer set designed for qPCR. RT-PCR showed that all seven onion lines were amplified into one band of 171 bp. qPCR was performed in six replicates by the SNP-3 primer, and the results were calculated using the ΔΔCt method. The gene expression level was normalized to the actin gene expression level and compared with the susceptible line S&P 7483 with the highest susceptibility (Figure 8). In the qPCR analysis, the expression level of the AKR gene was higher in the resistant group than in the susceptible group, with S&P 7522 showing the highest expression level and S&P 7130 showing the lowest expression level.

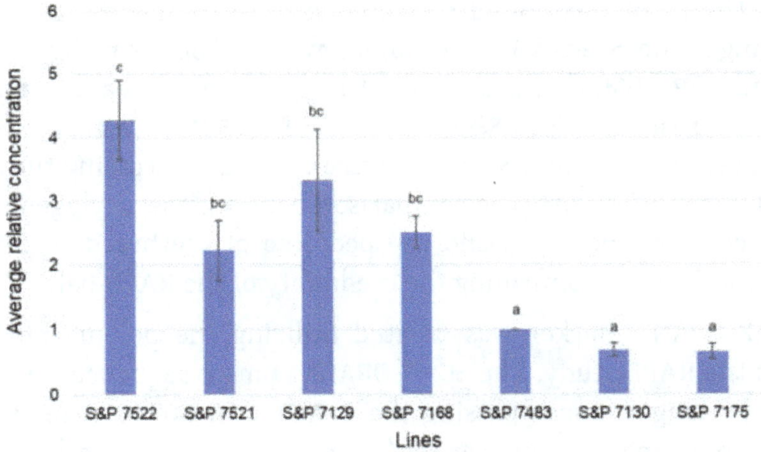

Figure 8. The gene expression level of the aldo-keto reductase (ARK) gene in the seven onion lines using the SNP-3 primer by the qPCR analysis. Four resistant lines (S&P 7522, S&P 7521, S&P 7129, and S&P 7168) and three susceptible lines (S&P 7483, S&P 7130, and S&P 7175) were analyzed. The qPCR analysis was performed in six replicates, and the standard errors are indicated as bars. Means with the same letter on the bar were not significantly different according to Duncan's multiple comparison test.

4. DISCUSSION

Molecular markers enable rapid and straightforward identification of genetic variations at the DNA level, without the need for phenotyping. RAPD markers can readily detect genetic variations by employing brief random primers, typically 10 base pairs long. Furthermore, RAPD is not dependent on prior knowledge of the target DNA sequence and is characterized by its cost-effectiveness, simplicity, speed, and user-friendly nature [25,26,27]. The RAPD data allow for the identification of minor genetic variations throughout extensive genomes. The genetic links among the onion lines can also be verified using the unweighted pair group technique with an arithmetic mean cluster dendrogram [15,16].

A phylogenetic analysis of lines resistant and susceptible to gray mold illness was undertaken in this work utilizing RAPD. The phylogenetic analysis derived from the RAPD revealed a close relationship between the resistant lines and other resistant lines, as well as between the susceptible lines and other susceptible lines. The varieties S&P 7522, S&P 7521, S&P 7129, and S&P 7168 are classified as belonging to the group of gray mold disease-resistant lines. On the other hand, the varieties S&P 7483, S&P 7130, and S&P 7175 are classified as belonging to the group of gray mold disease-susceptible lines. Furthermore, a strong genetic similarity was seen between the resistant S&P 7522 and S&P 7521 lines, as well as between the resistant S&P 7129 and S&P 7168 lines. In the phylogenetic study, it was observed that the susceptible lines S&P 7483, S&P 7130, and S&P 7175 had significant genetic similarities. The results exhibited a resemblance to the genealogy of the onion lines belonging to the 'Seeds & People' Company. Based on the pedigrees, two lines, S&P 7129 and S&P 7168, that are resistant to gray mold disease were developed by crossing the resistant lines S&P 7522 and S&P 7521. The gray mold disease-vulnerable lines S&P 7130 and S&P 7175 were created by crossing the susceptible lines S&P 7406 and S&P 7405. The comparison between the phylogenetic analysis results derived from the RAPD and the pedigree of the breeding lines showed consistent genetic links, confirming the credibility of the RAPD data.

The SCAR-OPAN1 marker was created utilizing the polymorphic fragment identified in the RAPD study. The SCAR-OPAN1 marker sequence was specifically designed to be longer, encompassing the 5′-ACT CCA CGT C-3′ sequence of the OPAN-1 random primer. The primer set for the SCAR- OPAN1 marker was created taking into account the annealing temperature, GC content, the potential for dimer formation, and the presence of hairpin loops. A BLAST analysis was conducted to examine the products that were amplified by the SCAR marker. The BLAST analysis of the DNA sequence derived from the amplified segment of the SCAR marker did not yield any significant matches with sequences from onions or other species. Consequently, further analysis is necessary. SCAR markers have been employed to differentiate between different cultivars and species [28,29]. SCAR markers are utilized for the purpose of selecting individuals with particular features and identifying individuals that are resistant to diseases. Reports have indicated the use of SCAR markers to select for male infertility-dominant Ms and -recessive ms in onions [30], anthrax resistance in grapes [31], and Acokita blight resistance in lentils [32]. The use of the SCAR-OPAN1 marker in the onion breeding lines of 'Seeds & People' Co. has been shown effective in selecting gray mold disease-resistant onions.

Furthermore, to authenticate the developed SCAR-OPAN1 marker, further examination was conducted on gray mold disease-resistant and -susceptible lines obtained from the 'Asia Seed' Co. During the examination of the 'Asia Seed' Co. lines for resistance to gray mold illness, it was verified that a 2Kb product was only detected in the lines that were resistant. Thus, it has been verified that the SCAR-OPAN1 marker, established in this work, not only enables the identification of gray mold-resistant onions but also expedites the selection process significantly.

The HRM marker was created to identify onion lines that are resistant to gray mold disease. This was achieved by analyzing the selected transcripts obtained through a DEG analysis from RNA sequencing. After doing an analysis of the RNA-seq data, we further examined 14 specific transcripts. To do this, we utilized the Arabidopsis Information Resource (TAIR) ID, which is obtained from Arabidopsis. The transcripts that were matched were subjected to analysis using GO annotation and primarily categorized into cellular components, molecular functions, and biological processes. From the available transcripts, those associated with disease resistance were chosen. These transcripts were categorized as 'reaction to biotic stimulation' and 'response to stress'.

Plant cells produce reactive oxygen species (ROS) and pathogenesis-related proteins (PR proteins) upon detecting the presence of pathogens. This triggers a hypersensitive response (HR) that inhibits the growth of pathogens. These resistance reactions can trigger a systemic acquired resistance (SAR) response, which is a resistance response that affects the entire plant. When the systemic acquired resistance (SAR) is triggered by pathogens including fungi, bacteria, and viruses, there is an accumulation of salicylic acid (SA), which is crucial for the development of SAR [33,34].

The SNP-10 transcript was used to target the SAR gene in this investigation. It revealed differentially expressed genes (DEGs) and variations between the groups of gray mold disease-resistant and -susceptible lines, as shown in Table 2. Following an analysis of the correlation between SA accumulation and SAR during the early 1990s, SARs have been detected in Arabidopsis thaliana, tobacco, and cucumbers. The immune response triggered by systemic acquired resistance (SAR) is initiated by pathogenesis-related (PR) proteins and has been extensively investigated for its ability to detect salicylic acid (SA) as a signaling molecule [35,36,37]. Confirmation of the resistance mechanism in tobacco against B. cinerea and Pseudomonas syringae occurred when it was observed that SAR, PR genes (PR-1 and PR-5), and SA were involved before the fabrication of complete plants [38].

Unlike SAR, induced systemic resistance (ISR) is another type of plant defense response that is triggered by the plant hormones jasmonic acid (JA) and ethylene, through the activation of different signaling pathways. Jasmonic acid (JA) is a phytohormone that plays a crucial role in the regulation of disease resistance in plants. The transcript specifically targeted by SNP-1 was identified as the lipoxygenase (LOX) 3 gene, as shown in Table 2. The LOX gene plays a role in the production of jasmonic acid (JA). Specifically, the expression of LOX3 was quickly increased when the pathogen B. cinerea was introduced into Arabidopsis thaliana. The upregulation of LOX was observed in LOX3 and LOX4, which is likely associated with the early reaction of oligogalacturonides functioning as damage-associated molecular patterns (DAMPs) [39].

Furthermore, the glutathione S-transferase (GST) gene, which is the specific gene affected by SNP-6, has been discovered in numerous research investigating disease resistance in plants (Table 2) [40,41,42]. Glutathione S-transferase (GST) functions as a significant antioxidant and exhibits resistance in plants when it comes to a hypersensitive reaction leading to cell death. The Nicotiana benthamiana plant underwent a genetic transformation that resulted in resistance to the fungal pathogens Colletotrichum destructivum and *C. orbiculare*. This resistance was achieved through the incorporation of the GST gene [43].

Additional transcripts have also verified that target genes possess diverse activities associated with plant resistance. One of the transcripts, SNP-3, which was identified as an HRM marker, was found to be an aldo-keto reductase gene (AKR) according to Table 2. The SNP-3 HRM marker associated with the AKR gene has been verified to be capable of identifying gray mold disease-resistant lines from both 'Seeds & People' Co. and 'Asia Seed' Co. The expression of the AKR gene was greater in the group of gray mold disease-resistant lines compared to the group of susceptible lines. Previous studies have discussed AKR and it is recognized that it enhances resistance when expressed at high levels [44,45].

Given that the SNP-3 HRM marker was a marker based on genes, a quantitative polymerase chain reaction (qPCR) investigation was performed to verify the level of expression of the AKR gene. The expression level of AKR was elevated in the resistant group compared to the susceptible group. The qPCR results were compared to the in silico data. The RNA-seq analysis revealed that the resistant group had a greater expression level of AKR compared to the susceptible group. Additionally, S&P7522 displayed the highest expression level among all the genes studied. The congruity of the AKR gene expression levels between the qPCR analysis and in silico RNA-seq results demonstrated the gene's correlation with resistance to gray mold sickness, hence enhancing the dependability of the in silico data employed in the development of molecular markers.

Furthermore, it has been determined that the AKR gene confers several stress tolerances in plants. Specifically, research has been conducted on abiotic challenges, including herbicide resistance, heat stress tolerance, and biotic stress, such as mildew [41,46,47]. The AKR gene group primarily participates in plant secondary metabolic pathways, specifically in the manufacture of flavonoids, during plant-microbial interactions to combat biotic stress caused by diseases like microbes. Thus, these AKRs primarily serve as plant defense mechanisms against biological stressors, such as an attack by pathogens. Although plant AKR has been identified as a possible target for generating plant species that are tolerant to abiotic and biotic stress, its importance has not been stressed yet, according to reports [46,48].

Multiple studies have demonstrated that AKR plays a role in relation to stress. This study also demonstrated variations in resistance and gene expression levels based on single nucleotide polymorphisms (SNPs) across the groups of gray mold-resistant and gray mold-susceptible lines. The AKR gene was found to influence the resistance of the group to gray mold sickness. Hence, additional research on AKR is necessary, and the SNP-3 HRM marker established in this investigation could be employed for the purpose of choosing onion lines that are resistant to gray mold illness.

5. CONCISE SUMMARY

Onions possess a considerable genome size, and the available data on the current DNA sequences is inadequate. This study aimed to create a molecular marker for the purpose of breeding onions that are resistant to gray mold illness. A dendrogram obtained from the RAPD analysis was used to conduct a phylogenetic analysis comparing the gray mold disease-resistant and -susceptible lines. The dendrogram that was created was compared to the pedigree of the gray mold disease-resistant and -susceptible breeding lines provided by 'Seeds & People' Co. The phylogenetic analysis results were in conformity with the genetic relationship of the breeding lines pedigree and demonstrated the credibility of the RAPD results. Subsequently, a SCAR-OPAN1 marker was created using the RAPD data, which revealed distinct pieces that differed between the resistant and susceptible lines. The SCAR-OPAN1 marker selectively amplified the resistant lines, producing a particular 2-Kb product, while no amplification was observed in the susceptible lines.

Furthermore, the RNA-seq analysis of onion lines that are resistant and vulnerable to gray mold illness was conducted using Next-Generation Sequencing (NGS) technology. The RNA-seq results were used to conduct DEG and GO analysis in order to discover variations, including SNPs and indels. Consequently, a selectable marker called SNP-3 HRM was created for the purpose of selecting gray mold-resistant lines. The SNP-3 HRM marker is used to select resistant lines. It consists of single nucleotide polymorphisms (SNPs) that are found in the AKR gene and are highly expressed in these lines. Therefore, in this research, the SCAR-OPAN1 and SNP-3 HRM markers were created to choose onion lines that are resistant to gray mold disease in breeding programs, with the aim of minimizing the harm caused by this illness. By utilizing these molecular markers, it is possible to reduce the breeding period of biennial onions by selecting an onion line that exhibits resistance to gray mold disease. This approach effectively mitigates the economic losses associated with onions.

AT A GLANCE

Onions, scientifically known as *Allium cepa* L., are a highly popular vegetable crop that is widely consumed globally. However, they are susceptible to many fungal diseases that can occur either in the field or during storage. Gray mold disease, caused by the necrotrophic pathogens *Botrytis cinerea* and Botrytis squamosa, is a condition that decreases the productivity and storage longevity of onions. Controlling gray mold disease in onions proves challenging when employing physical and chemical treatments. Developing onions with resistance to gray mold disease can effectively mitigate pathogen-induced damage, minimize the labor-intensive management measures, and alleviate environmental pollution resulting from fungicide usage. Onions possess a substantial genome size of 16Gb, which presents challenges in their analysis. Additionally, their biennial life cycle contributes to an extended breeding period. Thus, this study aimed to create markers that would reduce the duration of the onion breeding process. Initially, a technique called random amplified polymorphic DNA (RAPD) was utilized to verify the genetic correlation between the gray mold disease-resistant and -susceptible lines. This was accomplished by constructing a dendrogram. Furthermore, a marker called sequence characteristic amplified region (SCAR)-OPAN1 was established to identify resistant lines. This marker was created utilizing a polymorphic RAPD fragment. Next, the RNA-seq data from onion lines that are resistant and vulnerable to gray mold were evaluated using Next-Generation Sequencing (NGS) technology. Analyzed the RNA-seq results and did DEG and GO analyses to construct a selectable marker for the resistant line by studying the variants, such as SNPs and indels. This work devised the SNP-3 HRM marker to choose gray mold disease-resistant lines. This was accomplished by utilizing the single nucleotide polymorphisms (SNPs) found in the aldo-keto reductase (AKR) gene, which exhibited elevated expression levels in these lines. The SCAR-OPAN1 and SNP-3 HRM markers, which were created during this research, have the potential to be utilized in breeding programs to identify gray mold disease-resistant onions. This would help mitigate the harm caused by gray mold disease.

REFERENCES

1. Nam, K.H.; Baik, H.W.; Choi, T.Y.; Yoon, S.G.; Park, S.W.; Joung, H.J. Effects of ethanol extract of onion on the lipid profiles in patients with hypercholesterolemia. J. Nutr. Health 2007, 40, 242–248.
2. Kim, J.Y.; Seo, Y.J.; Noh, S.K.; Cha, Y.J. Concentrated onion extract lowers serum lipid levels in rats fed a high-fat diet. Korean J. Food Preserv. 2010, 17, 398–404.

3. Khosa, J.S.; McCallum, J.; Dhatt, A.S.; Macknight, R.C. Enhancing onion breeding using molecular tools. Plant. Breed. 2016, 135, 9–20.

4. Jo, J.; Purushotham, P.M.; Han, K.; Lee, H.R.; Nah, G.; Kang, B.C. Development of a genetic map for onion (*Allium cepa* L.) using reference-free genotyping-by-sequencing and SNP assays. Front. Plant. Sci. 2017, 8, 1606.

5. Havey, M.J. Onion Breeding. In Plant Breeding Review; Wiley: New Jersey, NJ, USA, 2018; Volume 42, pp. 39–85.

6. Chilvers, M.I.; du Toit, L.J. Detection and identification of Botrytis species associated with neck rot, scape blight, and umbel blight of onion. Plant. Health Prog. 2006, 7, 38.

7. Williamson, B.; Tudzynski, B.; Tudzynski, P.; van Kan, J.A.L. *Botrytis cinerea*: The cause of grey mould disease. Mol. Plant. Pathol. 2007, 8, 561–580.

8. Choquer, M.; Fournier, E.; Kunz, C.; Levis, C.; Pradier, J.M.; Simon, A.; Viaud, M. *Botrytis cinerea* virulence factors: New insights into a necrotrophic and polyphageous pathogen. FEMS Microbiol. Lett. 2007, 277, 1–10.

9. Carisse, O.; Tremblay, D.M.; McDonald, M.R.; Brodeur, L.; McRoberts, N. Management of botrytis leaf blight of onion: The Québec experience of 20 years of continual improvement. Plant. Dis. 2001, 95, 504–514.

10. Abbey, J.A.; Percival, D.; Abbey, L.; Asiedu, S.K.; Prithiviraj, B.; Schilder, A. Biofungicides as alternative to synthetic fungicide control of grey mould (*Botrytis cinerea*)—prospects and challenges. Biocontrol. Sci. Technol. 2018, 29, 207–228.

11. Howell, G.S. Sustainable grape productivity and the growth-yield relationship: A review. Am. J. Enol. Vitic. 2001, 52, 165–174.

12. Tripathi, P.; Dubey, N.K.; Shukla, A.K. Use of some essential oils as post-harvest botanical fungicides in the management of grey mould of grapes caused by *Botrytis cinerea*. World J. Microbiol. Biotechnol. 2008, 24, 39–46.

13. Fan, F.; Hamada, M.S.; Li, N.; Li, G.Q.; Luo, C.X. Multiple fungicide resistance in *Botrytis cinerea* from greenhouse strawberries in Hubei province, China. Plant. Dis. 2017, 101, 601–606.

14. Rupp, S.; Weber, R.W.S.; Rieger, D.; Detzel, P.; Hahn, M. Spread of *Botrytis cinerea* strains with multiple fungicide resistance in German horticulture. Front. Microbiol. 2017, 7, 2075.

15. Maniruzzaman, M.; Haque, M.E.; Haque, M.M.; Sayem, M.A.; Al-Amin, M. Molecular characterization of onion (*Allium cepa*) using RAPD markers. Bangladesh J. Agric. Res. 2010, 35, 313–322.

16. Sudha, G.S.; Ramesh, P.; Sekhar, A.C.; Krishna, T.S.; Bramhachari, P.V.; Riazunnisa, K. Genetic diversity analysis of selected onion (*Allium cepa* L.) germplasm using specific RAPD and ISSR polymorphism markers. Biocatal. Agric. Biotechnol. 2019, 17, 110–118.

17. Wang, Z.; Gerstein, M.; Snyder, M. RNA-Seq: A revolutionary tool for transcriptomics. Nat. Rev. Genet. 2009, 10, 57–63.

18. Han, J.; Thamllarasan, S.K.; Natarajan, S.; Park, J.I.; Chung, M.Y.; Nou, I.S. De novo assembly and transcriptome analysis of bulb onion (*Allium cepa* L.) during cold acclimation using contrasting genotypes. PLoS ONE 2016, 11, e0161987.

19. Graham, R.; Liew, M.; Meadows, C.; Lyon, E.; Wittwer, C.T. Distinguishing different DNA Heterozygotes by high-resolution melting. Clin. Chem. 2005, 51, 1295–1298.

20. Reed, G.H.; Kent, J.O.; Wittwer, C.T. High-resolution DNA melting analysis for simple and efficient molecular diagnostics. Pharmacogenomics 2007, 8, 597–608.

21. Doyle, J.J.; Doyle, J.L. Isolation of plant DNA from fresh tissue. Focus 1990, 12, 13–15.

22. Sokal, R.R.; Michener, C.D. A statistical method for evaluation systematic relationships. Univ. Kans. Sci. Bull. 1958, 28, 1409–1438.

23. Jaccard, P. Étude comparative de la distribution florale dans une portion des Alpes et des Jura. Bull. Soc. Vaud. Sci. Nat. 1901, 37, 547–579.

24. Wang, L.; Feng, Z.; Wang, X.; Wang, X.; Zhang, X. DEGseq: An R package for identifying differentially expressed genes from RNA-seq data. Bioinformatics 2010, 26, 136–138.

25. Welsh, J.; McClelland, M. Fingerprinting genomes using PCR with arbitrary primers. Nucleic. Acids Res. 1990, 18, 7213–7218.

26. Williams, J.G.K.; Kubelik, A.R.; Livak, K.J.; Rafalski, J.A.; Tingey, S.V. DNA polymorphisms amplified by arbitrary primers are useful as genetic markers. Nucleic. Acids Res. 1990, 18, 6531–6535.

27. Sidiq, Y.; Subiastuti, A.S.; Wibowo, W.A.; Daryono, B.S. Development of SCAR marker linked to begomovirus resistance in melon (*Cucumis melo* L.). Jordan J. Biol. Sci. 2020, 13, 145–151.

28. Baite, M.S.; Upadhyay, B.K.; Dube, S.C. Development of a sequence-characterized amplified region marker for detection of *Ascochyta rabiei* causing Ascochyta blight in chickpea. Folia Microbiol. 2020, 65, 103–108.

29. Mei, Z.; Khan, M.A. Genetic authentication of Eclipta prostrate (Asteraceae) from Penthorum chinense (Penthoraceae) by sequence characterized amplified region (SCAR) markers. Rev. Biol. Trop. 2020, 68, 180–188.

30. Yang, Y.Y.; Huo, Y.M.; Miao, J.; Liu, B.J.; Kong, S.P.; Gao, L.M.; Liu, C.; Wang, Z.B.; Tahara, Y.; Kitano, H.; et al. Identification of two SCAR markers co-segregated with the dominant Ms and recessive ms alleles in onion (*Allium cepa* L.). Euphytica 2013, 190, 267–277.

31. Kim, G.H.; Yun, H.K.; Choi, C.S.; Park, J.H.; Jung, Y.J.; Park, K.S.; Dane, F.; Kang, K.K. Identification of AFLP and RAPD markers linked to anthracnose resistance in grapes and their conversion to SCAR markers. Plant. Breed. 2008, 127, 418–423.

32. Chowdhury, M.A.; Andrahennadi, C.P.; Slinkard, A.E.; Vandenberg, A. RAPD and SCAR markers for resistance to acochyta blight in lentil. Euphytica 2001, 118, 331–337.

33. Hunt, M.D.; Ryals, J.A.; Reinhardt, D. Systemic acquired resistance signal transduction. Crit. Rev. Plant. Sci. 1996, 15, 583–606.

34. Durrant, W.E.; Dong, X. Systemic acquired resistance. Annu. Rev. Phytopathol. 2004, 42, 185–209.

35. Ward, E.R.; Uknes, S.J.; Williams, S.C.; Dincher, S.S.; Wiederhold, D.L.; Alexander, D.C.; Aphl-Goy, P.; Metraux, J.P.; Ryals, J.A. Coordinate gene activity in response to agents that induce systemic acquired resistance. Plant. Cell 1991, 3, 1085–1094.

36. Sticher, L.; Mauch-Mani, B.; Métraux, A.J. Systemic acquired resistance. Annu. Rev. Phytopathol. 1997, 35, 235–270.

37. Mauch-Mani, B.; Métraux, J.P. Salicylic acid and systemic acquired resistance to pathogen attack. Ann. Bot. 1998, 82, 535–540.

38. Frías, M.; Brito, N.; González, C. The *Botrytis cinerea* cerato-platanin BcSpl1 is a potent inducer of systemic acquired resistance (SAR) in tobacco and generates a wave of salicylic acid expanding from the site of application. Mol. Plant. Pathol. 2013, 14, 191–196.

39. Windram, O.; Madhou, P.; McHattie, S.; Hill, C.; Hickman, R.; Cooke, E.; Jenkins, D.J.; Penfold, C.A.; Baxter, L.; Breeze, E.; et al. Arabidopsis defense against *Botrytis cinerea*: Chronology and regulation deciphered by high-resolution temporal transcriptomic analysis. Plant. Cell 2012, 24, 3530–3557.

40. Poland, C.A.; Duffin, R.; Kinloch, I.; Maynard, A.; Wallace, W.A.; Seaton, A.; Stone, V.; Brown, S.; MacNee, W.; Donaldson, K. Carbon nanotubes introduced into the abdominal cavity of mice show asbestos-like pathogenicity in a pilot study. Nat. Nanotechnol. 2008, 3, 423–428.

41. Turóczy, Z.; Kis, P.; Török, K.; Cserháti, M.; Lendvai, Á.; Dudits, D.; Horváth, G.V. Overproduction of a rice aldo-keto reductase increases oxidative and heat stress tolerance by malondialdehyde and methylglyoxal detoxification. Plant. Mol. Biol. 2011, 75, 399–412.

42. Gullner, G.; Komives, T.; Király, L.; Schröder, P. Glutathione S-transferase enzymes in plant-pathogen interactions. Front. Plant Sci. 2018, 9, 1836.

43. Dean, J.D.; Goodwin, P.H.; Hsiang, T. Induction of glutathione S-transferase genes of Nicotiana benthamiana following infection by *Colletotrichum destructivum* and *C. orbiculare* and involvement of one in resistance. J. Exp. Bot. 2005, 56, 1525–1533.

44. Simpson, P.J.; Tantitadapitak, C.; Reed, A.M.; Mather, O.C.; Bunce, C.M.; White, S.A.; Ride, J.P. Characterization of two novel aldo—keto reductases from Arabidopsis: Expression patterns, broad substrate specificity, and an open active-site structure suggest a role in toxicant metabolism following stress. J. Mol. Biol. 2009, 392, 465–480.

45. Kanayama, Y.; Mizutani, R.; Yaguchi, S.; Hojo, A.; Ikeda, H.; Nishiyama, M.; Kanahama, K. Characterization of an uncharacterized aldo-keto reductase gene from peach and its role in abiotic stress tolerance. Phytochemistry 2014, 104, 30–36.

46. Sengupta, D.; Naik, D.; Reddy, A.R. Plant aldo-keto reductases (AKRs) as multi-tasking soldiers involved in diverse plant metabolic processes and stress defense: A structure-function update. J. Plant. Physiol. 2015, 179, 40–55.

47. Vemanna, R.S.; Vennapusa, A.R.; Easwaran, M.; Chandrashekar, B.K.; Rao, H.; Ghanti, K.; Sudhakar, C.; Mysore, K.S.; Makarla, U. Aldo-keto reductase enzymes detoxify glyphosate and improve herbicide resistance in plants. Plant. Biotechnol. J. 2017, 15, 794–804.

48. Wang, H.; He, Z.; Luo, L.; Zhao, X.; Lu, Z.; Luo, T.; Li, M.; Zhang, Y. An aldo-keto reductase, Bbakr1, is involved in stress response and detoxification of heavy metal chromium but not required for virulence in the insect fungal pathogen, *Beauveria bassiana*. Fungal Genet. Biol. 2018, 111, 7–15.

6 Chapter

GENERATION OF A HIGH-DENSITY GENETIC MAP OF PEPPER (*CAPSICUM ANNUUM* L.) BY SLAF-SEQ AND QTL ANALYSIS OF *PHYTOPHTHORA CAPSICI* RESISTANCE

1. INTRODUCTION: AN OVERVIEW

Pepper, scientifically known as *Capsicum annuum* L., is a widely used seasoning and a vegetable crop of great economic importance. Not only is it utilized in numerous culinary traditions, but it is also shown to possess numerous therapeutic attributes. According to data from the Food and Agriculture Organization (FAO), the global production of chilies and peppers in 2019 amounted to around 212.04 million tons. This production was spread across an area of about 49.31 million hectares. Pepper is vulnerable to several diseases, including CMV, TMV, Colletotrichum spp., and Phytophthora capsici (*P. capsici*) [1,2,3,4]. Phytophthora blight has a substantial negative impact on both the quantity and quality of pepper crops [5]. The disease is caused by the oomycete plant pathogen *P. capsici*, which first infects the roots and crown roots, and then spreads to all parts of the plant, including the roots, stems, fruits, and leaves [6]. Phytophthora blight is a serious disease that typically develops in warm (25–28 °C) and highly humid environments [7,8,9]. As of now, there are no known strategies that are both effective and safe for controlling Phytophthora blight, save for chemical control [10,11,12,13]. Thus, the use of resistant cultivars has emerged as a straightforward, efficient, and secure method for addressing the presence of Phytophthora blight in pepper plants. Plant breeders have also prioritized the selection of cultivars with elevated levels of resistance.

The three physiological races of *P. capsici*, referred to as “races 1-3,” have been identified based on their ability to infect four different pepper varieties: early calwonder (susceptible), PI201234 (resistant), PBC137 (partially resistant), and PBC602 (partially resistant) [14]. Prior research has shown other pepper varieties that exhibit resistance to *P. capsici*, such as PI123469, PI201232, PI201234, AC2258, and CM334 (Criollo de Morelos 334) [4,14,15,16,17,18]. The resistance to *P. capsici* is mostly controlled by a single dominant gene in PI201234 or by one dominant gene with modifiers [9,19,20,21]. AC2258, which is generated from PI201234, exhibits resistance to *P. capsici* [17,18]. Research has demonstrated that the resistance to *P. capsici* in CM334 is governed by at least two genes [22,23]. Furthermore, these findings have uncovered that the regulatory mechanism governing resistance to *P. capsici* in pepper is exceedingly intricate. Several studies have examined the impact of pepper QTLs on chromosomes that are linked to resistance against *P. capsici* [1 8,23,24,25,26,27,28,29,30,31]. Pc5.1 is a QTL (quantitative trait locus) located on chromosome 5 of CM334, PI201234, and Perennial. It is homologous, meaning it is similar or equivalent, in all three varieties. Pc5.1 has been linked to resistance to *P. capsici*, a specific pathogen. This association has been supported by research studies referenced as [23, 29, 31]. In their study, Mallard et al. (2013) conducted a meta-analysis and successfully discovered resistance quantitative trait loci (QTLs) within three meta-QTLs, namely MetaPc5.1, MetaPc5.2, and MetaPc5.3 [31]. In their study, Siddique et al. (2019) discovered three quantitative trait loci (QTLs) on chromosome P5, namely QTL5.1, QTL5.2, and QTL5.3. These QTLs were found to be linked to resistance to three different isolates of *P. capsici* (race 1, race 2, and race 3). The researchers employed a combination of standard QTL mapping and genome-wide association study (GWAS) techniques to establish this association [30]. Furthermore, some QTLs with small effects have been detected on several chromosomes [23, 27, 28, 32].

The utilization of next-generation sequencing (NGS) has led to the identification of extensive SNP markers, which have greatly facilitated the development of the pepper genetic map. SLAF-seq is an innovative method for sequencing that is both cost-effective and less complicated compared to high-quality reference genome libraries [33]. Furthermore, the SLAF-seq technique has been widely employed for the creation of dense genetic maps in several species and for the mapping of quantitative trait loci (QTL) [34,35,36,37,38,39,40,41,42]. This approach has also been effectively employed in the construction of high-density pepper genetic maps [40,43,44]. Guo et al. (2017) identified two potential CMV resistance genes on pepper chromosomes 2 and 11 utilizing SLAF-seq and BSA technologies [43]. Zhang et al. (2019) employed SLAF-seq to identify two significant QTLs that exhibited a substantial correlation with FFN [40].

In this study, we constructed a pepper linkage map with high density using SLAF-seq. Additionally, we found quantitative trait loci (QTLs) associated with resistance to *P. capsici* by analyzing F2 populations derived from a cross between two parental lines, 1287 (female, susceptible to *P. capsici*) and PI201234 (male, resistant to *P. capsici*). Ultimately, we examined the primary impact of Quantitative Trait Loci (QTLs) and identified specific genes of interest. Our findings have the potential to aid in understanding the genetic process that causes resistance to *P.*

capsici in pepper plants. This knowledge can serve as a basis for developing pepper cultivars that are highly resistant to the disease.

2. MATERIALS AND METHODS

2.1. Mapping Population

The sweet pepper line "1287" that is vulnerable to *P. capsici* was acquired from Zhongjiao808, whereas the P. capsici-resistant line "PI201234" was taken from Central America. The current investigation employed an F2 mapping population consisting of 150 individuals derived from the cross between female parent 1287 and male parent PI201234, which was then utilized as the mapping population. The parental lines and the F2 population were cultivated at the Chongqing experimental station of the Chongqing Academy of Agricultural Sciences, located in Chongqing, China. In 2018, pepper seeds were sown in 50-cell trays filled with a mixture of peat and vermiculite. The mixture was sterilized using an autoclave for 30 minutes.

2.2. Pathogen Preparation and Plant Inoculation with *P. capsici*

P. capsici isolate HT1 was used for *P. capsici* resistance identification in pepper. HT1 has been identified as physiological race 3 and was isolated from infected pepper fruit at the experimental station in Jiulongpo District, Chongqing, China. The isolate was cultured on V8 juice-agar medium at 28 °C in an incubator. To prepare the inoculums for disease screening, the cultures were soaked in 5 mL ddH2O and cultivated at 4 °C for 1 h and then set at room temperature for 1 h to promote sporulation. Spore density was determined using a hemocytometer and adjusted to 1 × 105 spores/mL in distilled water. Before inoculation, pepper plants were soaked in water. Then, 5 mL of suspension was injected into the root of each six- to seven-true-leaf stage pepper plant. The inoculated plants were then grown at 28 °C for 16 h/day and at %80 relative humidity.

2.3. Disease Evaluation

Seven days post inoculation, the plants were assessed for disease symptoms using the 0–5 scale of the Chinese standard NY/T 2060.1-2011 (Ministry of Agriculture of the People's Republic of China 2011), which consisted of the following: 0 = no disease symptoms; 1 = emergence of brown lesions in the roots and stems with no to slight wilting of leaves; 2 = extension of root and stem lesions by 1–2 cm, the leaves wilted and had fallen off; 3 = root and stem lesions exceed 2 cm and leaves clearly show wilting or defoliation; 4 = large brown lesions on stems are extended and dehydrated, with the exception of the uppermost leaves which have been lost; and 5 = plant death. According to the disease grade of each plant, the disease index (DI) of each identification material was calculated. The DI was calculated using the equation below:

$$\mathrm{DI} = \frac{\Sigma(\mathrm{s} \times \mathrm{n})}{\mathrm{N} \times \mathrm{S}} \times 100,$$

where s is the disease level ranging between 0 and 5; n is the number of plants with corresponding disease level; N is the number of plants investigated in each F2; and S is the representative value of the highest grade.

2.4. Statistical Analysis of Phenotypic Data

The laboratory study was conducted at the experimental station of the Chongqing Academy of Agricultural Sciences. The phenotypic data collected for the disease parameters were considered and analyzed as individual traits. The resistance traits were recorded for the F2 population and parents. The traits means were calculated using DPS 18.10 (DPS, China).

2.5. DNA Extraction, SLAF Library Construction, and High-Throughput Sequencing

An improved CTAB method was utilized to extract genomic DNA from the young leaves of two parental lines and 150 F2 individuals that were at the five- to six-leaf stage [**45**]. We employed an ND-1000 spectrophotometer (NanoDrop, Wilmington, DE, USA) and performed 1.0% agarose gel electrophoresis to respectively measure DNA concentration and quality. The SLAF-seq library was constructed as detailed previously by Sun et al. [**33**], with only a few small changes. The restriction enzyme *Hae*III (New England Biolabs, NEB, USA) was utilized for digestion of the genomic DNA of the parental lines and individuals of the F2 population. We added polyA tails to the 3′ ends of the digested fragments, which were then connected to duplex-labelled sequencing adapters and PCR amplified. PCR was performed with the diluted restriction-ligation DNA sample, Q5® High-Fidelity DNA polymerase (NEB), dNTPs, and PCR primers (forward, -'5AATGATACGGCGACCACCGA-3′ and reverse, 5′-CAAGCAGAAGACGGCATACG-3′). The PCR products were purified with Agencourt AMPure XP beads (Beckman Coulter, High Wycombe, UK) and then resolved on a 2% agarose gel. Fragments that were 314 to 364 bp in size were separated and purified with a QIAquick gel extraction kit (Qiagen, Hilden, Germany). SLAF-seq was then conducted on an Illumina High-Seq 2500 sequencing platform (Illumina, San Diego, CA, USA) at Beijing Biomarker Technologies Corp. (Beijing, China, **http://www.biomarker.com.cn**, accessed on: 8 January 2019). We employed the *Oryza sativa* L. genome as reference for quality control and conducted library construction and sequencing using similar settings as that for the pepper mapping population.

2.6. SLAF-seq Data Grouping and Genotyping

In this study, reads with a quality score below Q30 (quality score < 30e) were filtered out. After that, high-quality reads were mapped to the pepper reference genome utilizing BWA software, with the paired-end mapped reads at the identical position and >95% identity divided into a single SLAF locus. In each SLAF, a polymorphism locus was observed between the parents, of which most were SNPs. All of the polymorphism SLAF loci were then genotyped with consistency at SNP loci of the offspring and parents. SLAFs that consisted of more than eight SNPs were screened out, and then the parental SLAFs with a sequencing depth of <10-fold were discarded. A high-density linkage map was then created using polymorphic SLAFs showing parental homozygosis (aa × bb).

2.7. High-Density Linkage Map Construction

We quantified the modified logarithm of odds (MLOD) value between two adjacent markers and markers with MLOD values < 5 were filtered out. Then, the SLAF markers were assigned to chromosomes (Chr), and 12 Chr were obtained. Simultaneously, we analyzed the linear array of markers in every Chr using HighMap software [46] and then estimated the genetic distances between a pair of adjacent markers.

2.8. QTL Mapping of P. capsici Resistance and Candidate Gene Prediction

QTL analysis was identified by r/QTL software using CIM methods [**47,48**]. The LOD score thresholds for evaluating the statistical significance of the QTL effects were established using 1000 permutations ($p < 0.05$). The predicted genes within the target QTL interval were determined by comparison with the annotated Zunla1- and CM334 reference genomes (**http://peppersequence.genomics.cn**, accessed on: 20 January 2019). The function of genes identified in the candidate regions was manually determined by BLASTX (**https://blast.ncbi.nlm.nih.gov/**, accessed on: 20 January 2019). In addition, the predicted genes were further annotated based on KEGG (**https://www.kegg.jp/kegg/**, accessed on: 20 January 2019), COG (**http://www.ncbi.nlm.nih.gov/COG/**, accessed on: 20 January 2019), Swiss-Prot (**http://www.ebi.ac.uk/uniprot/** accessed on: 20 January 2019), and NR (**https://blast.ncbi.nlm.nih.gov/** accessed on: 20 January 2019) databases.

2.9. qRT-PCR Analysis

For expression analysis, we conducted qRT-PCR to investigate the expression pattern of five disease-resistant or defense-related genes for *P. capsici* resistance in pepper. Leaf samples were gathered from days 6 ,5 ,4 ,3 ,2 ,1 ,0, and 7 post inoculation with *P. capsici* in the resistant line "PI201234" and the susceptible line "Early calwonder." "Early calwonder" was defined as susceptible to three physiological races of *P. capsici*. Total RNAs were extracted utilizing the Plant RNA Kit (Tiangen DP441, China) as per the company's instructions. Subsequently, cDNAs were reverse-transcribed using TaKaRa Reverse Transcription Kit (Takara Biomedical Technology (Beijing) Co., Ltd., Beijing, China). Quantitative PCR was conducted on a 7500 Real-Time PCR System (Thermo Fisher Scientific, Waltham, MA, USA) using TB Green® Premix Ex TaqTM Kit (TaKaRa). The PCR program was as follows: Holding Stage Step 1: 95 °C 30 s, followed by 40 cycles of Step 1: 95 °C for 5 s, Step 2: 60 °C for 30 s, and 72 °C for 2 min. After the last cycle, the amplification was extended for 7 min at 72 °C. *AY572427* was used as internal control for qRT-PCR analysis. We employed the 2-ΔΔT method to determine relative expression levels of candidate genes, which were normalized to that of actin gene *(AY572427).* Each target sample was analyzed using three biological replicates. All values were reported as the mean ± standard deviation (n = 3), and the statistical significance of any differences was analyzed using a Student's *t*-test.

3. RESULTS

3.1. Sequencing and Genotyping Based on SLAF-seq

In this study, genotyping of 150 F2 individuals and their parents was performed using the SLAF-seq technology. The sequencing data generated in this work were sent to the NCBI SRA database (**http://www.ncbi.nlm.nih.gov/sra/** accessed on: 20 October 2020) as accession no. PRJNA669602. Approximately 76.22 GB of raw bases and 381.15 Mb of paired-end reads were generated, of which %94.37 achieved or exceeded quality score of 30 (Q30), and GC (guanine-cytosine) content was 38.86% (**Table 1**). *Oryza sativa* L. was used as control for evaluating the effectiveness of library construction. In addition, 12,250,440 reads representing 139,046 SLAFs with average depths of 63.83 were obtained from the male parent (PI201234), and 13,232,257 reads representing 141,584 SLAFs with average depths of 72.32 were obtained from the female parent (1287) (**Table 1**). In the offspring (F2 population), 2,371,153 reads that were representing 124,582 SLAFs with average depths of 14.66 were generated (**Table 1**).

Table 1. Specific-locus amplified fragment sequencing (SLAF)-seq data statistics of the Capsicum F2 population.

Samples	Total Read	Total Bases	Q30 (%)	GC Content (%)	SLAF Number	Total Depth	Average Depth(X)
PI201234	12,250,440	2,449,757,552	93.98	38.42	139,046	8,875,578	63.83
1287	13,232,257	2,646,335,364	94.71	38.15	141,584	10,239,208	72.32
Offspring	2,371,153	474,202,396	94.37	38.36	124,582	1,825,928	14.66
Total	381,155,587	76,226,452,308	94.37	38.86	405,212	/	/

After filtration of low-depth SLAF tags, approximately 174,193 high-quality SLAF markers were obtained, of which 19.77% (34,432) were polymorphic SLAFs (**Table 2**). In addition, 25,839 of the 34,432 polymorphic SLAFs were cultured into eight segregation patterns (aa×bb, ab×cc, ab×cd, cc×ab, ef×eg, hk×hk, lm×ll, and nn×np) (**Figure 1**). As the parents were homozygous (i.e., with genotype aa or bb), 21,069 SLAFs exhibited the aa×bb segregation pattern and were successfully selected for map construction.

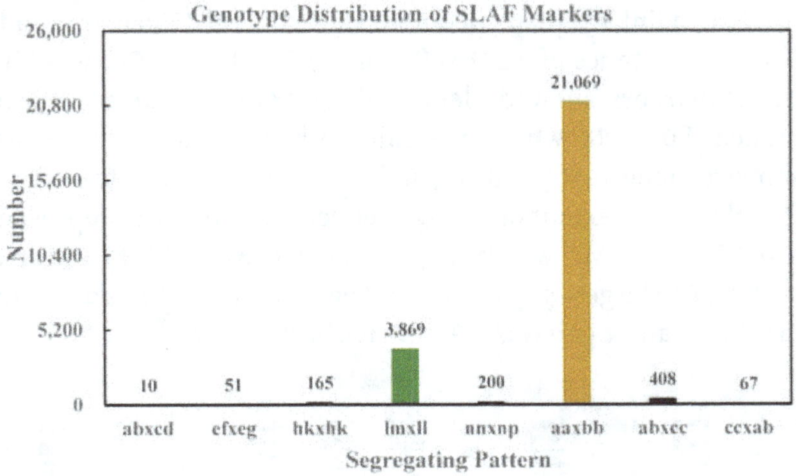

Figure 1. Specific-locus amplified fragment sequencing (SLAF) polymorphism analysis. Marker count in eight segregation patterns. The x-axis represents eight segregation patterns for the polymorphic SLAF markers, and the y-axis shows the number of markers.

Table 2. Description on basic characteristics of the 12 linkage groups.

Linkage Group	SLAF Number	Polymorphic
Chr01	16,109	3221
Chr02	9259	1626
Chr03	15,231	3159
Chr04	12,696	1569
Chr05	13,024	2986
Chr06	12,887	2640
Chr07	11,667	1907
Chr08	9426	1263
Chr09	14,507	3250
Chr10	11,356	1687
Chr11	11,890	3937
Chr12	13,089	2343
Other	23,052	4844
Total	174,193	34,432

3.2. Genetic Map Construction

After four-step filtering, our final map contained 5565 markers on 12 Chrs, which were designated Chr01-Chr12 using HighMap software and presented in Table 2 and Figure 2. The linkage map encompassed a total of 1535.69 cM and exhibited an average marker distance of 0.28 cM (Figure 2). The largest Chr was Chr03, which consisted of 444 markers, showed a length of 169.18 cM, and an average marker-to-marker distance of 0.38 cM, while the smallest Chr was Chr05 that consisted of 460 markers, showed a length of 99.98 cM, and an average marker-to-marker distance of 0.22 cM (Table 3). The extent of linkage between markers was represented by the percentage of "Gaps ≤ 5 cM," which ranged from 99.08% to 100%, and an average of 99.70% (Table 3). The largest gap on this linkage map was situated on Chr10 at 9.99 cM, whereas the smallest gap was 3.94 cM at Chr08 (Table 3).

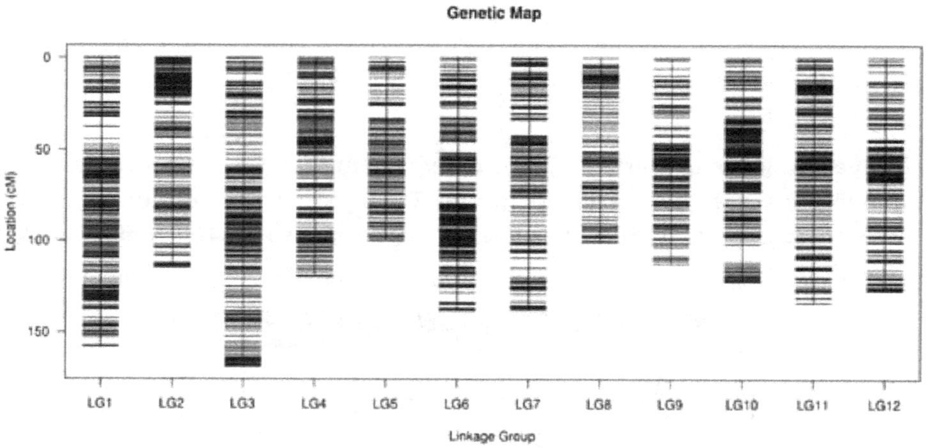

Figure 2. Distribution of SLAF markers across 12 pepper linkage groups. The black bar indicates a SLAF marker. The x-axis shows the linkage group number, whereas the y-axis represents genetic distance.

Table 3. Basic information of the 12 linkage groups. The closer the Spearman's rank correlation coefficient is to 1, the better the collinearity.

Linkage Group	Marker Number	Average Distance between Markers (cM)	Size (cM)	Gaps ≤ 5	Max Gap (cM)	Correlation Coefficient
Chr01	437	0.36	158.32	99.08%	6.77	0.9968
Chr02	415	0.28	114.98	100.00%	4.57	0.9273
Chr03	444	0.38	169.18	100.00%	4.63	0.9979
Chr04	336	0.36	120.20	99.70%	6.38	0.9948
Chr05	460	0.22	99.98	99.78%	7.29	0.9054
Chr06	722	0.19	137.92	99.58%	6.31	0.9977

Linkage Group	Marker Number	Average Distance between Markers (cM)	Size (cM)	Gaps ≤ 5	Max Gap (cM)	Correlation Coefficient
Chr07	517	0.27	137.66	99.22%	8.71	0.9712
Chr08	373	0.27	100.81	100.00%	3.94	0.9568
Chr09	414	0.27	112.88	99.52%	5.82	0.9875
Chr10	532	0.23	122.63	99.81%	9.99	0.9963
Chr11	458	0.29	133.80	100.00%	4.70	0.9974
Chr12	457	0.28	127.33	99.56%	6.38	0.9803
Maximum	722	0.38	169.18	100.00%	9.99	0.9979
Minimum	336	0.19	99.98	99.08%	3.94	0.9054
Total	5565	0.28	1535.69	/	/	/
Average	463.75	/	127.97	99.70%	/	0.9758

3.3. Quality and Accuracy of the Genetic Map

The quality and accuracy of the genetic map were assessed based on collinearity between the genetic and physical maps. The average integrity of each marker was 99.91% (**Figure 3**). Furthermore, among the 12 linkage groups, Chr03 showed the highest collinearity, with a correlation coefficient of 0.9979, and the average Spearman's rank correlation coefficient was 0.9758 (**Table 3**). On average, the coverage of these markers was 108.74-fold in PI201234 (male parent), 126.25-fold in 1287 (female parent), and 22.73-fold in every F2 individual (**Table 4**), thereby indicating genotyping accuracy. Furthermore, collinearity with the physical map was utilized to examine the quality of the genetic map. **Figure 4** shows that most of the genetically mapped loci were collinear with their physical positions on the reference genome sequence of *C. annuum* cv. Zunla1- v**45**] 2.0]. Every correlation coefficient of 12 linkage groups was also assessed. The correlation coefficients of the 12 linkage groups were all close to 1, which indicated relatively high collinearity between linkage groups and the pepper reference genome (**Figure 3**).

Table 4. Details on the depth of mapped markers.

Samples	Marker Numbers	Total Depth(X)	Average Depth(X)
PI201234	5565	605,153	108.74
1287	5565	702,568	126.25
Offspring	5513	125,293	22.73

3.4. Phenotypic Analysis of *P. capsici* Resistance

In 2008, the disease indices (DIs) of 150 F2 populations were determined (**Table 5**). The highest DI value was recorded in the susceptible 1287 (84.3), while the lowest ID value was recorded in the resistance parent PI201234 (7.9). The DI values of the F2 population varied between 0.00 and 100. The average DI value of the F2 population was 47.2. The variation was 0.75, and skewness and kurtosis value of the DI in the F2 population was small, indicating that the population was suitable for QTL identification.

Table 5. Descriptive statistics of disease index and the whole population of the parents.

Traits	F_2 Population								Parent		
	Min	Max	Range	Average	Standard Error	Var	Skew	Kurt	PI201234	1287	Midparent
Disease index (100%)	0	1	0-1	0.4772	0.029	0.75	-0.17	-1.43	7.9	84.3	46.1

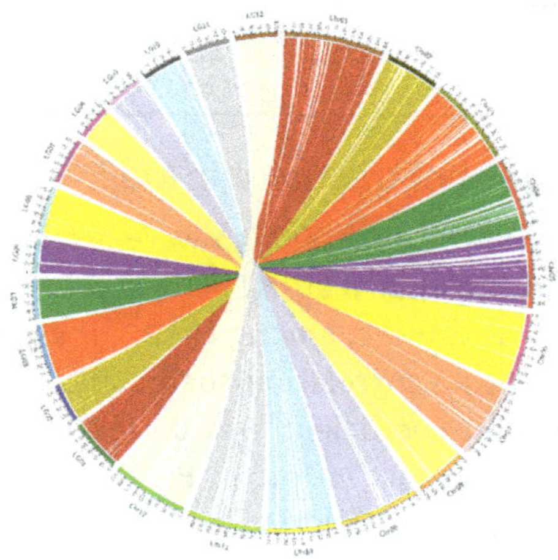

Figure 3. Collinearity between genetic and physical maps. The correlation between the pepper chromosomes (Chr) and the linkage group (LG) of the genetic map is illustrated.

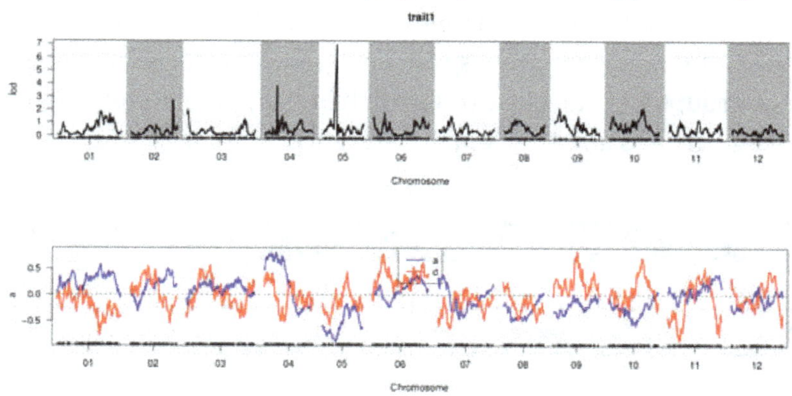

Figure 4. Quantitative trait locus (QTL) analysis of the P. capsici resistance trait of pepper. The x-axis indicates linkage group in pepper, and the y-axis presents LOD values. The blue line indicates the additive effect, and the red line represents the dominant effect.

3.5. QTL Mapping of *P. capsici* Resistance

In present study, the maximum LOD value of 6.972 was used as the threshold to determine the existence of QTL. Based on the high-density genetic map, a single major QTL for the *P. capsici* resistance trait was identified in the F2 population and designated as *CQPc5.1* (**Table 6, Figure 4**), which explained 11.76% of the observed phenotypic variance. *CQPc5.1* was localized in 19.4–17.9 Mb on Chr05, which encompassed a genetic distance of about 0.35 cM, as well as a physical distance of about 1.47 Mb on Chr05 (**Table 6**).

Table 6. Quantitative trait locus (QTL) analysis of the P. capsici resistance trait in F2 populations.

QTL	LOD Threshold	Chr ID	Physical Distance Interval (bp)	Genetic Distance Interval (cM)	Max LOD	ADD	DOM	PVE (%)
CQPc5.1	6.125	05	17,967,630–19,446,349	33.103–33.448	6.972	-0.897	0.079	11.758
Note: LOD, logarithm of odds. Maximum LOD score (QTL peak). ADD, additive effects. DOM, dominance effects. PVE, phenotypic variance explained.								

3.6. Candidate Gene Prediction and qRT-PCR Analysis

According to the annotations of the *C. annuum* cv. Zunla1- v2.0 genome, 23 predicted candidate genes were determined in the physical interval of *CQPc5.1* on Chr05 (**Table 7**). Among these, nine candidate genes were identified in the COG database, including 11 genes with KEGG annotations and 12 genes with Swiss-Prot annotations. Furthermore, 5 of the 23 genes were related to disease resistance or defense, and thus might be involved in *P. capsici* resistance in pepper; the *Capana05g000595* gene was annotated as disease resistance protein, RPP-8like; two genes (*Capana05g000596* and *Capana05g000597*) were annotated as disease resistance protein, RPP13-like; *Capana05g000598* was annotated as likely LRR receptor-like serine/threonine-protein kinase; and *Capana05g000604* was annotated as an F-box/LRR-repeat protein. These five genes were then analyzed by qRT-PCR. The primer sequences are listed in **Table 8**. The results showed that three genes (i.e., *Capana05g000595*, *Capana05g000596*, and *Capana05g000597*) were up-regulated in "PI201234," and expression levels peaked 2–3 days after pathogen inoculation (**Figure 5**). Five genes were up-regulated in "Early calwonder" after pathogen inoculation, and expression levels peaked at 5 days. In "Early calwonder," the expression of *Capana05g000604* gradually increased over time; however, it was expressed at a markedly lower level in "PI201234."

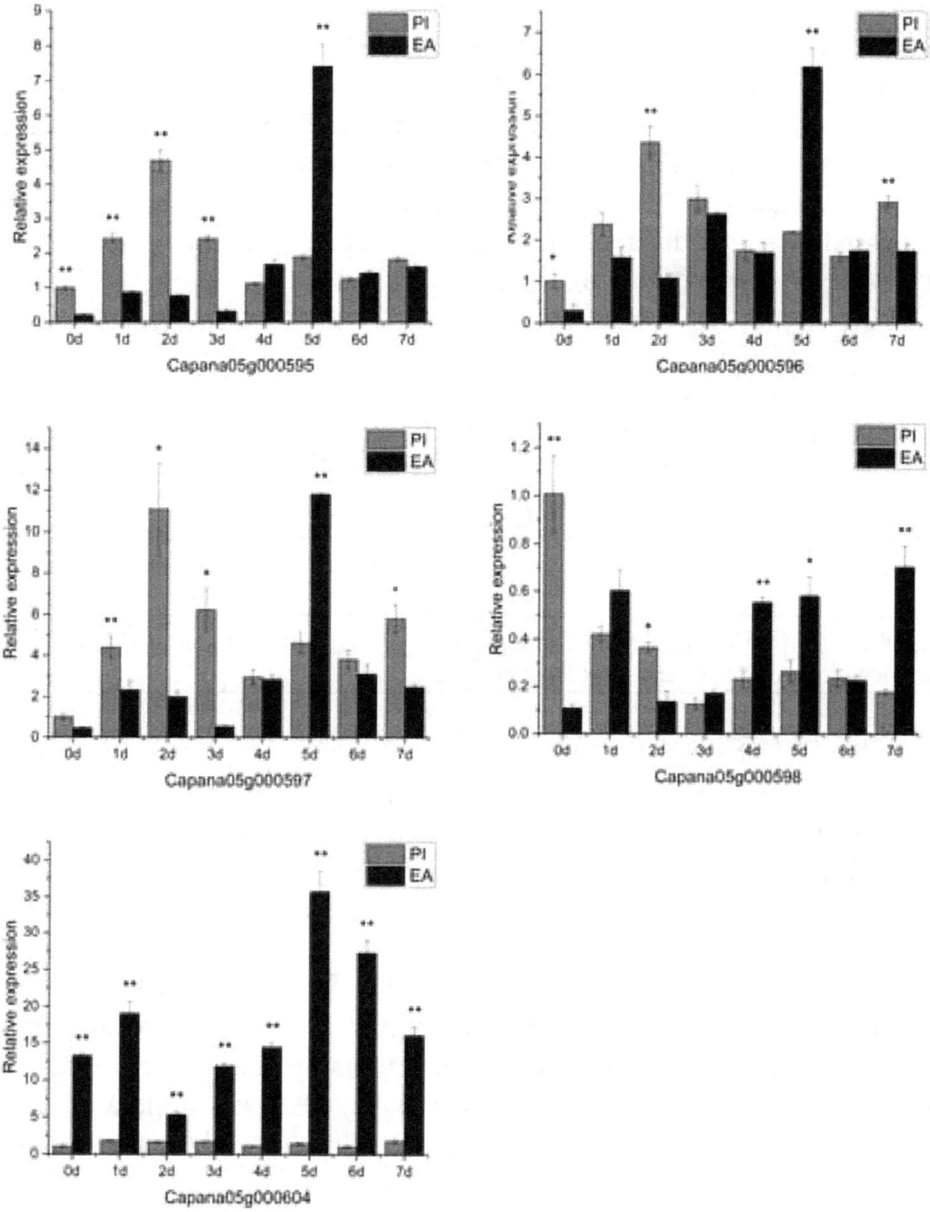

Figure 5. qPCR analysis of five genes in PI201234 and Early calwonder that are related to disease resistance or defense. PI: PI201234 inoculated using P. capsici zoospore suspension; EA: Early calwonder inoculated using P. capsici zoospore suspension. The x-axis shows the time points of sample collection, d: days post-inoculation. The y-axis shows the relative expression quantity of genes. Gene expression was normalized to that of actin, and the data were expressed as the mean ± standard deviation of two biological replicates and three technical replicates. A Student's t-test was used to analyze statistical significance of differences. * 0.05 level of significance; ** 0.01 level of significance.

Table 7. Details of the annotated candidate genes.

Gene	Start	Stop	COG	KEGG	Swiss-Prot	Nr
Capana05g000592	18,024,840	18,035,001	–	K17550 (protein phosphatase 1 regulatory subunit 7)	Protein phosphatase 1 regulatory inhibitor subunit PPP1R7 homolog	PREDICTED: protein phosphatase 1 regulatory subunit pprA-like
Capana05g000594	18,317,198	18,327,621	–	K13099 (CD2 antigen cytoplasmic tail-binding protein 2)	–	PREDICTED: CD2 antigen cytoplasmic tail-binding protein 2
Capana05g000595	18,357,315	18,357,857	–	–	Disease resistance RPP8-like protein	Hypothetical protein T459_14155
Capana05g000596	18,358,568	18,359,167	General function prediction only	–	Putative disease resistance RPP13-like protein	Hypothetical protein T459_14156
Capana05g000597	18,359,457	18,359,987	–	–	Disease resistance protein RPP13	Hypothetical protein BC332_12877
Capana05g000598	18,387,075	18,390,244	Transcription	–	Probable LRR receptor-like serine/threonine-protein kinase At3g47570	PREDICTED: probable LRR receptor-like serine/threonine-protein kinase At3g47570
Capana05g000599	18,390,817	18,396,058	Carbohydrate transport and metabolism	K05298 (glyceraldehyde-3-phosphate dehydrogenase (NADP+))	Glyceraldehyde-3-phosphate dehydrogenase A, chloroplastic (Fragment)	PREDICTED: glyceraldehyde-3-phosphate dehydrogenase A, chloroplastic
Capana05g000600	18,401,537	18,405,948	General function prediction only	–	Protein high chlorophyll fluorescent 107	PREDICTED: protein high chlorophyll fluorescent 107
Capana05g000601	18,410,078	18,412,165	Post-translational modification, protein turnover, chaperones	K00587 (protein-S-isoprenylcysteine O-methyltransferase)	Protein-S-isoprenylcysteine O-methyltransferase B	PREDICTED: protein-S-isoprenylcysteine O-methyltransferase A-like isoform X1

Gene	Start	Stop	COG	KEGG	Swiss-Prot	Nr
Capana05g000602	18,413,946	18,415,745	–	–	Pentatricopeptide repeat-containing protein At2g13600 OS = Arabidopsis thaliana	PREDICTED: pentatricopeptide repeat-containing protein At2g13600
Capana05g000603	18,424,439	18,427,748	Coenzyme transport and metabolism	–	FAD synthetase 1, chloroplastic	PREDICTED: FAD synthetase 1, chloroplastic-like
Capana05g000604	18,449,805	18,454,056	Transcription	K10268 (F-box and leucine-rich repeat protein)	F-box/LRR-repeat protein 4	PREDICTED: F-box/LRR-repeat protein 20
Capana05g000605	18,709,353	18,710,879	–	–	UPF0481 protein At3g47200	PREDICTED: putative UPF0481 protein At3g02645 isoform X1
Capana05g000607	18,749,830	18,750,435	–	–	–	PREDICTED: uncharacterized protein LOC107870380 isoform X1
Capana05g000608	18,775,198	18,776,266	Cell wall/ membrane/ envelope biogenesis	K18819 (inositol 3-alpha-galactosyltransferase)	Galactinol synthase 2	Galactinol synthase 2
Capana05g000609	18,823,996	18,827,956	–	–	–	Hypothetical protein CQW23_12126
Capana05g000611	18,835,570	18,835,953	–	K13495 (cis-zeatin O-glucosyltransferase)	Zeatin O-xylosyltransferase	PREDICTED: zeatin O-glucosyltransferase
Capana05g000612	18,851,478	18,852,197	–	K13495 (cis-zeatin O-glucosyltransferase)	Zeatin O-glucosyltransferase	PREDICTED: zeatin O-xylosyltransferase-like
Capana05g000613	18,857,046	18,857,423	–	K13495 (cis-zeatin O-glucosyltransferase)	Zeatin O-glucosyltransferase	Hypothetical protein T459_14175
Capana05g000614	18,879,064	18,879,435	–	–	Putative cis-zeatin O-glucosyltransferase	PREDICTED: zeatin O-xylosyltransferase-like

Gene	Start	Stop	COG	KEGG	Swiss-Prot	Nr
Capana05g000615	18,879,694	18,880,182	–	K13495 (cis-zeatin O-glucosyltransferase)	Zeatin O-xylosyltransferase	Hypothetical protein T459_14174
Capana05g000617	19,193,439	19,194,854	–	K13495 (cis-zeatin O-glucosyltransferase)	Zeatin O-glucosyltransferase	PREDICTED: zeatin O-glucosyltransferase-like
Capana05g000618	19,354,157	19,355,500	Transcription	–	Receptor-like protein Cf-9	Hypothetical protein T459_14173
Total			9	11	20	23
Note: COG, Clusters of Orthologous Groups. KEGG, Kyoto Encyclopedia of Genes and Genom.						

Table 8. Information on genes employed in qRT-PCR analysis.

Gene Name	Primer Sequence (5'-3')	PCR Product Size (bp)	TM (°C)
Capana05g000595	F:AAGGAGGCATTTAGCCGCAA R:TGTCTCAAGGCGAGCAACAT	115	59.0
Capana05g000596	F:CTGCAAGAAAGCGTGTCAGG R:AGCCTCCACATCTTTCCACC	98	59.0
Capana05g000597	F:CAATCCCTCAAGCGACGAGT R:CCAGGTCGGACCGATTGTTA	121	55.0
Capana05g000598	F:ACCTTCCGTGGTGAAATCCC R:CGATCCGCGTAACAGGTTTG	190	55.0
Capana05g000604	F:TTAGCTGTTGCTGAGGGGTG R:GCTTGCGTCCAGAGAGACAAA	163	59.0
Actin (AY572427)	F:AGCAACTGGGACGATATGGAGAAG R:AAGAGACAACACCGCCTGAATAGC	198	50.0

4. DISCUSSION

4.1. Genetic Map Constructed of Pepper

Phytophthora blight, caused by the pathogen P. capsici, is a highly detrimental disease in pepper plants. It leads to a substantial decrease in both crop yield and quality [49]. Despite extensive genetic study on the resistance of pepper to P. capsici over the course of several decades, no cultivars that are resistant to the pathogen have been developed up until now. Currently, genetic linkage maps serve as a crucial tool not only in plant genetics but also in identifying genomic areas associated with agronomic and qualitative features via QTL mapping. SLAF-seq has recently been employed to generate genetic linkage maps of pepper, resulting in the successful production of several high-density genetic maps. Zhu et al. (2019) discovered six QTLs related to flowering time and number of flowers per node in pepper. They used a molecular genetic linkage map created through SLAF-seq. The map consisted of 9038 markers spaced an average of 0.18 cm apart, distributed across 12 linkage groups. The total distance covered by the markers was 1586.78 cM [44]. In 2019, Zhang and colleagues discovered two significant pepper QTLs (Ffn2.1 and Ffn2.2) that were highly associated with FFN. They used a high-density genetic map consisting of 9328 SLAF markers from 12 linkage groups, which had a total genetic distance of 2009.69 cM and an average distance of 0.22 cM [40]. In their study, Sun et al. (2020) identified two Quantitative Trait Loci (QTLs), namely Rmpas-1 and Rmprp-1, which were found to be associated with aphid survival and reproduction, respectively. These QTLs were identified using a genetic linkage map that consisted of 167 Single Nucleotide Polymorphism (SNP) markers [50]. We utilized the SLAF-seq method to create a genetic map based on an F2 population in this study. The map had 5565 markers that were allocated to 12 linkage groups, with a total length of 1535.69 cM, and displayed an average genetic distance of 0.28 cM. The genetic map demonstrated sufficient coverage of the polymorphic markers in the regions of interest, and the mapped QTLs displayed precise positioning.

4.2. Identification QTL with the Resistance to *P. capsici* Traits

Previous studies have showed that the major QTLs related to resistance to *P. capsici* are situated on Chr05, despite the use of various resistant lines, pepper populations, or *P. capsici* isolates [**23,26,27,32,51**]. Mallard et al. (2013) utilized published pepper genome information and identified three major QTLs, namely, *Pc5.1*, *Pc5.2*, and *Pc5.3*, which were localized to the 22.4–24.6, 53.0–162.6, and 9.7–13.3 Mb regions on Chr05, respectively [**31**]. Siddique et al. (2019) reported three major QTLs on Chr05, namely, *QTL5.1* (19.5–18.7 Mb), *QTL*29.2–27.3) *5.2* Mb), and *QTL*37–34.6) *5.3* Mb) that were related to resistance to three *P. capsici* isolates on using combined traditional QTL mapping with GWAS [**30**]. Here, we performed *P. capsici* resistance QTL analysis of pepper. We detected a major QTL *CQPc5.1* based on a high-density linkage map of F2 plants. *CQPc5.1* was localized to the 19.4–17.9 Mb region on Chr05, with a genetic distance of 33.448–33.103 cM. In an earlier study, Collard et al. (2005) documented that a QTL is only described as "major" when it accounts for >10% of the PVE [**52**]. In the present study, the phenotypic variance of *CQPc5.1* was %11.58. In addition, the position of *CQPc5.1* on Chr05 differs from *Pc5.1*, *Pc5.2*, and *Pc5.3*, yet the location of *CQPc5.1* that was identified in this work coincides with that of the earlier determined locus *QTL***30**] *5.1*]. However, the physical location of *CQPc5.1* on the chromosome is closer than that of *QTL5.1*, so we infer that *CQPc5.1* represents a more accurate mapping of resistance to *P. capsici* in pepper.

4.3. Candidate Gene Prediction

Here, we identified five genes that are related to disease resistance in the *CQPc5.1* QTL region. We identified three genes annotated as disease-resistance protein RPP-13like; *RPP13* was a singleton NBS-LRR gene located in *CQPc5.1* on Chr05. *Capana05g000595* gene was identified as disease resistance protein RPP8-. Two genes (*Capana05g000596* and *Capana05g000597*) were annotated to be disease resistance protein RPP13-like. *RPP13* is a CC (coiled-coil)-NBS-LRR domain-containing *R* gene that controls resistance to *Peronospora parasitica* oomycete pathogen in *Arabidopsis thaliana* [**53,54**]. These two candidate genes encode *RPP13*-like NBS-LRR proteins and serve as potential candidates for *P. capsici* resistance in pepper. *Capana05g000598* was annotated as a probable LRR receptor-like serine/threonine-protein kinase. *Capana05g000604* was annotated to be an F-box/LRR-repeat protein. Several LRR domain proteins have been determined to participate in defense responses to infiltrating pathogens [**12,55,56,57**].

4.4. Candidate Gene qRT-PCR Analysis

The results of our quantitative real-time polymerase chain reaction (qRT-PCR) show that the expression levels of three genes (Capana05g000595, Capana05g000596, and Capana05g000597) are increased in both the resistant "PI201234" and susceptible "Early calwonder" lines following pathogen inoculation. The gene Capana05g000598 exhibited a decrease in expression levels when subjected to pathogen inoculation in the "PI201234" variety, while it showed an increase in expression levels in the "Early calwonder" variety. Based on our analysis, it appears that Capana05g000598 may function as a negative regulator of resistance to P. capsici in PI201234. The gene Capana05g000604 showed increased expression in the "Early calwonder" variety after infection, whereas its expression level remained consistently lower in the "PI201234" variety. Curiously, the level of expression of five genes in the "PI201234" variety reached its highest point 2-3 days after being infected by a pathogen. In contrast, the expression in the "Early calwonder" variety reached its peak 5 days after being infected by a pathogen. Hence, we inferred that Capana05g000595, Capana05g000596, and Capana05g000597 could potentially be associated with resistance against P. capsici. These three genes have a strong correlation with CQPc5.1, although there is no evidence of their functional validation. Hence, it is imperative to carry out a functional investigation of these genes in order to validate their molecular roles in conferring resistance to P. capsici in pepper plants. The outcome of this investigation would furnish data for other research stages, including gene functional analysis, pyramiding breeding, and marker-assisted selection (MAS) as well.

AT A GLANCE

Pepper, scientifically known as *Capsicum annuum* L., is a very important crop and spice with global economic significance. The oomycete plant pathogen, Phytophthora capsici (P. capsici), can greatly diminish the crop's productivity. A high-density genetic map was constructed using a mapping panel of F2 populations derived from 150 individuals of parental lines PI201234 and 1287. Specific-locus amplified fragment sequencing (SLAF) was employed to identify loci associated with resistance to P. capsici. The genetic map has a sequencing depth of 108.74-fold for the male parent, 126.25-fold for the female parent, and 22.73-fold for the offspring. A pepper genetic map with a high level of detail was created, containing 5565 markers and 12 linkage groups. This map covers a distance of 1535.69 cM, with an average distance of 0.28 cM between markers. A significant quantitative trait locus (QTL) named CQPc5.1 was discovered on Chr05, which accounted for 11.758% of the reported phenotypic variance in P. capsici resistance. Within the QTL CQPc5.1 interval, a total of 23 candidate genes were found. These genes include Capana05g000595, which encodes the RPP8-like protein, as well as Capana05g000596 and Capana05g000597, which encode an RPP13-like protein. The results of quantitative reverse-transcription PCR (qRT-PCR) showed elevated expression levels of Capana05g000595, Capana05g000596, and Capana05g000597 in P. capsici resistance accessions, indicating a potential link between these genes and resistance to P. capsici in pepper.

REFERENCES

1. Chaim, A.B.; Grube, R.C.; Lapidot, M.; Jahn, M.; Paran, I. Identification of quantitative trait loci associated with resistance to cucumber mosaic virus in *Capsicum annuum*. *Theor. Appl. Genet.* 2001, *102*, 1213–1220.
2. Cai, W.-Q.; Fang, R.-X.; Shang, H.-S.; Wang, X.; Zhang, F.-L.; Li, Y.-R.; Zhang, J.-C.; Cheng, X.-Y.; Wang, G.-L.; Mang, K.-Q. Development of CMV-and TMV-resistant chili pepper: Field perfermance and biosafety assessment. *Mol. Breed.* 2003, *11*, 25–35.
3. Hong, J.K.; Yang, H.J.; Jung, H.; Yoon, D.J.; Sang, M.K.; Jeun, Y.-C. Application of Volatile Antifungal Plant Essential Oils for Controlling Pepper Fruit Anthracnose by *Colletotrichum gloeosporioides*. *Plant Pathol. J.* 2015, *31*, 269–277.
4. Bosland, P.W. A Seedling Screen for Phytophthora Root Rot of Pepper, *Capsicum annuum*. *Plant Dis.* 1991, *75*, 1048.
5. Parra, G.; Ristaino, J.B. Resistance to Mefenoxam and Metalaxyl Among Field Isolates of *Phytophthora capsici* Causing Phytophthora Blight of Bell Pepper. *Plant Dis.* 2001, *85*, 1069–1075.
6. Oelke, L.M.; Bosland, P.W.; Steiner, R. Differentiation of Race Specific Resistance to Phytophthora Root Rot and Foliar Blight in *Capsicum annuum*. *J. Am. Soc. Hortic. Sci.* 2003, *128*, 213–218.
7. Lefebvre, V.; Palloix, A. Both epistatic and additive effects of QTLs are involved in polygenic induced resistance to disease: A case study, the interaction pepper-*Phytophthora capsici* Leonian. *Theor. Appl. Genet.* 1996, *93*, 503–511.
8. Foster, J.M.; Hausbeck, M.K. Resistance of Pepper to Phytophthora Crown, Root, and Fruit Rot Is Affected by Isolate Virulence. *Plant Dis.* 2010, *94*, 24–30.
9. Barksdale, T.H.; Papavizas, G.C.; Johnston, S.A. Resistance to foliar blight and crown rot of pepper caused by *Phytophthora capsici*. *Plant Dis.* 1984, *68*, 506–509.
10. Flett, S.; Ashcroft, W.; Jerie, P.; Taylor, P. Control of Phytophthora root rot in processing tomatoes by metalaxyl and fosetyl-Al. *Aust. J. Exp. Agric.* 1991, *31*, 279–283.
11. Polizzi, G.; Agosteo, G.E.; Cartia, G. Soil solarization for the control of *Phytophthora capsici* on pepper. *Acta Hortic.* 1994, *366*, 331–338.
12. Lehmann, P. Structure and evolution of plant disease resistance genes. *J. Appl. Genet.* 2002, *43*, 403–414.
13. Stieg, J.R.; Walters, S.A.; Bond, J.P.; Babadoost, M. Effects of fungicides and cultivar resistance for *Phytophthora capsici* control in bell pepper production. *HortScience* 2006, *41*, 1076.
14. Black, L. Studies on Phytophthora blight in pepper. In *Talekar NS (ed) AVRDC Report 1998*; Asian Vegetable Research and Development Center: Shanhua, Taiwan, 1999; pp. 25–27.

15. Ortega, R.G.; Espanol, C.P.; Zueco, J.C. Genetics of Resistance to *Phytophthora capsici* in the Pepper Line 'SCM334-'. *Plant Breed.* 1991, *107*, 50–55.

16. Kim, B.S. Characteristics of bacterial spot resistant lines and Phytophthora Blight resistant lines of *Capsicum pepper*. *Hortic. Environ. Biotechnol.* 1988, *29*, 247–252.

17. Smith, P.G.; Kimble, K.A.; Grogan, R.G.; Millett, A.H. Inheritance of resistance in peppers to Phytophthora root rot. *Phytopathology* 1967, *57*, 377–379.

18. Sugita, T.; Yamaguchi, K.; Kinoshita, T.; Yuji, K.; Sugimura, Y.; Nagata, R.; Kawasaki, S.; Todoroki, A. QTL analysis for resistance to Phytophthora Blight (*Phytophthora capsici* Leon.) using an intraspecific Doubled-Haploid population of *Capsicum annuum*. *Breed. Sci.* 2006, *56*, 137–145.

19. Kim, B.S.; Kwon, Y.S. Inheritance of resistance to Phytophthora Blight and to bacterial spot in pepper. *J. Korean Soc. Hortic. Sci.* 1990, *7*, 17–24.

20. Saini, S.S.; Sharma, P.P. Inheritance of resistance to fruit rot (*Phytophthora capsici* Leon.) and induction of resistance in bell pepper (*Capsicum annuum* L.). *Euphytica* 1978, *27*, 721–723.

21. Wang, P.; Wang, L.; Guo, J.; Yang, W.; Shen, H. Molecular mapping of a gene conferring resistance to *Phytophthora capsici* Leonian race 2 in pepper line PI201234 (*Capsicum annuum* L.). *Mol. Breed.* 2016, *36*, 1–11.

22. Reifschneider, F.J.B.; Boiteux, L.S.; Vecchia, P.T.D.; Poulos, J.M.; Kuroda, N. Inheritance of adult-plant resistance to *Phytophthora capsici* in pepper. *Euphytica* 1992, *62*, 45–49.

23. Thabuis, A.; Palloix, A.; Pflieger, S.; Daubèze, A.-M.; Caranta, C.; Lefebvre, V. Comparative mapping of Phytophthora resistance loci in pepper germplasm: Evidence for conserved resistance loci across Solanaceae and for a large genetic diversity. *Theor. Appl. Genet.* 2003, *106*, 1473–1485.

24. Pflieger, S.; Palloix, A.; Caranta, C.; Blattes, A.; Lefebvre, V. Defense response genes co-localize with quantitative disease resistance loci in pepper. *Theor. Appl. Genet.* 2001, *103*, 920–929.

25. Minamiyama, Y.; Tsuro, M.; Kubo, T.; Hirai, M. QTL Analysis for Resistance to *Phytophthora capsici* in Pepper Using a High Density SSR-based Map. *Breed. Sci.* 2007, *57*, 129–134.

26. Kim, H.-J.; Nahm, S.-H.; Lee, H.-R.; Yoon, G.-B.; Kim, K.-T.; Kang, B.-C.; Choi, D.; Kweon, O.Y.; Cho, M.-C.; Kwon, J.-K.; et al. BAC-derived markers converted from RFLP linked to *Phytophthora capsici* resistance in pepper (*Capsicum annuum* L.). *Theor. Appl. Genet.* 2008, *118*, 15–27.

27. Truong, H.T.H.; Kim, K.T.; Kim, D.W.; Kim, S.; Chae, Y.; Park, J.H.; Oh, D.G.; Cho, M.C. Identification of isolate-specific resistance QTLs to phytophthora root rot using an intraspecific recombinant inbred line population of pepper (*Capsicum annuum*). *Plant Pathol.* 2011, *61*, 48–56.

28. Rehrig, W.Z.; Ashrafi, H.; Hill, T.; Prince, J.; Deynze, A.V. CaDMR1 Co-segregates with QTL Pc5.1 for resistance to *Phytophthora capsici* in pepper (*Capsicum annuum*). *Plant Genome* 2014, *7*, 1–12.

29. Kim, N.; Kang, W.H.; Lee, J.; Yeom, S.I. Development of clustered resistance gene analogs-based markers of resistance to *Phytophthora capsici* in chili pepper. *BioMed Res. Int.* 2019, *2019*, 1–12.

30. Siddique, M.I.; Lee, H.Y.; Ro, N.Y.; Han, K.; Venkatesh, J.; Solomon, A.A.-O.; Patil, A.A.-O.; Changkwian, A.; Kwon, J.K.; Kang, B.C. Identifying candidate genes for *Phytophthora capsici* resistance in pepper (*Capsicum annuum*) via genotyping-by-sequencing-based QTL mapping and genome-wide association study. *Sci. Rep.* 2019, *9*, 9962.

31. Mallard, S.; Cantet, M.; Massire, A.; Bachellez, A.; Ewert, S.; Lefebvre, V. A key QTL cluster is conserved among accessions and exhibits broad-spectrum resistance to *Phytophthora capsici*: A valuable locus for pepper breeding. *Mol. Breed.* 2013, *32*, 349–364.

32. Bonnet, J.; Danan, S.; Boudet, C.; Barchi, L.; Sage-Palloix, A.-M.; Caromel, B.; Palloix, A.; Lefebvre, V. Are the polygenic architectures of resistance to *Phytophthora capsici* and *P. parasitica* independent in pepper? *Theor. Appl. Genet.* 2007, *115*, 253–264.

33. Sun, X.; Liu, D.; Zhang, X.; Li, W.; Liu, H.; Hong, W.; Jiang, C.; Guan, N.; Ma, C.; Zeng, H.; et al. SLAF-seq: An Efficient Method of Large-Scale De Novo SNP Discovery and Genotyping Using High-Throughput Sequencing. *PLoS ONE* 2013, *8*, e58700.

34. Li, B.; Tian, L.; Zhang, J.; Huang, L.; Han, F.; Yan, S.; Wang, L.; Zheng, H.; Sun, J. Construction of a high-density genetic map based on large-scale markers developed by specific length amplified fragment sequencing (SLAF-seq) and its application to QTL analysis for isoflavone content in Glycine max. *BMC Genom.* 2014, *15*, 1086.

35. Zhang, Z.; Shang, H.; Shi, Y.; Huang, L.; Li, J.; Ge, Q.; Gong, J.; Liu, A.; Chen, T.; Wang, D. Construction of a high-density genetic map by specific locus amplified fragment sequencing (SLAF-seq) and its application to Quantitative Trait Loci (QTL) analysis for boll weight in upland cotton (*Gossypium hirsutum*). *BMC Plant Biol.* 2016, *16*, 79.

36. Qi, Z.; Huang, L.; Zhu, R.; Xin, D.; Liu, C.; Han, X.; Jiang, H.; Hong, W.; Hu, G.; Zheng, H.; et al. A High-Density Genetic Map for Soybean Based on Specific Length Amplified Fragment Sequencing. *PLoS ONE* 2014, *9*, e104871.

37. Zhang, Y.; Wang, L.; Xin, H.; Li, D.; Ma, C.; Ding, X.; Hong, W.; Zhang, X. Construction of a high-density genetic map for sesame based on large scale marker development by specific length amplified fragment (SLAF) sequencing. *BMC Plant Biol.* 2013, *13*, 141.

38. Xu, X.; Lu, L.; Zhu, B.; Xu, Q.; Qi, X.; Chen, X. QTL mapping of cucumber fruit flesh thickness by SLAF-seq. *Sci. Rep.* 2015, *5*, 15829.

39. Zhu, Y.; Yin, Y.; Yang, K.; Li, J.; Sang, Y.; Huang, L.; Fan, S. Construction of a high-density genetic map using specific length amplified fragment markers and identification of a quantitative trait locus for anthracnose resistance in walnut (*Juglans regia* L.). *BMC Genom.* 2015, *16*, 1–13.

40. Zhang, X.-F.; Wang, G.-Y.; Dong, T.-T.; Chen, B.; Du, H.-S.; Li, C.-B.; Zhang, F.-L.; Zhang, H.-Y.; Xu, Y.; Wang, Q.; et al. High-density genetic map construction and QTL mapping of first flower node in pepper (*Capsicum annuum* L.). *BMC Plant Biol.* 2019, *19*, 167.

41. Hu, X.H.; Zhang, S.Z.; Miao, H.R.; Cui, F.G.; Shen, Y.; Yang, W.Q.; Xu, T.T.; Chen, N.; Chi, X.Y.; Zhang, Z.M.; et al. High-Density Genetic Map Construction and Identification of QTLs Controlling Oleic and Linoleic Acid in Peanut using SLAF-seq and SSRs. *Sci. Rep.* 2018, *8*, 5479.

42. Wang, L.; Yang, X.; Cui, S.; Zhao, N.; Li, L.; Hou, M.; Mu, G.; Liu, L.; Li, Z. High-density genetic map development and QTL mapping for concentration degree of floret flowering date in cultivated peanut (*Arachis hypogaea* L.). *Mol. Breed.* 2020, *40*, 1–14.

43. Guo, G.; Wang, S.; Liu, J.; Pan, B.; Diao, W.; Ge, W.; Gao, C.; Snyder, J.C. Rapid identification of QTLs underlying resistance to Cucumber mosaic virus in pepper (*Capsicum frutescens*). *Theor. Appl. Genet.* 2016, *130*, 41–52.

44. Zhu, Z.; Sun, B.; Wei, J.; Cai, W.; Huang, Z.; Chen, C.; Cao, B.; Chen, G.; Lei, J. Construction of a high density genetic map of an interspecific cross of *Capsicum chinense* and *Capsicum annuum* and QTL analysis of floral traits. *Sci. Rep.* 2019, *9*, 1–14.

45. Murray, M.G.; Thompson, C.L.; Wendel, J.F. Rapid isolation of high molecular weight plant DNA. *Nucleic Acids Res.* 1980, *8*, 4321–4325.

46. Li, R.; Li, Y.; Kristiansen, K.; Wang, J. SOAP: Short oligonucleotide alignment program. *Bioinformatics* 2008, *24*, 713–714.

47. Peichel, C.L.; Nereng, K.S.; Ohgi, K.A.; Cole, B.L.E.; Colosimo, P.F.; Buerkle, C.A.; Schluter, D.; Kingsley, D.M. The genetic architecture of divergence between threespine stickleback species. *Nat. Cell Biol.* 2001, *414*, 901–905.

48. Wang, W.; Huang, S.; Liu, Y.; Fang, Z.; Yang, L.; Hua, W.; Yuan, S.; Liu, S.; Sun, J.; Zhuang, M.; et al. Construction and analysis of a high-density genetic linkage map in cabbage (*Brassica oleracea* L. var. capitata). *BMC Genom.* 2012, *13*, 523.

49. Burdon, J.J.; Thrall, P.H. Coevolution of Plants and Their Pathogens in Natural Habitats. *Science* 2009, *324*, 755–756.
50. Sun, M.; Voorrips, R.E.; Westende, W.V.; van Kaauwen, M.; Visser, R.G.F.; Vosman, B. Aphid resistance in Capsicum maps to a locus containing LRR-RLK gene analogues. *Theor. Appl. Genet.* 2020, *133*, 227–237.
51. Ogundiwin, E.A.; Berke, T.F.; Massoudi, M.; Black, L.L.; Huestis, G.; Choi, D.; Lee, S.; Prince, J.P. Construction of 2 intra-specific linkage maps and identification of resistance QTLs for *Phytophthora capsici* root-rot and foliar-blight diseases of pepper (*Capsicum annuum* L.). *Genome* 2005, *48*, 698–711.
52. Collard, B.C.Y.; Jahufer, M.Z.Z.; Brouwer, J.B.; Pang, E.C.K. An introduction to markers, quantitative trait loci (QTL) mapping and marker-assisted selection for crop improvement: The basic concepts. *Euphytica* 2005, *142*, 169–196.
53. Rose, L.E.; Bittner-Eddy, P.D.; Langley, C.H.; Holub, E.B.; Michelmore, R.W.; Beynon, J.L. The Maintenance of Extreme Amino Acid Diversity at the Disease Resistance Gene, RPP13, in *Arabidopsis thaliana*. *Genetics* 2004, *166*, 1517–1527.
54. Serra, H.; Choi, K.; Zhao, X.; Blackwell, A.R.; Kim, J.; Henderson, I.R. Interhomolog polymorphism shapes meiotic crossover within the Arabidopsis RAC1 and RPP13 disease resistance genes. *PLoS Genet.* 2018, *14*, e1007843.
55. Alder, M.N.; Rogozin, I.B.; Iyer, L.M.; Glazko, G.V.; Cooper, M.D.; Pancer, Z. Diversity and Function of Adaptive Immune Receptors in a Jawless Vertebrate. *Science* 2005, *310*, 1970–1973.
56. Shanmugam, V. Role of extracytoplasmic leucine rich repeat proteins in plant defence mechanisms. *Microbiol. Res.* 2005, *160*, 83–94.
57. Zhang, X.S.; Choi, J.H.; Heinz, J.; Chetty, C.S. Domain-Specific Positive Selection Contributes to the Evolution of Arabidopsis Leucine-Rich Repeat Receptor-Like Kinase (LRR RLK) Genes. *J. Mol. Evol.* 2006, *63*, 612–621.

7 Chapter

DEVELOPMENT OF HIGH-THROUGHPUT SSR MARKERS AND ANALYSIS OF GENETIC DIVERSITY IN CAPSICUM FRUTESCENSE

1. INTRODUCTION: AN OVERVIEW

Pepper is a vital condiment and a significant horticultural crop that is grown worldwide. It has its origins in South America and is classified within the genus Capsicum (Solanaceae) [1,2,3]. This genus has a wide array of cultivated and wild species. Nevertheless, the contemporary literature typically refers to only five species as domesticated and culinary species. The species included are *C. annuum, C. chinense, C. frutescens, C. baccatum,* and *C. pubescens* [4]. Out of all the species, *C. annuum* is the most common one. It consists of numerous commercial cultivars that differ significantly in the size, shape, color of the fruit, and especially the level of spiciness. Nevertheless, because to extensive artificial selection, ongoing cultivation, and domestication, the performance of pepper has become increasingly varied, while the genetic backgrounds have gotten more limited. As a result, the decreased genetic diversity index has caused difficulties in the generation of spicy pepper cultivars. In order to enhance the qualitative qualities, the process has involved searching for and recovering powerful characteristics from related wild species or wild pepper variants [5]. In addition, the increased genetic variety found in the other four species, as well as the remainder of the wild species, has greatly enhanced the potential for enhancing C. annuum through the use of new technologies [6]. The genetic diversity studies of the indicated accessions have yielded valuable research resources, allowing for the utilization of their benefits and the possibility to allocate parent traits effectively, ensuring improved breeding efficiency in the future [7].

Capsicum frutescens is a cultivated shrub crop that can be either an annual or perennial. It produces small, upright fruit that resemble wheat grains and are known for their intense spiciness. The immature fruit display a green to pale yellow coloration, which transitions to a crimson hue during the process of ripening. This species exhibits numerous wild variants, with one of the most frequently documented being the wild pepper species found in China. These variants are primarily found in tropical regions, specifically in the Yunnan and Hainan provinces of China. The plant species *C. frutescens* is commonly referred to as 'Xiaomijiao' in the surrounding area, as documented by the Kunming Institute of Botany, Chinese Academy of Sciences in 1979 [8]. Furthermore, this particular plant species has a notable capacity to withstand a wide range of both living and non-living challenges, as indicated by previous studies [8,9,10,11]. Furthermore, it possesses the capacity to endure pathogenic assaults [9]. C. frutescens is distinguished by its great productivity, exquisite aroma, and significant ethnomedicinal value. Thus, studying the genetic variety of the germplasm resources of C. frutescens has the potential to serve as a foundation for enhancing the quality of farmed pepper in China.

Simple sequence repeats (SSRs), also known as microsatellites, are a collection of DNA sequences that consist of one to six nucleotide units that are repeated in tandem. These can be found everywhere in the genetic material of both simple and complex animals [12,13]. SSRs, or simple sequence repeats, are widely acknowledged genetic markers that are commonly used in plant breeding. They are valued for their distinct characteristics of co-dominant inheritance, multi-allelic nature, broad genome coverage, high abundance, and particularly good repeatability [12,14]. Polymorphism arising from SSRs can be identified by amplifying genes using primers that surround the repetitive motifs [15]. SSR markers are extensively employed for studying various aspects of plants and animals, including population characteristics analysis [16], functional diversity assessment [17,18,19], linkage map construction [20,21], DNA fingerprinting [22,23,24,25], and assisted breeding approaches [26,27,28]. Numerous SSR markers have been designed and effectively utilized. Nevertheless, it is crucial to cultivate a set of robust and highly variable SSR markers to meet the requirements of complex investigations such as high-density genetic mapping, genome comparative mapping, and genome-wide association analysis.

Examining the diversity of the pepper germplasm through the evaluation of morphological traits is not only rigorous but also frequently leads to misleading conclusions. In addition, environmental variables might also have an equal impact on these traits. Nevertheless, molecular markers offer numerous benefits in effectively reflecting vast quantities of data, as well as demonstrating improved stability and great efficiency in analyzing the genetic diversity of plant germplasm resources. Thus, molecular markers are extensively utilized in studies on genetic diversity [29]. Recently, molecular markers have been widely used to study the genetic diversity of pepper germplasm resources. Furthermore, out of the several categories of molecular markers, SSRs are particularly favored for the purpose of assessing diversity because of their significant polymorphism, extensive repeatability, and co-dominance characteristics [30]. These traits have been corroborated by numerous previous reports. In the study conducted by Gu [31] on pepper diversity, a total of 1904 pepper materials were classified into two categories using 29 pairs of polytropic SSR markers. In a similar manner, Zhang [32] using SSR labeling technology to examine 372 pepper specimens and subsequently classify them into three distinct groups. The results were determined to be in accordance with the botanical features. Chen [5] discovered that when evaluating the effectiveness of SRAP and SSR for analyzing the genetic diversity of eight pepper germplasm, SSRs demonstrated greater efficiency, displaying higher levels of polymorphism and detection capability. In a separate study, Li [33] utilized 14 pairs of SSR polymorphic primers to examine the genetic diversity of 169 pepper samples. Upon categorizing them into seven groups and confirming the findings, it was noted that the results perfectly aligned with the fundamental classification of the pepper species.

Pepper germplasm resources serve as the fundamental material foundation for the breeding and cultivation of pepper crops. Nevertheless, the genetic diversity of the pepper germplasm resources in China is limited. Thus, it is recommended to prioritize the acquisition of substantial data resources through the efficient application of collected wild germplasm resources. 'Xiaomijiao' is the sole indigenous *C. frutescens* plant discovered in China among the existing resources. Consequently, this study aimed to investigate the creation of SSR molecular markers for *C. frutescens* using the 'Xiaomijiao' sequence data (which has not been published). Subsequently, the genetic diversity of the collected germplasm resources was analyzed at the genome level using SSR markers. The newly developed polymorphic microsatellite markers serve as the foundation for future population studies. This research examines the genetic diversity of 147 *C. frutescens* germplasm from 25 countries. The goal is to gain insights into the genetic relationships and composition of different accessions. This information will serve as a foundation for more efficient usage of these germplasm resources in the future.

2. MATERIALS AND METHODS

2.1. Plant Materials and DNA Extraction

This study utilized a total of 147 pepper (*C. frutescens*) accessions, which were gathered from 25 different nations. A summary is provided that includes the English names, cultivation zones, DNA concentrations (ng/uL), and the selected species type (wild/cultivated) for the samples. After the sample procedures, newly-formed leaves were promptly frozen using liquid nitrogen and moved to a temperature of -80 °C for subsequent DNA extraction. Subsequently, the genomic

DNA extraction was carried out using the CTAB (Cetyl Trimethyl Ammonium Bromide) procedures as outlined by Murray and Thompson [34]. Furthermore, the purity of the isolated DNA was determined using a NanoDropOneC Microvolume UV-Vis Spectrophotometer. Ultimately, the DNA concentrations were modified to a range of 10 to 35 ng/μL in order to be utilized in the subsequent polymerase chain reactions (PCR).

2.2. Source of Genic Sequences, SSR Identification and Primer Design

In this study, the wild pepper 'Xiaomijiao' (*C. frutescens*) was sequenced at Beijing Nuohe Zhiyuan Technology Co., Ltd. The genome sequencing of 'Xiaomijiao' (*C. frutescens*) was performed with Illumina HiSeq4000 (300x coverage) and PacBio Sequel (30x coverage). The assembled sequences, which totaled 2.95 G bases, were used in this study to characterize the distribution of microsatellites in the pepper genome. The completeness of the 2.95 Gbp assembly is supported by the mapping of over 99% of ~3 million EST reads (generated using HiSeq4000 technology) from 'Xiaomijiao' (*C. frutescens*) leaf, stem and root tissues. A MISA (MIcroSAtellite) SSR identification tool program was employed for the sequence identification [**35**]. It was confirmed that 2 to 6 nucleotide motifs could be considered for identifying the presence of microsatellites. The minimum repeating units for the dinucleotides, trinucleotides, tetranucleotides, pentanucleotides, and hexanucleotides were defined as 6, 5, 4, 4, and 4, respectively. We allowed up to 5 nucleotide mismatches at the 5' end of the primer, but no mismatches at the 3' end, and a minimum of 80% overall match homology. For a given primer pair, we considered that a specific amplicon was generated if both forward and reverse primers were mapped to the same chromosomes/scaffold. Then, based on the MISA results, Primer 5 software was employed for designing the SSR primers. They generated amplicon sizes of 100 to 300 bp with the following criteria: 22 to 25 bp lengths with 40 to 70% GC content levels; 45 to 65 °C melting temperature (Tm); and the remaining parameters used the program's default values. Eight pepper cultivars, referred to as GRIF 9194, GRIF 9316, PI 439309, PI 439489, PI 631142, VI029462, VI062180, and LJ091, respectively, were selected for validating the primers via PCR and electrophoretic techniques. A total number of 240 SSR markers, which were distributed among 12 linkage groups, were screened. Finally, 33 markers were identified as being evenly distributed along the linkage groups, which produced clear bands with high polymorphism. These markers were further used to analyze all of the accessions, as detailed in **Table 1.**

Table 1. Details of 33 SSR primers used in the study.

No.	Primers	Forward Primer Sequence (5'-3')	Reverse Primer Sequence (5'-3')	Repeat Unit	Repeat Number
1	Chr1SSR8	GACTATAGCAAACATGTCCCCAG	CCAGATTTTGTGGATCCTATTGA	TCA	4
2	Chr1SSR12	ATTCAAAGAGGGCATGATGTAGA	TGTCAACTTAGTGGGTGGGTTAG	GGTAGG	6
3	Chr1SSR18	TTAGTGTTGTCAAAATACCCGTG	CAATAAACATATCACACGTGCAAC	ATAG	6
4	Chr2SSR12	GAAAGATCCGTCTAAGCAAACAA	AAATCTCATGATTTTCGGTGATG	AAT	4
5	Chr2SSR14	TGAATTTCGGAGTGTTACGTAGAG	GCGAAGGTGAGTCTGTTCAATTA	TAA	4
6	Chr2SSR15	AATTTGAAGAGCGTGCATAAAAA	ACCATACCATACCATGGAAACAA	ATGGT	4
7	Chr3SSR5	ATAAAATAGACCCCACCCACTTC	GGGCTGCTGTACTAAAGAAGAAGA	TTC	4
8	Chr3SSR6	CTCTCTAGAATGAAAAGTGCCGA	TTGTCGATTTGTTCTTCTTCCAT	CAC	4
9	Chr3SSR13	ACCCTAAAAGCTATGGAGTTGCT	CCCACCATTCTTCTAAACTTTCC	AC	6
10	Chr4SSR11	CCTACAAGTGAGGTCTGAGGAGA	CAGCAAGTTGGAGAAACTAATGC	ACCCT	4
11	Chr4SSR14	AGTGGAAAGTGCTGTTACGATGT	GCGTGATAATTTTTCCACAAGAA	TTTA	5
12	Chr4SSR20	AAAACAACACGACACACAGCTTA	TATATTTTTCTTGGGAACGAGCA	TATAT	6
13	Chr5SSR13	TAAAACCTTGTCACATGTACCCC	AAATGAATAAATCCTCTGCATGG	TGCA	4
14	Chr5SSR14	TAAGTTGTTCGAGAAATCAGACG	CTGTCTCGTATTGAAGGGTGTTT	CGGTGA	4
15	Chr5SSR16	CACTGTGAGAGCAACTTTCTGTG	CACTATTTTCTCATGCACTTTACCA	ACGGGC	4
16	Chr6SSR12	TTCGACCTCCGTATCACTATCAG	CGTACTCTATCGCTTGTTGCTTT	ACTC	7
17	Chr6SSR17	TCTGAAAAATCCTCGGCTAAAGT	ACTGTCCCACCTTACATCCCTAT	AAAAG	5
18	Chr6SSR19	CAGAGGCAGTTAGGTAGTAGCGA	TGCTAAACCCACCTTCAGTCTTA	GGATTC	4
19	Chr6SSR20	TCTGCTTTCCCACAGTATCTCTC	TCAACAGATAAGCGTCAAGTGAA	TCACCA	4
20	Chr7SSR2	CTGCTTAAAAAGTTGAAGATGAGAA	TTGATACGCTAATAAAATTGTTGAAA	AAT	4
21	Chr7SSR15	GCCTGGCATGTTTTGTATTTCTA	TTTGGTGACCGACAATATAAAGG	ATT	4

No.	Primers	Forward Primer Sequence (5′-3′)	Reverse Primer Sequence (5′-3′)	Repeat Unit	Repeat Number
22	Chr8SSR6	TGAGTCAAGAAAACTTGCAGAAA	TTTGAAAATAATTGAAGTTCCGC	AAAGCA	4
23	Chr8SSR13	TAGATGTTGAACCCCTATTGGAA	GGTAGAGGGTAGAGTGTACGCAG	TACCAC	4
24	Chr8SSR15	GGTGTCATGCGTAAGCTCATAGT	GGTGTCATGCGTAAGCTCATAGT	GAGTTG	4
25	Chr9SSR16	CCCCACCGATGAATTTAGTAGA	TGATGATGTGTCATGGTGTATGA	CCCCCA	4
26	Chr9SSR17	ATTCCCCATATCGAAACTTCTTC	AAATCAAGGAGACGATTGTTGAA	CTGGTG	5
27	Chr10SSR12	ACGAGAGTTTGCTTTCTTTTCCT	TCAGAGGTAGAGGTATGGACTGC	AAC	4
28	Chr10SSR14	TTTTTCAGGCTTTTGTGTATGTAAA	CGGAAACAACCTCTCTACTTCAG	CAA	5
29	Chr11SSR4	AATTTTCAACAACAAACTCCACG	CAGCAGTGAGGATGAAAAAGTTTA	GAG	4
30	Chr11SSR14	TTTTAACTTTTTGTAATTTCGTGTCA	ATCCAATTTTGTTAGGCCTATTG	ATCAAA	4
31	Chr11SSR16	ACGAGTGAACACGTACATGAAAA	GGTAGGAGCTAGGAGTAGTGACG	TTA	6
32	Chr12SSR10	AAAAGGGACTTGTTTTCCTATCA	AAATGGGAATGCGATTATCTAAA	AAAAAT	4
33	Chr12SSR19	TCTTTCCTTCGGATTAAGTTTCC	GAGGCAAGAAATAAGAGATGCCT	CAAATC	4

2.3. PCR Amplification and Polyacrylamide Gel Electrophoresis

PCR were carried out in the final volume of the 10 μL reaction mixture (1 μL DNA, 1 μL forward primer, 1 μL reverse primer (100 ng/μL), 5 μL of 2 × T5 Super PCR Mix, and 2 μL nuclease-free water) on a thermocycler (Applied Biosystems, StepOnePlus ABI7500) with the following reaction conditions: initial denaturation at 94 °C for 3 min, then 30 cycles at 94 °C for 30 s; annealing at 55 °C for 30 s, extension at 72 °C for 1 min; and final extension at 72 °C for 5 min. Electrophoretic analysis was completed in order to assess the PCR amplicons, employing 8% polyacrylamide gel and 0.5 × TBE buffer at a constant voltage of 180 V, 150 mA, and 50 W for three hours, along with a 100 bp DNA ladder. When the electrophoresis was completed, the gel was carefully retrieved, rinsed with sterile water, and kept incubated with 1% silver nitrate (AgNO3) (1 L) for 20 min under shaking conditions. Following the incubation, the gel was washed 3 times with sterile water and immersed in 1 L of developing solution (1.5% sodium hydroxide (NaOH) and 4 mL formaldehydes (CHHO)) until the bands were clearly visible (approximately 5 min). Then, based on the number of clearly visible bands, the alleles in each pepper variety were visually determined.

2.4. Data Statistics and Analysis Results

The scoring was given as 1 (presence) and 0 (absence) for the amplified fragments in each microsatellite loci, and data matrixes were constructed accordingly. Then, employing Popgen (version 1.32) software, several indices were calculated, such as the observed number of alleles (Na), effective number of alleles (Ne), observed heterozygosity (Ho), expected heterozygosity (He), and the Shannon information index (I) [36]. The major allele frequency (MAF), polymorphism information content (PIC), and gene diversity index were calculated using PowerMarker (version 3.0) software [37]. In addition, the cluster analysis of the germplasms was based on the Nei genetic distance [38] and a neighbor joining (NJ) method was used to construct a dendrogram via PowerMarker (version 3.0) software. The dendrogram tree was visualized and edited using MEGA7 (version 7.0) [39].

3. RESULTS

3.1. Distribution of the SSRs in the Capsicum Frutescens Genome

The search results of the genome sequences of *C. frutescens* resulted in a total number of 278,425 SSR loci being identified. (AT)n(TTG)n(AAAT)n(AAATA) n(TATAGA)n is the most common repeat motifs corresponding to the dinucleotide to hexanucleotide repeat, respectively. The SSR repeat types were found to be different. For example, the dominant amongst the 1638 SSR repeats were the trinucleotides and dinucleotides, which accounted for 158,967) %57.1) and 94,916) %34.1), respectively. The remaining was occupied by tetranucleotide, pentanucleotide, and hexanucleotide repeat motifs, accounting for 5.9% (16,351), 1.6% (4531), and 1.3% (3660), respectively. Taken together, it was found that the majority of the SSR repeat motifs along the entirety of genome sequences were trinucleotide repeats, and the hexanucleotide repeats were the fewest, as illustrated in **Figure 1**A.

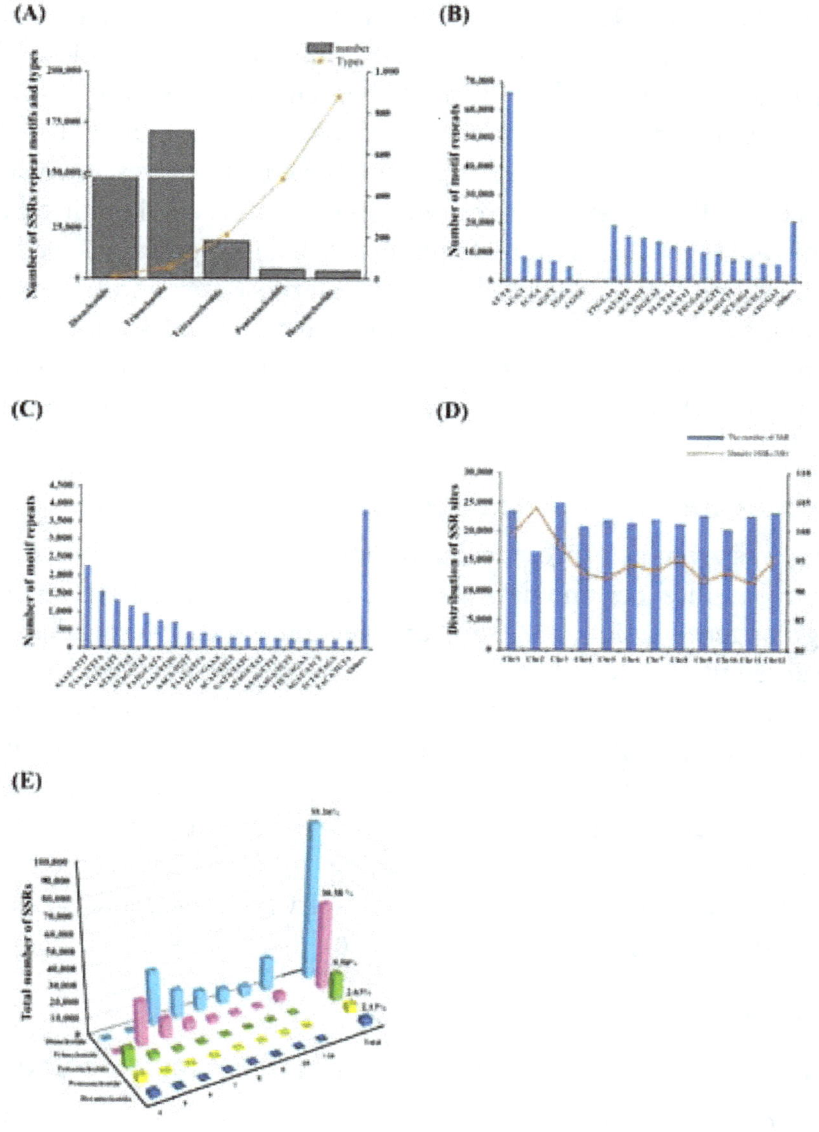

Figure 1. Total number, distribution, frequency, and distinct motif types of SSRs in the entire genome of C. frutescens: (A) Distribution of the SSR in the 'Xiaomijiao' genome; (B) Number of dinucleotide and trinucleotide repeat motifs in the genome; (C) Number of tetranucleotide repeat motifs in the genome; (D) Distribution and frequency of the SSR loci in the genome; (E) Relative frequency (%) of the SSR types with different repeat numbers in the 'Xiaomijiao' genome.

The frequencies of each of SSR motif types along the entire *C. frutescens* genome were also detected. Among the dinucleotide motifs, AT/TA was observed to be the most common (66,356 ;%69.91). This was followed by AC/GT (8671 ;%9.14) and TC/GA (7476 ;%7.88). Meanwhile, the CG/GC motif repeats were rarely observed (%0.05). The trinucleotide repeat motif consisted of 30 different types. The predominant motifs were TTG/CAA and AAT/ATT, which accounted for 12.41 and %10.02, respectively (**Figure 1**B). In addition, AAAT/ATTT (13.85%) were the predominantly found tetranucleotide repeats (**Figure 1**C).

The statistical data showed that the SSR loci were widely distributed on all 12 chromosomes of the *C. frutescens* genome. These were mainly found in Chr3 25,014)), followed by Chr23,644) 1), Chr23,240) 12), Chr22,761) 9), Chr22,617) 11), Chr22,066) 7), Chr21,976) 5), Chr21,448) 6), Chr21,286) 8), Chr20,939) 4), Chr10 20,333)), and Chr16,579) 2), respectively. In addition, 16,522 SSR loci were unable to allocate in the chromosomes. It was found that, while analyzing the distribution frequency of the SSR loci/Mb, the results revealed that the number of SSR loci/Mb on each chromosome ranged from 91.27 to 104.13. However, the number of SSR loci on Chr3 was the highest. The density of the SSR loci on Chr3 was found to be the third highest, with an average of 97.45 SSR loci/Mb. Moreover, although the number of SSR loci on Chr2 was found to the fewest, the density of the SSR loci was the highest overall (104.13 SSR loci/Mb), as shown in **Figure 1**D.

3.2. Analysis of the SSR Repeat Motif Types and Frequencies

As shown in **Figure 1**E, the distributions of the SSRs were also examined from the aspect of the number of repeat units. It was observed that for all the SSR types, the SSR frequencies decreased as the number of repeat units increased. Meanwhile, the change rates became more gradual for the dinucleotides when compared with the longer repeat motif types. The dominant numbers of repeats in the pepper SSR loci ranged between 4 and 10, with the exception of a few (more than 10). At the same time, the majority of the observed repeat times were 6, accounting for 26.20% (45,076). The dinucleotides were found to be the most abundant number of repeats, accounting for 55.16% (94,916), whereas the number of repeats for remaining trinucleotide, tetranucleotide, pentanucleotide, and hexanucleotide were determined to be 30.58% (52,614), 9.50% (16,351), 2.63% (4531), and 2.13% (3660), respectively, as shown in **Figure 1**E.

3.3. Primer Design of the Pepper Plant Genomic SSR Markers

For the designing of the primers in the current study, a total of 240 SSRs, which were distributed on different chromosomes, were selected. Then, the reliability was evaluated on eight pepper cultivars. It was determined that out of the total primer sets tested, 41 were successfully amplified showing full length polymorphisms.

The remaining 199 primers were found to be either non-polymorphic, non-specific amplification with ambiguous bands, or not amplified, as evidenced from the gel results. Of the 41 amplified primer sets, only 33 (13.75%) were found to have generated both polymorphic and unambiguous bands on the gel. Therefore, those primer sets were selected for further analysis, as detailed in Table 1. Among the 33 SSR loci, 1 was observed to be dinucleotides, 11 were trinucleotides, 4 were tetranucleotides, 4 were pentanucleotides, and 13 were hexanucleotides, respectively.

3.4. Polymorphism Analysis with SSR

The results of polyacrylamide gel electrophoresis of several highly polymorphic SSR markers are shown in **Figure 2**C. In total, 91 alleles were obtained with the aforementioned 33 amplified SSR primers. Among those markers, the Number of Alleles (Na) per locus ranged from 2 (Chr1SSR8, Chr2SSR12, Chr2SSR14, Chr3SSR5, Chr3SSR6, Chr4SSR11, Chr5SSR13, Chr5SSR14, Chr6SSR12, Chr6SSR20, Chr7SSR2, Chr8SSR6, Chr9SSR16, Chr10SSR14, Chr11SSR14, and Chr11SSR16) to 6 (Chr4SSR20), with an observed average of 2.8 alleles. The Effective Number of Allele (Ne) per locus ranged from 1.0288 (Chr7SSR2) to 3.6226 (Chr7SSR15), with an observed average of 1.7055 alleles. The major allele frequency (MAF) ranged from a low of 0.3231 (Chr7SSR15) to a high reaching 0.9858 (Chr7SSR2), with an average of 0.7547. In addition, the Observed Heterozygosity (Ho) ranged from 0.000 (Chr1SSR8, Chr4SSR11, Chr6SSR17, Chr7SSR2, Chr10SSR12, Chr11SSR4, and Chr11SSR14) to 0.9863 (Chr5SSR16), with an average of 0.0989 observed. It was also determined that the Expected Heterozygosity (He) ranged from 0.0281 (Chr7SSR2) to 0.7264 (Chr7SSR15), with an average of 0.3313. Also, the Shannon information index (I) ranged between 0.0744 and 1.3273, with an average of 0.5758, and the polymorphic information content (PIC) ranged from 0.0276 to 0.6718, with a mean value of 0.2893. In the present study, the calculation of the mean gene diversity confirmed it to be 0.3300 for all of the 147 types of material, as shown in **Table 2**. It was observed that the different markers had displayed different polymorphism. For example, primer Chr7SSR15 was found to be the most informative (PIC value: 0.6718), whereas primer Chr7SSR2 was the least informative (PIC value: 0.0276). Therefore, this study concluded that when considered altogether, the performances of the selected SSR markers were very effective in detecting genetic diversity.

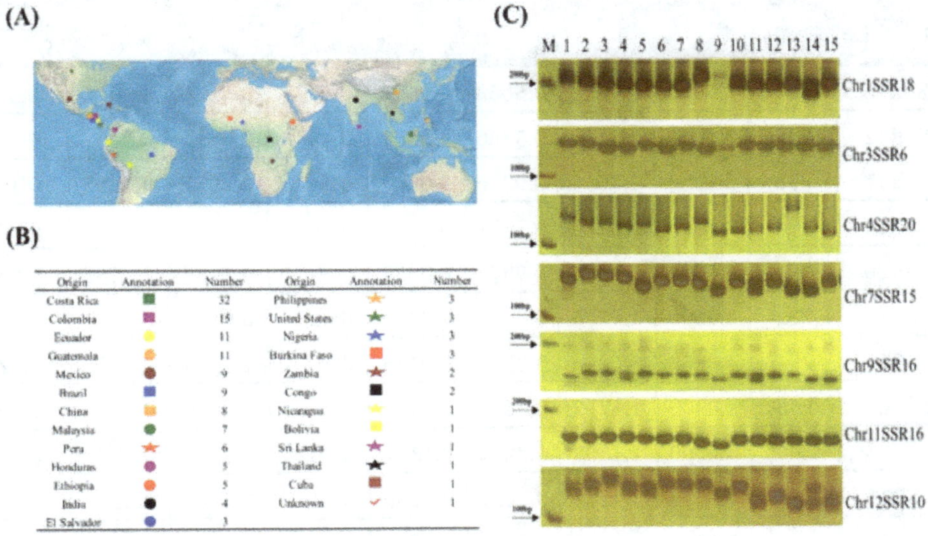

Figure 2. Regional distribution of the 147 pepper (C. frutescens) cultivars, and the detailed information material collected from these countries. (A) Regional distribution of the 147 pepper cultivars in this study, mainly collected from Central and South America, Africa, and Asia; (B) Labeling of 25 countries on the map, and information material collected from these countries (including an unidentified source); (C) Polyacrylamide gel electrophoresis analysis of 7 SSR markers with high polymorphism on samples 1 to 15.

Table 2. Polymorphism analysis of the 147 C. frutescens accessions with SSR primers.

Primer	Allele Size	Na	Ne	MAF	Ho	He	I	PIC	Gene Diversity
Chr1SSR8	134	2.0000	1.0615	0.9701	0.0000	0.0584	0.1342	0.0562	0.0579
Chr1SSR12	294	3.0000	1.2132	0.9048	0.1837	0.1763	0.3680	0.1665	0.1757
Chr1SSR18	278	4.0000	2.7999	0.4532	0.0360	0.6452	1.1545	0.5758	0.6428
Chr2SSR12	286	2.0000	1.1104	0.9476	0.0350	0.0997	0.2057	0.0945	0.0994
Chr2SSR14	294	2.0000	1.1764	0.9184	0.0272	0.1504	0.2827	0.1387	0.1499
Chr2SSR15	278	3.0000	1.3743	0.8417	0.2014	0.2733	0.4998	0.2452	0.2723
Chr3SSR5	286	2.0000	1.1259	0.9406	0.1189	0.1122	0.2254	0.1056	0.1118
Chr3SSR6	290	2.0000	1.3729	0.8379	0.0345	0.2725	0.4431	0.2347	0.2716
Chr3SSR13	272	3.0000	1.4543	0.8162	0.2059	0.3135	0.5784	0.2837	0.3124
Chr4SSR11	290	2.0000	1.1012	0.9517	0.0000	0.0922	0.1934	0.0877	0.0919
Chr4SSR14	204	3.0000	2.1276	0.5392	0.0588	0.5326	0.8242	0.4252	0.5300
Chr4SSR20	292	6.0000	2.3599	0.5822	0.0137	0.5782	1.0946	0.5175	0.5763
Chr5SSR13	282	2.0000	1.4302	0.8156	0.3404	0.3019	0.4780	0.2556	0.3008
Chr5SSR14	278	2.0000	1.0516	0.9748	0.0216	0.0493	0.1176	0.0479	0.0491

Primer	Allele Size	Na	Ne	MAF	Ho	He	I	PIC	Gene Diversity
Chr5SSR16	292	3.0000	2.1992	0.5068	0.9863	0.5472	0.8578	0.4426	0.5453
Chr6SSR12	248	2.0000	1.9541	0.5766	0.0081	0.4902	0.6814	0.3691	0.4883
Chr6SSR17	288	3.0000	1.1656	0.9236	0.0000	0.1426	0.2931	0.1338	0.1421
Chr6SSR19	294	3.0000	1.3276	0.8571	0.0136	0.2476	0.4375	0.2195	0.2467
Chr6SSR20	280	2.0000	1.0894	0.9571	0.0857	0.0823	0.1769	0.0787	0.0820
Chr7SSR2	282	2.0000	1.0288	0.9858	0.0000	0.0281	0.0744	0.0276	0.0280
Chr7SSR15	294	4.0000	3.6226	0.3231	0.0476	0.7264	1.3273	0.6718	0.7240
Chr8SSR6	288	2.0000	1.5732	0.7604	0.0486	0.3656	0.5506	0.2980	0.3644
Chr8SSR13	294	3.0000	1.4308	0.8265	0.1361	0.3021	0.5775	0.2795	0.3011
Chr8SSR15	286	4.0000	3.4282	0.4021	0.0699	0.7108	1.3023	0.6566	0.7083
Chr9SSR16	250	2.0000	1.9678	0.5640	0.0560	0.4938	0.6849	0.3709	0.4918
Chr9SSR17	294	3.0000	1.6031	0.7721	0.0748	0.3775	0.6873	0.3430	0.3762
Chr10SSR12	270	4.0000	3.5014	0.3556	0.0000	0.7171	1.3058	0.6606	0.7144
Chr10SSR14	292	2.0000	1.5026	0.7877	0.0411	0.3356	0.5170	0.2785	0.3345
Chr11SSR4	292	3.0000	1.3060	0.8699	0.0000	0.2351	0.4723	0.2206	0.2343
Chr11SSR14	76	2.0000	1.2321	0.8947	0.0000	0.1909	0.3365	0.1706	0.1884
Chr11SSR16	294	2.0000	1.7306	0.6973	0.0884	0.4236	0.6132	0.3331	0.4222
Chr12SSR10	278	3.0000	2.5179	0.4928	0.1007	0.6050	0.9921	0.5226	0.6028
Chr12SSR19	294	4.0000	1.3399	0.8571	0.2313	0.2546	0.5140	0.2366	0.2537
Mean	271	2.7576	1.7055	0.7547	0.0989	0.3313	0.5758	0.2893	0.3300

3.5. Genetic Diversity Analysis

The genetic diversity and phylogenetic relationships were determined using 147 collected pepper cultivars from 26 different countries around the world (**Figure 2 1**). This study adopted Nei genetic distance and neighbor-joining methods, and a dendrogram was constructed based on the genotypes detected by the newly developed SSR markers (**Figure 3**). These were clustered into seven main groups (designated in this study as Groups I, II, III, IV, V, VI, and VII), which were comprised of 18, 37, 32, 20, 21, 5, and 14 members, respectively.

The dendrogram not only reflected the phylogenetic relationships of the cultivars, but was also consistent with their places of origin. Remarkably, the dendrogram revealed that Group I had become a unique branch. Furthermore, Group II consisted of 37 accessions, majority of which were from Latin America, with the exception of 3 accessions (PI 281,419 and PI 281,420 from the Philippines, and PI 281,347 from India). Of plant material collected from 15 countries, 32 types comprised Group III. All of the materials collected in Africa were found to be clustered in that group. In addition, 4 Chinese cultivars were also assigned to Group III. Group IV consisted of 20 accessions, among which 7 were from Guatemala; 7 were from South America; 3 were from United States; 2 were from Mexico; and 1 was from an unknown country, which this study speculated may have been of North American origin. Group V was composed of 21 accessions which were from various geographical origins, such as North America, South America, and other Asian areas. The five members which were clustered in Group VI included four cultivars derived from Costa Rica, and one derived from El Salvador. In addition, 20 accessions derived from Latin America were clustered together in Group VII (**Figure 3**). These results suggested that the newly developed SSR markers were both stable and suitable for assessing the genetic relationships among *C. frutescens* cultivars.

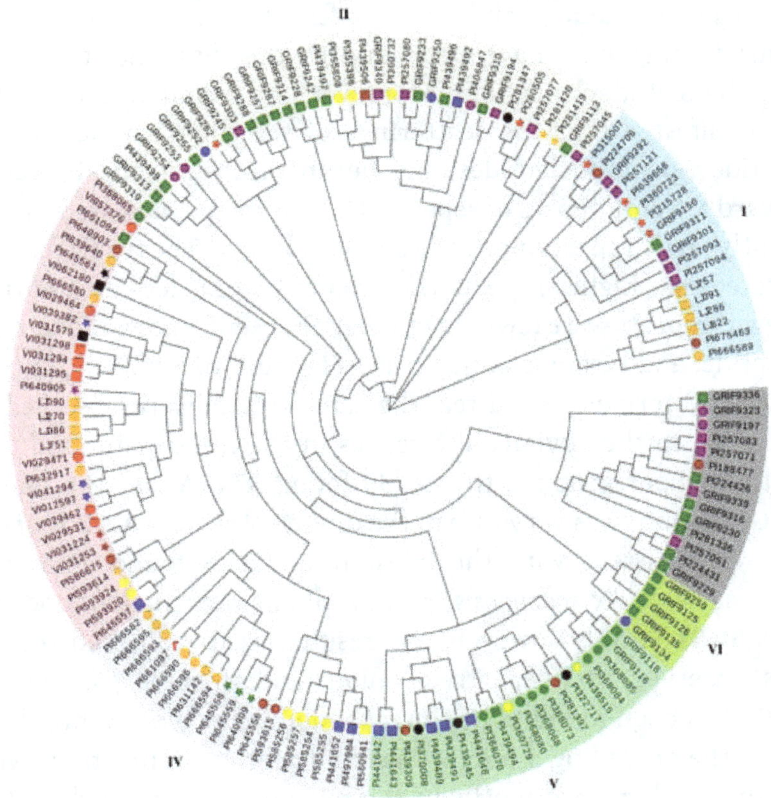

Figure 3. Dendrogram was constructed based on the genotypes from 33 SSR markers using neighbor-joining methods, and the icons indicated the information of material source, which was shown in Figure 2B.

4. DISCUSSION

The genome wide analysis of SSRs could provide the opportunity to decipher the optimal functions of these repeats in the regulation and organization of a genome.

Also, the potential uses of these markers, such as diversity and population analyses, evolutionary history, and genome and comparative mapping, are currently being explored [**19,20,21,40,41,42**]. In previous studies, based on the results of sequencing analyses, assessments of the high-quality genome sequences of *C. frutescens* were made possible, thereby accrediting the opportunity to develop suitable SSR primers. In the present study, it was observed that the dinucleotide and trinucleotide motif repeats were the most abundant, accounting for %85.74. The remainder of the repeats (%14.26) were contributed by the tetranucleotides, pentanucleotides, and hexanucleotides, as illustrated in **Figure 1**E. This phenomenon has also been reported in other plants, such as *Radix codonopsis*, *Anthuriumand raeanum*, and *Camellia sinensis*, respectively [**43,44,45**]. Nevertheless, a few reports have found high abundance of tetranucleotide repeats in some plants, such as *Cucumis sativus*, *Medicago truncatula*, and *Vitis vinifera* [**46,47,48**]. The differences in the previous finding may have been due to the dissimilarities in paradigms adopted for the SSR identifications. Moreover, the examination of the SSR motif frequency manifested that the scatterings of the dinucleotides, trinucleotides, tetranucleotides, pentanucleotides, and hexanucleotides repeats were generally skewed toward fewer numbers of repeats. These findings indicated that there were predominantly fewer repeats along the pepper genome. The results obtained in this study indicated that higher repeats were found in dinucleotide and trinucleotide SSRs. However, repeats were fewer in number or absent among the tetranucleotide, pentanucleotide, and hexanucleotide SSRs (**Figure 1**A). Similarly, in other plants, such as citrus, watermelon, and tea, the same trends were observed [**18,19,45**]. This may be due to the obvious differences in the frequencies and types of SSR motifs. In the pepper genome, this study found AT/TA was the most common, while the CG/GC motif was very rare in the dinucleotide repeats (**Figure 1**B). These findings were consistent with the motif frequencies found among cucumber, strawberry, maize, *Radix codonopsis*, potato, plum, watermelon, and horseradish [**19,43,48,49,50,51,52,53**]. However, our results greatly differed from the motif frequencies observed in rice, citrus, onion, and *Atremisia frigida* [**18,54,55,56**], where AG/CT has been found to be the most dominant type. Similarly, TTG, AAT, and ATT were the prevailing motifs of the trinucleotide in this study, with CCG and AGG being the predominant motifs in the monocotyledons, such as barley, rice, and corn. It was found that the number of SSRs, along with their structure and repeat motifs, will greatly differ compared with those in plant species.

It has been found that SSR markers are very much beneficial in population genetics and molecular breeding. However, their effectiveness mainly relies on the marker quality and the accuracy of the experimentation [25]. In the present study, 240 selected SSR loci markers were scrutinized, resulting in 33 unique markers. It was found that when evaluating 147 pepper cultivars, these markers demonstrated remarkable and unambiguous amplification bands (Table 1). The screened SSR polymorphism primers accounted for 13.8% of the total. Previously, Li et al. obtained 17 pairs of SSR polymorphism primers with clear bands and high polymorphism from 152 pairs of SSR primers covering 12 chromosomes [33]. These accounted for 11.2% of the total number of SSR primers, which was slightly lower than that obtained in this result. Liu et al. evaluated 85 pairs of SSR polymorphic primers and 12 pairs were scrutinized [57]. These accounted for 14.1% of the total, which was similar to this study's research results. Wu et al. used three different peppers as templates to select 65 pairs of SSR polymorphism primers from 153 pairs of SSR primers [42]. These accounted for 42.5% of the total number, which was substantially higher than that obtained in this study. Therefore, it was determined that the proportion of SSR polymorphism primers screened in this study was relatively low, which may have been attributed to the small differences existing in these peppers However, although the proportion of polymorphic primers was low, the results could still be used to analyze the genetic diversity of the pepper population. In regard to pepper, many SSR markers have been developed and mapped to linkage groups [58,59,60], which provide a key basis for analyzing pepper genetic diversity. However, such factors as the number, size, and types of SSR markers, frequencies of the SSR motifs, as well as the various sampling schemes, apparently result in differences in genetic diversity [25,61].

It is of major significance for the collection and efficient utilization of germplasm resources to continue to carry out genetic diversity evaluations. Highly polymorphic, as well as stable markers, are the prerequisites for studying genetic relationships and diversity. Nevertheless, it was found that the SSR loci showed less diversity, as evidenced by low polymorphism information index and gene diversity when compared to earlier reports. For example, in the study conducted by Nicolaï [62], a PIC of 0.67 and a gene diversity of 0.7 were reported, which were 0.38% and 0.37% higher than those of the current study (PIC: 0.29; Gene Diversity: 0.33). However, this may have been due to the volume and types of test sampling. In the aforementioned study, 1352 accessions from 89 countries were utilized, including 11 species of *Capsicum*. However, this study only examined a single species, *C. frutescens*. Therefore, compared with Li's report [33], the amount of accessions used was approximate. The PIC was approximate (slightly higher than this study), but the number of markers was different. Therefore, it was considered that the hypothesis that the difference in genetic diversity was influenced by the number of SSR markers had been confirmed.

Previously, researchers reported that genetic diversity of some peppers (*Capsicum* spp.) accessions, including *Capsicum chinense*, *Capsicum annuum*. In 2016, 71 *C. chinense* accessions from different Brazilian geographic regions, using fruit morphological descriptors and AFLP molecular markers, were analyzed [**63**]. The results found no association between the morphological descriptors and AFLP markers [**63**]. In the same year, the researchers investigated patterns of molecular diversity using a transcriptome-based 48 single nucleotide polymorphisms (SNPs) in a large germplasm collection comprising 3821 accessions. Among the 11 species examined, *Capsicum annuum* showed the highest genetic diversity (HE=0.44, I=0.69), whereas the wild species *C. galapagoense* showed the lowest genetic diversity (HE=0.06, I=0.07). The *Capsicum* germplasm collection was divided into 10 clusters (cluster 1 to 10) based on population structure analysis, and five groups (group A to E) based on phylogenetic analysis [**64**]. The dendrogram constructed in this study from 147 pepper accessions using NJ methods, indicated that the genetic relatedness of the pepper cultivars clustered in the majority of the groups were in good agreement with their geographic origins. This study's analysis results were also consistent with the previous findings reported by Luo et al. and Jia et al. [**7**,**65**]. Moreover, the geographical sources of the Group III and Group V materials were found to be diverse, and not only attributed to Asian and African countries, but also to Latin American countries. It was observed that the pepper plants of the same geographical origin were not strictly divided into the same groups. For example, the eight pepper materials from China were grouped into Group I and Group III. These findings suggested that numerous complex migrations had occurred in the pepper genotype as the result of human migration, which had led to their adoption, acclimatization, and local selection.

5. CONCISE SUMMARY

A total of 278,425 simple sequence repeats (SSRs) were discovered by a thorough examination of the genomic sequences of C. frutescens. The AT/TA, TTG/CAA, and AAAT/ATTT were found to be the predominant repeat motifs in the dinucleotide, trinucleotide, and tetranucleotide repeats, respectively. Dinucleotides were the most prevalent type of repetitions, comprising 55.16% of the total. This research study examined the genetic diversity of C. frutescens germplasms by utilizing 33 SSR markers that were uniformly dispersed across all chromosomes. The study utilized 147 experimental materials consisting of wild peppers that exhibited a wide range of genetic variability. Their prominent characteristics, such as resistance to disease, high productivity, and others, have the potential to offer valuable genes for the development of new pepper breeding varieties. Additionally, these findings may serve as a foundation for enhancing the current cultivated pepper species, while also playing a crucial role in broadening the limited genetic diversity of pepper breeding in China. Furthermore, the genome-wide detection and creation of SSR markers were deemed valuable and could potentially offer valuable insights into multiple research domains related to C. frutescens in the future, including high-density genetic mapping, genome comparative mapping, and genome-wide association analyses.

AT A GLANCE

Capsicum frutescens, a cultivated pepper species found globally, is considered highly beneficial for its robust resistance to plant diseases, high yield, and powerful fragrance. Nevertheless, the scarcity of molecular markers hampers the effectiveness and precision of genetic breeding for pepper. The recently established advanced next-generation sequencing technology allows for the creation of genome sequences of C. frutescens. These sequences can now be utilized for the identification of SSR markers through the process of data mining. The pepper genome was analyzed using MISA software, resulting in the detection of 278,425 SSRs. The primary repeat motif was found to be trinucleotides. Subsequently, the dinucleotide, tetranucleotide, pentanucleotide, and hexanucleotide repeat types were observed. The repeat motifs (AT)n, (TTG)n, (AAAT)n, (AAATA)n, and (TATAGA) n are recognized as the most prevalent patterns of dinucleotide to hexanucleotide repetitions, respectively. Furthermore, a comprehensive set of 240 SSR primers, which were evenly dispersed throughout all 12 chromosomes, were specifically constructed and tested against 8 different cultivars of C. frutescens. A total of 33 SSR markers with high polymorphism were examined in 147 accessions from 25 countries. The dendrogram grouped these accessions into seven distinct clusters. The groups were determined to be congruent with their respective origins. The investigation yielded SSR molecular markers and valuable information on the genetic diversity of C. frutescens.

REFERENCES

1. Moscone, E.A.; Scaldaferro, M.; Grabiele, M.; Cecchini, N.; García, Y.S.; Jarret, R.; Daviña, J.; Ducasse, D.; Barboza, G.; Ehrendorfer, F. The evolution of chili peppers (*Capsicum*-solanaceae): A cytogenetic perspective. *Acta Hortic.* 2007, *745*, 137–169.
2. Olmstead, R.G.; Bohs, L.; Migid, H.A.; Santiago-Valentin, E.; Garcia, V.F.; Collier, S.M. A molecular phylogeny of the solanaceae. *TAXON* 2008, *57*, 1159–1181.
3. Qin, C.; Yu, C.; Shen, Y.; Fang, X.; Chen, L.; Min, J.; Cheng, J.; Zhao, S.; Xu, M.; Luo, Y.; et al. Whole-genome sequencing of cultivated and wild peppers provides insights into *Capsicum* domestication and specialization. *Proc. Natl. Acad. Sci. USA* 2014, *111*, 5135–5140.
4. Nee, M. Peppers, the domesticated *Capsicums*. *Brittonia* 1985, *37*, 218.
5. Chen, X.; Zhou, K.; Zong, H.; Fang, R. Genetic diversity of *Capsicum frutescens* in China as revealed by SRAP and SSR markers. *Acta Bot. Boreali-Occident. Sin.* 2012, *32*, 2201–2205.
6. Zhao, H.; Cao, Y.; Zhang, Z.; Zhang, B.; Bai, R.; Zhao, Y.; Wang, L. Analysis and evaluation of genetic diversity of pepper (*Capsicum* spp.) Core Germplasm Resources in China. *China Veg.* 2018, *1*, 25–34.
7. Jia, H.; Wei, X.; Yao, Q.; Yuan, Y.; Wang, Z.; Jiang, J.; Yang, S.; Zhao, Y.; Wang, B.; Zhang, X. Genetic diversity analysis of *Capsicum* genus germplasm resources using SSR markers. *Mol. Plant Breed.* 2017, *15*, 353–363.

8. Deng, M.; Wen, J.; Zhu, H.; Zou, X. The hottest pepper variety in China. *Genet. Resour. Crop Evol.* 2009, *56*, 605–608.

9. Liu, S.; Li, W.; Wu, Y.; Chen, C.; Lei, J. De Novo Transcriptome Assembly in Chili Pepper (*Capsicum frutescens*) to Identify Genes Involved in the Biosynthesis of Capsaicinoids. *PLoS ONE* 2013, *8*, e48156.

10. Wei, J.; Zheng, J.; Yu, J.; Zhao, D.; Cheng, Y.; Ruan, M.; Ye, Q.; Yao, Z.; Wang, R.; Zhou, G.; et al. Production and identification of interspecific hybrids between pepper (*Capsicum annuum* L.) and the wild relative (*Capsicum frutescens* L.). *J. Agric. Sci. Technol.* 2019, *21*, 761–769.

11. Wei, J.; Li, J.; Yu, J.; Cheng, Y.; Ruan, M.; Ye, Q.; Yao, Z.; Wang, R.; Zhou, G.; Deng, M.; et al. Construction of high-density bin map and QTL mapping of horticultural traits from an interspecific cross between *Capsicum annuum* and Chinese wild *Capsicum frutescens*. *Biotechnol. Biotechnol. Equip.* 2020, *34*, 549–561.

12. Tautz, D.; Renz, M. Simple sequences are ubiquitous repetitive components of eukaryotic genomes. *Nucleic Acids Res.* 1984, *12*, 4127–4138.

13. Tóth, G.; Gáspári, Z.; Jurka, J. Microsatellites in different eukaryotic genomes: Survey and analysis. *Genome Res.* 2000, *10*, 967–981.

14. Powell, W.; Machray, G.C.; Provan, J. Polymorphism revealed by simple sequence repeats. *Trends Plant Sci.* 1996, *1*, 215–222.

15. Chen, C.; Zhou, P.; Choi, Y.A.; Huang, S.; Gmitter, F.G. Mining and characterizing microsatellites from citrus ESTs. *Theor. Appl. Genet.* 2006, *112*, 1248–1257.

16. Cheng, Z.; Huang, H. SSR fingerprinting Chinese peach cultivars and landraces (*Prunus persica*) and analysis of their genetic relationships. *Sci. Hortic.* 2008, *120*, 188–193.

17. Mccouch, S.; Teytelman, L.; Xu, Y.; Lobos, K.; Clare, K.; Walton, M.; Fu, B.; Maghirang, R.; Li, Z.; Xing, Y.; et al. Development and mapping of 2240 new SSR markers for rice (*Oryza sativa* L.). *DNA Res.* 2002, *9*, 199–207.

18. Liu, S.; Li, W.; Long, D.; Hu, C.; Zhang, J. Development and characterization of genomic and expressed SSRs in citrus by genome-wide analysis. *PLoS ONE* 2013, *8*, e75149.

19. Zhu, H.; Song, P.; Dal-Hoe, K.; Guo, L.; Li, Y.; Sun, S.; Weng, Y.; Yang, L. Genome wide characterization of simple sequence repeats in watermelon genome and their application in comparative mapping and genetic diversity analysis. *BMC Genom.* 2016, *17*, 557–583.

20. Tan, L.; Wang, L.; Wei, K.; Zhang, C.; Wu, L.; Qi, G.; Cheng, H.; Zhang, Q.; Cui, Q.; Liang, J. Floral transcriptome sequencing for SSR marker development and linkage map construction in the tea plant (*Camellia sinensis*). *PLoS ONE* 2013, *8*, e81611.

21. Bali, S.; Mamgain, A.; Raina, S.N.; Yadava, S.K.; Bhat, V.; Das, S.; Pradhan, A.K.; Goel, S. Construction of a genetic linkage map and mapping of drought tolerance trait in Indian beveragial tea. *Mol. Breed.* 2015, *35*, 1–20.

22. Li, L.; He, W.; Ma, L.; Liu, P.; Xu, H.; Xu, J.; Zheng, X. Construction Chinese cabbage (*Brassica rapa* L.) core collection and its EST-SSR finger print database by EST-SSR molecular markers. *Genom. Appl. Biol.* 2009, *28*, 76–88.

23. Sarao, N.K.; Vikal, Y.; Singh, K.; Joshi, M.A.; Sharma, R.C. SSR marker-based DNA fingerprinting and cultivar identification of rice (*Oryza sativa* L.) in Punjab state of India. *Plant Genet. Resour.* 2009, *8*, 42–44.

24. Tan, L.; Peng, M.; Xu, L.; Wang, L.; Chen, S.; Zou, Y.; Qi, G.; Cheng, H. Fingerprinting 128 Chinese clonal tea cultivars using SSR markers provides new insights into their pedigree relationships. *Tree Genet. Genomes* 2015, *11*, 1–12.

25. Liu, S.; Liu, H.; Wu, A.; Hou, Y.; An, Y.; Wei, C. Construction of fingerprinting for tea plant (*Camellia sinensis*) accessions using new genomic SSR markers. *Mol. Breed.* 2017, *37*, 1–14.

26. Ercisli, S.; Ipek, A.; Barut, E. SSR marker-based DNA fingerprinting and cultivar identification of olives (*Olea europaea*). *Biochem. Genet.* 2011, *49*, 555–561.

27. Hameed, U.; Pan, Y.B.; Muhammad, K.; Afghan, S.; Iqbal, J. Use of simple sequence repeat markers for DNA fingerprinting and diversity analysis of sugarcane (*Saccharum* spp.) cultivars resistant and susceptible to red rot. *Genet. Mol. Res.* 2012, *11*, 1195–1204.

28. Lanteri, S.; Portis, E.; Acquadro, A.; Mauro, R.; Mauromicale, G. Morphology and SSR fingerprinting of newly developed *Cynara cardunculus* genotypes exploitable as ornamentals. *Euphytica* 2012, *184*, 311–321.

29. Zhebentyayeva, T.; Reighard, G.; Gorina, V.; Abbott, A. Simple sequence repeat (SSR) analysis for assessment of genetic variability in apricot germplasm. *Theor. Appl. Genet.* 2003, *106*, 435–444.

30. Grover, A.; Sharma, P.C. Development and use of molecular markers: Past and present. *Crit. Rev. Biotechnol.* 2016, *36*, 290–302.

31. Gu, X.; Cao, Y.; Zhang, Z.; Zhang, B.; Zhao, H.; Zhang, X.; Wang, H.; Li, X.; Wang, L. Genetic diversity and population structure analysis of *Capsicum* germplasm accessions. *J. Integr. Agric.* 2019, *18*, 1312–1320.

32. Zhang, X.; Zhang, Z.; Gu, X.; Mao, S.; Li, X.; Chadoeuf, J.; Palloix, A.; Wang, L.; Zhang, B. Genetic diversity of pepper (*Capsicum* spp.) germplasm resources in China reflects selection for cultivar types and spatial distribution. *J. Integr. Agric.* 2016, *15*, 1991–2001.

33. Li, Y.; Zhao, H.; Wang, Y.; Jiang, J.; Meng, X.; Wei, X.; Li, J. Genetic diversity analysis of 169 accessions of *Capsicum*. *J. Henan Agric. Sci.* 2018, *47*, 91–97.

34. Murray, M.G.; Thompson, W.F. Rapid isolation of high molecular weight plant DNA. *Nucleic Acids Res.* 1980, *8*, 4321–4325.

35. Thiel, T.; Michalek, W.; Varshney, R.; Graner, A. Exploiting EST databases for the development and characterization of gene-derived SSR-markers in barley (*Hordeum vulgare* L.). *Theor. Appl. Genet.* 2003, *106*, 411–422.

36. Yeh, F.; Boyle, T.B. Population genetic analysis of co-dominant and dominant markers and quantitative traits. *Belg. J. Bot.* 1997, *129*, 157.

37. Liu, K.; Muse, S.V. PowerMarker: An integrated analysis environment for genetic marker analysis. *Bioinformatics* 2005, *21*, 2128–2129.

38. Nei, M. Estimation of average heterozygosity and genetic distance from a small number of individuals. *Genetics* 1978, *89*, 583–590.

39. Kumar, S.; Stecher, G.; Tamura, K. MEGA7: Molecular evolutionary genetics analysis version 7.0 for bigger datasets. *Mol. Biol. Evol.* 2016, *33*, 1870–1874.

40. Zhao, D.; Yang, J.; Yang, S.; Kenji, K.; Luo, J. Genetic diversity and domestication origin of tea plant *Camellia taliensis* (Theaceae) as revealed by microsatellite markers. *BMC Plant Biol.* 2014, *14*, 14.

41. Göl, Ş.; Göktay, M.; Allmer, J.; Doğanlar, S.; Frary, A. Newly developed SSR markers reveal genetic diversity and geographical clustering in spinach (*Spinacia oleracea*). *Mol. Genet. Genom.* 2017, *292*, 847–855.

42. Wu, J.; Cheng, F.; Cai, C.; Zhong, Y.; Jie, X. Association mapping for floral traits in cultivated *Paeonia rockii* based on SSR markers. *Mol. Genet. Genom.* 2017, *292*, 187–200.

43. Wang, D.; Cao, L.; Gao, J. Data mining of simple sequence repeats in *Codonopsis pilosula* transcriptome. *Chin. Tradit. Herb. Drugs* 2014, *45*, 2390–2394.

44. Yu, Y.; Tian, D.; Pan, X.; Jin, L.; Ge, Y. Mining and developing SSR molecular markers based on transcriptome sequences of *Anthurium*. *Mol. Plant Breed.* 2015, *13*, 1349–1354.

45. Liu, S.; An, Y.; Li, F.; Li, S.; Liu, L.; Zhou, Q.; Zhao, S.; Wei, C. Genome-wide identification of simple sequence repeats and development of polymorphic SSR markers for genetic studies in tea plant (*Camellia sinensis*). *Mol. Breed.* 2018, *38*, 1–13.

46. Mun, J.; Kim, D.; Choi, H.; Gish, J.; Debéllé, F.; Mudge, J.; Denny, R.; Endré, G.; Saurat, O.; Dudez, A.; et al. Distribution of microsatellites in the genome of *Medicago truncatula*: A resource of genetic markers that integrate genetic and physical maps. *Genetics* 2006, *172*, 2541–2555.

47. Jaillon, O.; Aury, J.; Noel, B.; Policriti, A.; Clépet, C.; Casagrande, A.; Choisne, N.; Aubourg, S.; Vitulo, N.; Jubin, C.; et al. The grapevine genome sequence suggests ancestral hexaploidization in major angiosperm phyla. *Nature* 2007, *449*, 463–467.

48. Cavagnaro, P.F.; Senalik, D.A.; Yang, L.M.; Simon, P.W.; Harkins, T.T.; Kodira, C.D.; Huang, S.W.; Weng, Y.Q. Genome-wide characterization of simple sequence repeats in cucumber (*Cucumis sativus* L.). *BMC Genom.* 2010, *11*, 569–587.

49. Dong, Q.; Wang, X.; Zhao, M.; Song, C.; Ge, A.; Wang, J. Development of EST-derived SSR markers and their application in strawberry genetic diversity analysis. *Sci. Agric. Sin.* 2011, *44*, 3603–3612.

50. Xu, J.; Liu, L.; Xu, Y.; Chen, C.; Rong, T.; Ali, F.; Zhou, S.; Wu, F.; Liu, Y.; Wang, J.; et al. Development and characterization of simple sequence repeat markers providing genome-wide coverage and high resolution in maize. *DNA Res.* 2013, *20*, 497–509.

51. Gong, L.; Cheng, Y.; Gan, X.; Chen, Y.; Nie, F.; Zhang, L.; Shi, L.; Song, Y.; Guo, Z.; Wang, F. Mining EST-SSRs based on potato transcriptome and analyzing their polymorphism. *Mol. Plant Breed.* 2015, *13*, 1535–1544.

52. Fang, Z.; Ye, X.; Zhou, D.; Jiang, C.; Pan, S. Analysis on SSR information in 'Furongli' plum transcriptome and development of molecular markers in *Prunus salicina* Lindl. *J. Fruit Sci.* 2016, 33, 416–424.

53. Zhang, Y.; Wei, X.; Wang, X.; Pan, S.; Ye, X. SSR loci in transcriptome of *Moringa oleifera* lam. *Fujian J. Agric. Sci.* 2017, *32*, 955–958.

54. Zhang, Z.; Deng, Y.; Tan, J.; Hu, S.; Yu, J.; Xue, Q. A Genome-wide microsatellite polymorphism database for the indica and japonica Rice. *DNA Res.* 2007, *14*, 37–45.

55. Li, M.; Zhang, S.; Deng, P.; Hou, X.; Wang, J. Analysis on SSR information in transcriptome of onion and the polymorphism. *Acta Hortic. Sin.* 2015, *42*, 1103–1111.

56. Yue, C.; Chen, C.; Guo, F.; Li, H.; Sun, H.; Pei, D.; Ma, X.; Chen, F.; Yang, H.; Li, Q. Data mining of simple sequence repeats in transcriptome sequences of mongolia medicinal plant artemisia frigida willd. *J. Agric. Sci. Technol.* 2016, *18*, 31–43.

57. Liu, Z.; Yang, Y.; Sun, J.; Liu, Z.; Cao, Z. Purity identification of Rela No.2 pepper and genetic diversity analysis of excellent pepper inbred lines. *Chin. J. Trop. Crop.* 2014, *35*, 847–853.

58. Rivera, A.; Monteagudo, A.B.; Igartua, E.; Taboada, A.; García-Ulloa, A.; Pomar, F.; Riveiro-Leira, M.; Silvar, C. Assessing genetic and phenotypic diversity in pepper (*Capsicum annuum* L.) landraces from North-West Spain. *Sci. Hortic.* 2016, *203*, 1–11.

59. Yuan, X.; Zhou, K.; Wu, Y.; Fang, R.; Chen, X. Genetic Diversity and population structure analysis of pepper core collections. *Mol. Plant Breed.* 2019, *17*, 3090–3104.

60. Uncu, A.T. Genome-wide identification of simple sequence repeat (SSR) markers in *Capsicum chinense* Jacq. with high potential for use in pepper introgression breeding. *Biologia* 2019, *74*, 119–126.

61. Yao, M.; Ma, C.; Qiao, T.; Jin, J.; Chen, L. Diversity distribution and population structure of tea germplasms in China revealed by EST-SSR markers. *Tree Genet. Genomes* 2012, *8*, 205–220.

62. Nicolaï, M.; Cantet, M.; Lefebvre, V.; Sage-Palloix, A.M.; Palloix, A. Genotyping a large collection of pepper (*Capsicum* spp.) with SSR loci brings new evidence for the wild origin of cultivated *C. annuum* and the structuring of genetic diversity by human selection of cultivar types. *Genet. Resour. Crop Evol.* 2013, *60*, 2375–2390.

63. Baba, V.Y.; Rocha, K.; Gomes, G.P.; Ruas, C.F.; Ruas, P.; Rodrigues, R.; Gonçalves, L. Genetic diversity of *Capsicum chinense* accessions based on fruit morphological characterization and AFLP markers. *Genet. Resour. Crop Evol.* 2015, *63*, 1371–1381.

64. Lee, H.; Ro, N.; Jeong, H.; Kwon, J.; Jo, J.; Ha, Y.; Jung, A.; Han, J.; Venkatesh, J.; Kang, B. Genetic diversity and population structure analysis to construct a core collection from a large *Capsicum* germplasm. *BMC Genet.* 2016, *17*, 142–154.

65. Luo, Y.; Li, J.; Li, M. Analysis of genetic diversity of *Capsicum* germplasm resources by using SSR markers. *Biotechnol. Bull.* 2006, *S1*, 337–341.

8
Chapter

AN ASSESSMENT WAS CONDUCTED ON 130 GENOTYPES OF EGGPLANT (*SOLANUM MELONGENA* L.) TO DETERMINE THEIR SUITABILITY FOR FUTURE BREEDING PROGRAMS

1. INTRODUCTION: AN OVERVIEW

Eggplant, also known as brinjal (*Solanum melongena* L.; 2n = 2x = 24), is classified as a highly diverse member of the Solanaceae family, which encompasses around 1300 different species. It may thrive in a wide range of climatic conditions seen in different biological locations. The species has a great number of different types with significant capacity for adapting to different environments, which has made it the most economically important vegetable. Eggplant is a collective term used to refer to many species of Solanum that are grown for their fruits. These species include the East Asian aubergine (*S. melongena* L.) and two African native eggplants, Scarlet (*S. aethiopicum* L.) and Gboma (*S. macrocarpon* L.) [1]. The health-promoting characteristics of eggplant have led to its increased prominence. Hence, eggplant and its related plants has a multitude of medical uses, encompassing 77 unique medicinal characteristics [2]. Eggplant is a widely cultivated vegetable that is grown in various regions, ranging from the subtropics to the Mediterranean. It is particularly popular in Asia, Africa, and the southern United States. In 2019, global output of eggplant reached a substantial amount of 55.15 million tons. Asia accounts for almost 90% of the world's eggplant production, with 87% of the total growing area [3,4]. In Bangladesh, the vegetable that is produced the second most, after potatoes, is ranked. It covers approximately 15% of the overall area dedicated to vegetable farming and contributes to around 8% of the total vegetable output [5]. Nevertheless, wild varieties coexist with landraces and cultivars in the specific regions where they originated and were domesticated. The exchange of genetic material between wild and cultivated organisms, along with the influence of both natural and human-driven selection, has led to the emergence of intermediate characteristics that are associated with certain traits seen in the wild. The contribution of traits that are of significant value for breeding to diversity is not evenly distributed. The primary factors that caused the differences between accessions in Indian landraces of S. melongena were the yield per plant, fruit breadth, number of long-styled flowers per plant, flowering earliness, total phenolic content, and ascorbic acid content [6]. However, the outcomes are contingent upon the size of the sample utilized. Thus far, there has been no comprehensive study conducted on a sizable and representative sample that encompasses the entirety of the phenotypic variations found in all farmed eggplants. Eggplant cultivation is widespread in Bangladesh, however the yield is inadequate due to the absence of improved and desired varieties and the significant infestation of insect pests. Nevertheless, the analysis of the physical characteristics of eggplants has been valuable in examining the correlation and variety within different eggplant cultivars. The European Eggplant Genetic Resources Network (EGGNET) established the morphological characterization for eggplant [7], which has been verified and employed in the assessment of eggplant breeding materials in several research investigations [8,9,10]. Hence, the process of generating diversity through mutation, hybridization, and biotechnology techniques is both costly and time-consuming [8]. Therefore, it is necessary to characterize the collected germplasm (populations) in order to identify lines that are suitable for the generation of new varieties [11]. Plant breeders are intrigued by genetic diversity studies that focus on qualitative and quantitative features due to their ability to be efficiently and effortlessly evaluated utilizing cost-effective techniques.

Several papers have shown the phenotypic variety of fruits, plants, and other features in Solanum or *S. aethiopicum*, as well as in multiple eggplant species [12,13,14,15,16,17,18,19]. In addition, there have been numerous studies that extensively examine the phenotypic variety of eggplants using either Mendelian or quantitative heredity patterns to analyze specific qualities of interest [20, 21, 22]. Several scientists [23,24,25] conducted comprehensive research on the genetic diversity, heritability, and genetic advancement in several eggplant genotypes. As a result, breeders encounter a difficulty in choosing genotypes that have high yields in numerous characteristics, necessitating a dependable decision assistance tool. Adopting a rigorous selection method in plant breeding research can significantly reduce the consumption of time and resources.

The Smith-Hazel Index (SH index) is commonly used in plant breeding as a selection index that considers multiple traits [26]. To calculate the SH index, it is necessary to invert a phenotypic covariance matrix and a vector of economic weights. The presence of multicollinearity can lead to poorly conditioned matrices and biased index coefficients, which in turn might impact genetic gain calculations [26]. In order to address the problem of multicollinearity in multi-trait indexes, the utilization of a mixture of multivariate methodologies seems to be successful in surpassing their limitations. FAI-BLUP is a methodology that uses factor analysis to establish scores for each ideotype based on desirable and unwanted features [26]. Subsequently, the geographic likelihood is determined based on the genotype-ideotype distance, which enables the ranking of genotypes. Olivoto and Nardino [27] introduced a novel multi-trait genotype-132 ideotype distance index (MGIDI). This index, along with the exercise computations, is integrated into the R-Metan package. The R-Metan package encompasses all the necessary functions for genotype selection in plant breeding programs.

There are many different types of eggplant available in Bangladesh. The Genetic Resource Center (PGRC) of the Bangladesh Agricultural Research Institute (BARI Plant) is responsible for the collection and conservation of various types of eggplant germplasm from across Bangladesh. Assessing the degree of available diversity in a crop development program is a vital step in enhancing crop quality. This can be achieved by gathering and analyzing germplasm. Hence, this study aimed to assess the natural variation in local eggplant germplasm in order to select the most promising germplasm that displays genetic diversity for the purpose of enhancing crop improvement efforts using advanced multi-disciplinary methodology.

2. MATERIALS AND METHODS

2.1. Experimental Site

The experiment was executed at PGRC of BARI at Gazipur, Bangladesh during winter (Rabi season) 2019–2020 at 23.988929 N latitude, 90.412393 E longitudes and 8.40 m above sea level. The soil in the test field was silty clay with a pH of 6.

2.2. Experimental Materials, Treatments, Design and Procedures

A total of 130 genotypes (126 locally collected germplasm and Bangladesh has developed and released four Bt brinjal varieties expressing Cry1Ac gene (Bt brinjal), viz., BARI Bt brinjal-1,2,3 and 4 variety as a check) were used in the experiment (**Table S1**).

All 130 genotypes were arranged in an augmented randomized complete block design (augmented RCBD) with four check varieties and seven blocks was followed in this study. All check varieties received seven replications, giving a total of 154 experimental plots. The plot size was 3 m × 2.1 m. Each genotype was implanted in three rows per plot. The spacing was 70 × 60 cm.

Direct seeding was completed with inside the well-organized seedbeds on 12 November 2019. Thirty-day-old seedlings were transplanted in the organized pits of the predominant experimental plot on 22 December 2019.

Fertilizer doses were 10 tons ha^{-1} Cowdung, 210 kg ha^{-1} Urea, 33 kg ha^{-1} triple superphosphate (TSP), 200 kg ha^{-1}, MP and 5 kg ha^{-1} Borax ([**28**]. The total requirement of Cow-dung, TSP and Borax was applied during final land preparation about one week before transplanting. Urea and MP were supplied in the three equal splits [**29**]. Four times weeding and mulching were carried out in the first 25 days of mid-December. Sumithion 60 EC at 2.5 ml L^{-1}, Sevin 75 WP at 0.1 g pit^{-1} and Vertimac 18 EC at 1.2 ml L^{-1} were sprayed for controlling insect and mite, respectively. The data was noted as per the descriptor developed by IBPGR, 1990.

2.3. Data Recorded

Fourteen qualitative traits and ten quantitative traits (Table 1) were considered during the morphological characterization based on EGGNET [7] and IBPGR descriptors [30].

Table 1. Fourteen qualitative traits and ten quantitative traits.

Sl. No.	Trait	Code	Sl. No.	Trait	Code
	Qualitative Traits			**Quantitative Traits**	
1.	Plant Growth Habit	PGH	15.	Days to First Flowering	DFF (Day)
2.	Leaf Blade Lobing	LBL	16.	Plant Height	PH (cm)
3.	Leaf Blade Tip Angle	LBTA	17.	Fruit Diameter	FD (cm)
4.	Leaf Prickles	LP	18.	Fruit Length	FL (cm)
5.	Leaf Hairs	LH	19.	Fruit Weight	FW (g)
6.	Corolla Color	CC	20.	Normalized Difference Vegetation Index	NDVI
7.	Fruit Calyx Prickles	FCP	21.	Single Leaf Area	SLA (cm2)
8.	Fruit Color Distribution	FCD	22.	Soil Plant Analyses Development	SPAD
9.	Fruit Curvature	FC	23.	Total Number of Fruits	TF
10.	Fruit Apex Shape	FAS	24.	Yield Per Plant	YPP (kg)
11.	Fruit Cross Section	FCS			
12.	Fruit Color at Ripening	FCR			
13.	Fruit Flesh Density	FFD			
14.	Fruit Position	FP			

Quantitative Traits Measuring

Data on the number of days required from planting to the first opening of the flower (for example early genotypes took <91 days), optimum genotypes took 91–105 days, and late genotypes took >105 days).

Data on plant height was recorded from five randomly selected plants at the edible fruiting stage (for example, short (~30 cm), intermediate (~60 cm), and tall (61–100 cm).

The normalized difference vegetation index (NDVI) was measured by hand green seeker (Trimble) and green seeker RT100 (Agri Optics). Quantification of two natural light sources (near infrared-NIR and red light) is measured by NDVI. These two natural lights have individual mechanisms on vegetation. For example, NIR is reflected by the vegetation, whereas red light is absorbed by the vegetation.

The NDVI formula is:

$$\mathrm{NDVI} = \frac{\mathrm{NIR} - \mathrm{Red}}{\mathrm{NIR} + \mathrm{Red}}$$

The chlorophyll content in plant leaves was determined with a SPAD meter (Model: SPAD-502). The SPAD value was carried from the middle portion of the leaf of the tagged main shoot at the first flowering stage through using a self-calibrating Minolta chlorophyll meter. Measurements at each experimental plot consisted of an average of five readings [**31**].

Leaf area was calculated with an Automatic Leaf Area Meter (Leaf area meter-LICOR-3300, USA) at the first flowering stage of a single leaf. The leaf area index (LAI) of the crop at different growth stages was calculated using the equation as described by [**32**].

- Data on fruit length, fruit diameter, mean number of edible fruits, and mean weight of edible fruits were measures from 10 randomly selected fruits.
- Length of fruit was measured from base of calyx to tip of fruit (for example, very short (<1 cm), short (3–5 cm), Intermediate (6–10 cm), long (11–20 cm) and very long (>20 cm))
- Fruit diameter was estimated as small (2–3 cm), intermediate (4–5 cm), large (6–10 cm), and very large (>10 cm)
- The mean number of edible fruits was very low (<7), low (7–12), intermediate (13–18), high (19–24), very high (>25)).
- The mean weight of edible fruit was low (<30 g), intermediate (30–60 g) and high (>60 g)).

The multi-trait index based on factor analysis and genotype-ideotype distance (MGIDI) proposed by [**27**] was used to select the novel donors with high performing under optimum and low nitrogen conditions. We also compared the result of the MGIDI index with the result of the Smith–Hazel (SH) index proposed by Smith (29) and Hazel (30) multiple trait index based on factor analysis and ideotype-design (FAI-BLUP) index proposed by [**33**]. The MGIDI [**33**] was computed as follows:

$$MTSI_i = \left[\sum_{j=1}^{f} (\gamma_{ij} - \gamma_j)^2 \right]^{0.5}$$

where MGIDI*i* is the distance index of multi-trait genotype-ideotype for the *i*th genotype, γij is the score of the ith genotype in the jth factor I = 1, 2, ..., g; j = 1, 2, ..., f), g and f are the number of genotypes and factors, respectively, and Fj is the ideotype's jth score. This means that the genotype with the lowest MGIDI is closest to the ideotype, and as a result, it has the ideal values for all the analyzed features.

2.4. Statistical Analysis

Statistical analyses were performed under the R-statistics platform (software version 4.0.2) [**34**]. Analyses of variance (ANOVA) for each trait were assessed by using the R package 'augmented RCBD' [**35**]. Phenotypic and genetic variance (PV and GV), along with genotypic and phenotypic coefficients of variation (GCV and PCV), were calculated using the formula provided by [**36**]. Broad sense heritability (hBS), genetic advance (GA) calculation as formula elucidated in [**37**]. Components of the phenotypic variance of each trait were estimated using restricted maximum likelihood methods. For the estimation of variance components of linear mixed-effect "lmer", lme4 package was employed. R package ggplot2, scales and GGally were used for heatmap analysis.

The hierarchical clustering was performed using Spearman's rank correlation algorithm. Principal component analysis (PCA) was performed using R package ggplot2, ggfortify, usethis, devtools, plyr, scales and grid. Using a two-way matrix of 10 characteristics and 130 genotypes, a G × T biplot was constructed. The first two PCs were plotted. Genotypes were schemed according to scores on each PC, and traits were plotted based on the eigenvectors on each PC. The genotypic, phenotypic variance and broad-sense heritability were estimated using Agricola R-package [**38**] with "metan" package. Mathematic figures were plotted using the ggplot2 package [**39**].

3. RESULTS

3.1. Qualitative Traits

To determine the variability of the examined germplasm, different qualitative traits were evaluated (**Table 2** and **Table 3**).

Table 2. Variability in growth and foliage traits in eggplant germplasm.

Trait Names	Descriptor State	No. of Germplasm	Germplasm (%)
Plant Growth Habit	Upright	23	17.69
	Intermediate	64	49.23
	Prostrate	43	33.08
Leaf Blade Lobing	Weak	59	45.38
	Intermediate	43	33.08
	Strong	23	17.69
	Very strong	5	3.85
Leaf Blade Tip Angle	Very acute	6	4.62
	Acute	28	21.54
	Intermediate	64	49.23
	Obtuse	32	24.62
Leaf Prickles	Very few	13	10.00
	Few	38	29.23
	Intermediate	72	55.38
	Many	5	3.85
	Very many	2	1.54
Leaf Hairs	Very many	128	98.46
	Few	2	1.54
CorollaColor	Light Violet	41	31.54
	Pale Violet	84	64.62
	White	5	3.85

Table 3. Variability in fruit traits in eggplant germplasm.

Descriptor Name	Descriptor State	No. of Germplasm	Germplasm (%)
Fruit Calyx Prickles	Very few	3	2.31
	Few	31	23.85
	Intermediate	49	37.69
	Many	32	24.62
	Very many	15	11.54
Fruit Color Distribution	Uniform	18	13.85
	Mottled	21	16.15
	Netted	18	13.85
	Striped	73	56.15

Descriptor Name	Descriptor State	No. of Germplasm	Germplasm (%)
Fruit Curvature	None (fruit straight)	68	52.31
	Slightly curved	37	28.46
	Curved	25	19.23
Fruit Apex Shape	Rounded	39	30.00
	Depressed	91	70.00
Fruit Cross Section	Circular (no grooves)	130	100.00
Fruit Color at Ripening	Milky white	2	1.54
	Lilac grey	7	5.38
	Purple	32	24.62
	Green with mottled at the distal end	51	39.23
	Green with a yellowish stripe	37	28.46
	Purple with light green at the distal end	3	2.31
Fruit Flesh Density	Very loose (spongy)	74	56.92
	Loose (crumbly)	24	18.46
	Average density	21	16.15
	Dense	7	5.38
	Very dense	4	3.08
Fruit Position	Pendant	130	100.00

3.2. Variability in Growth and Foliage

All the traits studied related to plant growth habits and foliage exhibited noticeable variation among the germplasm except leaf hairs and corolla Color (Table 2). Plant growth habit was observed in upright (17.69%), intermediate (49.23%) and prostrate (33.08%). Leaf-blade lobing were exhibited four categories as Weak (45.38%), Intermediate (33.08%), Strong (17.69%), and very strong (3.85%). The leaf blade tip angles were also exhibited as four categories such as very acute (4.62%), Acute (21.54%), Intermediate (49.23%) and Obtuse (24.62%). Leaf prickles were found in maximum variation such as very few (10.0%), few (29.23%), intermediate (55.38%), many (3.58%) and very many (1.54%) in the studied germplasm. Leaf hairs and showed minimum variation. Corolla Colors were found light violet (31.54%), pale violet (64.62%) and white (3.85%). All the fruits traits displayed distinctive variation among the genotypes except fruit cross-section and fruit position (Table 3). The maximum variation was identified in overall leaf prickles, fruit calyx prickles, fruit color at ripening and fruit flesh density.

3.3. The Analysis of Variance and Frequency Distribution of Quantitative Traits

The germplasm panel, consisting of 126 accessions with four check varieties, was tested for characterized through different morphological traits including 14 qualitative and 10 quantitative traits. Analysis of variance (ANOVA) showed high significant variation among the accessions with a check for all the investigated traits. ANOVA for all the traits also revealed highly significant differences among the check, accession by check interaction as well as among the accession (Table 4) excluding normalized difference vegetation index (NDVI), plant height (PH), single leaf area (SLA), and yield per plant (YPP) for check varieties. Adjusted blocks are insignificant for all the traits. All the traits matched with normal distribution except for some traits skewed left and some were right (Figure 1 and Figure 2).

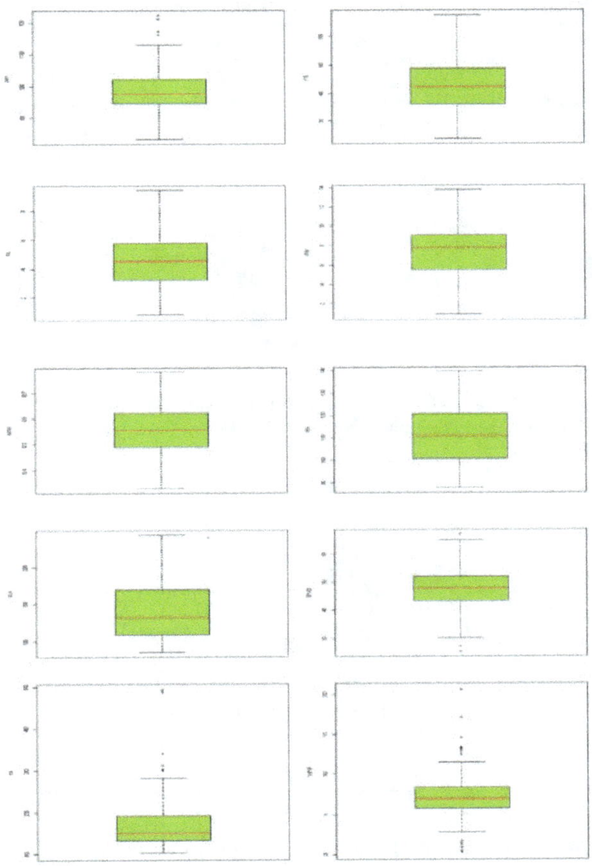

Figure 1. Box plots showing the pattern of the measured traits of germplasm. DFF = Days to first flowering (day), FD = Fruit diameter (cm), FL = Fruit length (cm), FW = Fruit weight (g), YPP =Yield plant-1 (kg), NDVI = Normalized difference vegetation index, PH = Plant height (cm), SLA = Single leaf area (cm2), SPAD = Soil plant analyses development, TF = Total number of fruits.

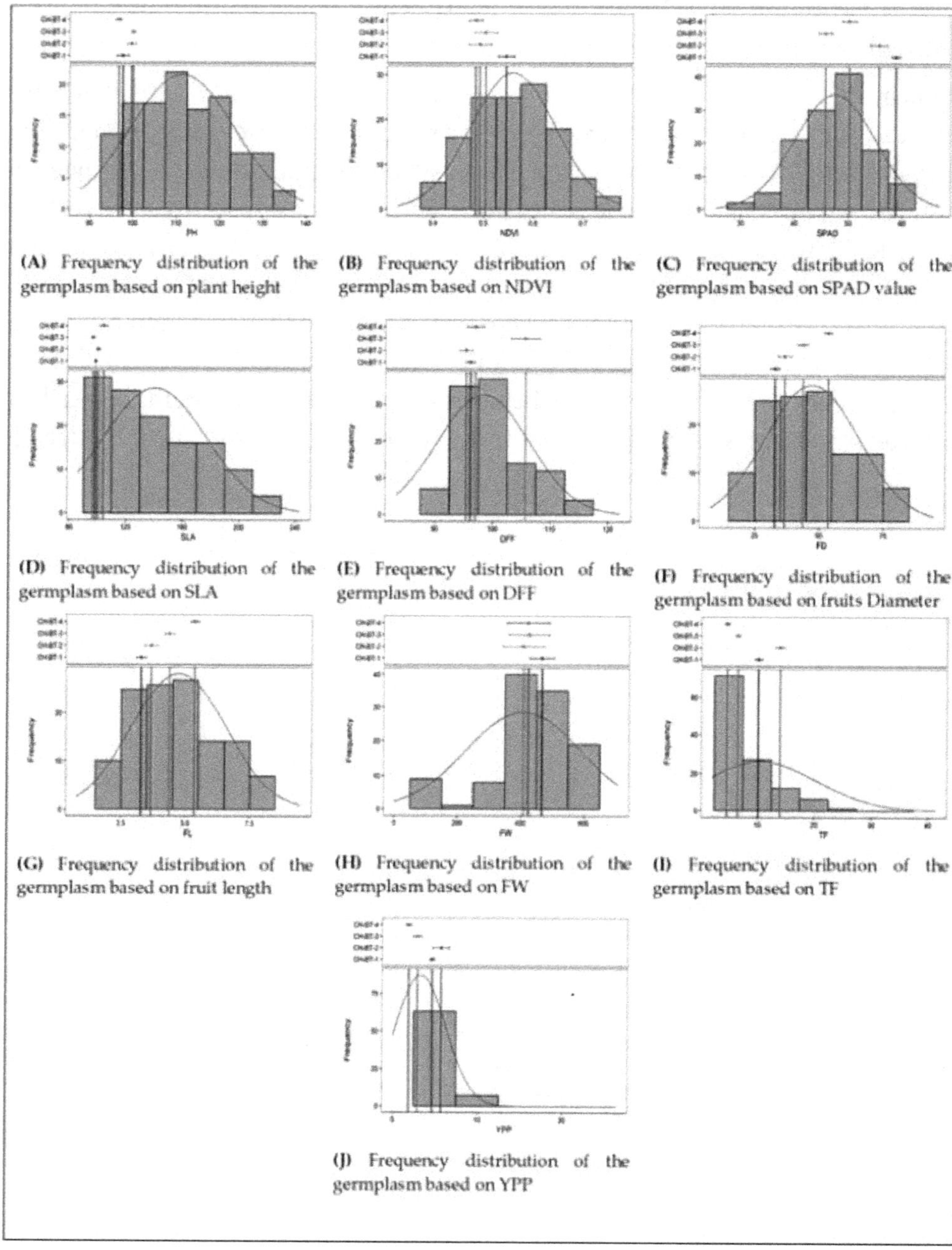

(A) Frequency distribution of the germplasm based on plant height

(B) Frequency distribution of the germplasm based on NDVI

(C) Frequency distribution of the germplasm based on SPAD value

(D) Frequency distribution of the germplasm based on SLA

(E) Frequency distribution of the germplasm based on DFF

(F) Frequency distribution of the germplasm based on fruits Diameter

(G) Frequency distribution of the germplasm based on fruit length

(H) Frequency distribution of the germplasm based on FW

(I) Frequency distribution of the germplasm based on TF

(J) Frequency distribution of the germplasm based on YPP

Figure 2. Frequency distribution of the germplasm based on quantitative traits: (A) Plant height, (B) NDVI, (C) SPAD value, (D) SLA, (E) DFF, (F) fruits diameter, (G) fruit length, (H) FW, (I) TF and (J) YPP of eggplant.

Table 4. Analysis of variance of the tested quantitative traits.

Traits	Source of Variation					
	Accession (G) with C (df = 129)	Check (C) (df = 3)	Accession (G) vs. C (df = 1)	Accession (G) (df = 125)	Adjusted Block (B) (df = 6)	Residuals (df = 18)
DFF	59.61 **	161.62 **	0.7 ns	57.63 **	20.31 ns	16.67
FD	390.72 **	586.56 **	766.74 **	383.01 **	31.44 ns	27.71
FL	3.91 **	5.87 **	7.67 **	3.83 **	0.31 ns	0.28
FW	8.63 **	49.04 **	0.36 ns	7.72 **	1.67 ns	1.27
NDVI	0.01 **	0.01 ns	0.07 **	0.01 **	0.0023 ns	0.0028
PH	169.94 **	18.57 ns	4014.75 **	142.81 **	3.14 ns	7.57
SLA	1572.19 **	71.58 ns	40431.19 **	1297.33 **	8.38 ns	22.74
SPAD	59.63 **	235.78 **	637.78 **	50.78 **	17.9 ns	9.96
TF	104.55 **	128.42 **	68.89 **	104.26 **	2.54 ns	4.25
YPP	9.59 **	0.9 ns	3.89 ns	9.84 **	0.97 ns	1.12

ns, non-significant at $p > 0.05$; **, significant at $p <= 0.01$, DFF = Days to first flowering (day), FD = fruit diameter (cm), FL = Fruit length (cm), FW = Fruit weight (g), YPP = Yield plant^{-1} (kg), NDVI = Normalized difference vegetation index, PH = Plant height (cm), SLA = Single leaf area (cm^2), SPAD = Soil plant analyses development, TF = Total number of fruits.

3.4. Descriptive Statistic of the Traits

For morpho-physiological traits evaluated in this study, their descriptive statistics including means and standard error (std.Error), standard deviations (std.Dev), minimum (Min), maximum (Max), skewness and kurtosis are summarized in Table 5. All the phenotypic values are shown a wide range of variability. Plant height (PH) ranged from 87.82 to 139.57. The NDVI value ranged from 0.33 to 0.78, SPAD ranged from 25.37 to 66.97, single leaf area (SLA) ranged from 86.12 to 242.85 cm2, DFF ranged from 70.21 to 80.5 days, FD ranged from 0.77 to 9.47 cm, FL ranged from 3.71 to 24.93 cm, FW ranged from 10.00 to 650.40 g, TF ranged from 10.21 to 49.21, and YPP ranged from 0.24 to 10.57 kg.

Skewness is a measure of the asymmetry and kurtosis is a measure of 'peakedness' of a distribution. The skewness and kurtosis were non-significant for all the traits except PH, TF and YPP indicating all the traits fitted with a normal distribution (Figure 2). The traits SLA, DFF, FD, TF and YPP were significant, and the distribution is positively skewed, which means that more accessions are below the mean than expected in a normal distribution. Only the trait SPAD was non-significant, and the distribution is negatively skewed, which means that more accessions above the mean than expected in a normal distribution (Figure 2). The traits PH, TF and YPP were significant and positive for kurtosis which means heavily leptokurtic distributions (Figure 2).

Table 5. Descriptive statistics of measured traits.

Trait	Mean	Std.Error	CV	Min	Max	Skewness	Kurtosis
PH	111.4	1.05	2.52	87.82	139.57	0.21 ns	2.25 *
NDVI	0.56	0.01	9.55	0.33	0.78	0.11 ns	2.7 ns
SPAD	47.5	0.66	6.54	25.37	66.97	-0.07 ns	3.37 ns
SLA	140.53	3.18	3.55	86.12	242.85	0.56 **	2.37 ns
DFF	98.4	0.7	4.15	83.32	122.07	0.46 *	3.27 ns
FL	47.25	1.6	11.35	3.71	24.73	0.47 *	2.71 ns
FD	4.72	0.16	11.35	0.77	9.47	0.47 *	2.71 ns
FW	7.55	0.23	39.52	10.00	650.40	0.14 ns	2.69 ns
TF	18.65	0.86	16.27	10.21	49.21	2.25 **	7.16 **
YPP	7.12	0.28	46.14	0.24	10.57	0.54 *	5.41 **

ns, non-significant; *, ** indicate the significance at 5% and 1% level of probability.

Table 6. Estimation of statistical and genetic parameters of yield and its contributing traits of different eggplant germplasm.

Traits	PV	GV	GCV	GCV	PCV	PCV	hBS	hBS	GA	GAM	GAM
PH	142.81	135.24	10.44	Medium	10.73	Medium	94.7	High	23.35	20.96	High
NDVI	0.01	0.0047	12.26	Medium	15.46	Medium	62.92	High	0.11	20.07	High
SPAD	50.78	40.82	13.45	Medium	15	Medium	80.38	High	11.82	24.88	High
SLA	1297.3	1274.6	25.4	High	25.63	High	98.25	High	73	51.95	High
DFF	57.63	40.96	6.5	Low	7.71	Low	71.07	High	11.13	11.31	Medium
FD	383.01	355.3	39.9	High	41.42	High	92.77	High	37.45	79.27	High
FL	3.83	3.55	39.9	High	41.42	High	92.77	High	3.75	79.27	High
FW	7.72	6.46	33.65	High	36.8	High	83.61	High	4.79	63.47	High
TF	104.26	100.01	53.64	High	54.76	High	95.92	High	20.21	108.37	High
YPP	9.84	8.72	41.49	High	44.07	High	88.63	High	5.74	80.58	High

GV = Genotypic variance, PV = Phenotypic variance, GCV = Genotypic coefficients of variation, PCV = Phenotypic coefficients of variation, hBS = broad-sense heritability, GA = Genetic advance at 5% selection intensity, GAM = Genetic advance as the percentage of the mean at 5% selection intensity, DFF = Days to first flowering (day), FD = Fruit diameter (cm), FL = Fruit length (cm), FW = Fruit weight (g), YPP =Yield Plant^{-1} (kg), NDVI = Normalized difference vegetation index, PH = Plant height (cm), SLA = Single leaf area (cm^2), SPAD = Soil plant analyses development, TF = Total number of fruits.

The results pertaining to genetic parameters viz., phenotypic coefficient of variation (PCV), genotypic coefficient of variation (GCV), broad-sense heritability (h2 BS), and genetic advance as percent of the mean (GAM) for all the 10 traits are summarized in **Table 6.** TF had the highest PCV (54.76%) and GCV (53.64%), followed by FD, FL, FW, and YPP. High PCV and high GCV suggesting that these traits were under the influence of genetic control. The traits PH, NDVI, and SPAD were noted for moderate magnitudes of both PCV and GCV respectively. The traits DFF recorded for low magnitudes of both PCV and GCV, respectively.

3.5. Variability in Fruit Traits

All the traits studied related to the fruit of eggplant showed distinct variation among the germplasm except fruit cross-section and fruit position (**Table 6**). The maximum variation was found in 'fruit Color at ripening stage'. Six categories of fruit Color at ripening such as milky white (1.54%), lilac grey (5.38%), Purple (24.62%), green with mottled at the distal end (39.23%), and Green with yellowish stripe (28.46%) and purple with light green at the distal end (2.31%) were observed. The next higher variation was found in 'fruit flesh density'. Very loose (spongy), loose (crumbly), average density, dense, and very dense type of fruit flesh density was found where the majority of the germplasm exhibited average density type. On the other hand, fruit calyx prickles were found as very few (1.6%), few (13.9%), intermediate (36.5%), many (23.0%), and very many (25.0%). Fruit Color distribution was exhibited as uniform (50.8%), mottled (13.1%), netted (16.8%), and striped (27.9%). No fruits were found curved in 80.56% of the germplasm. Only 8.33 % germplasm showed curved and 11.11% showed slightly curved. Fruit apexes were exhibited as two categories as rounded (40.28%) and depressed (59.72%).

3.6. Analysis of Correlation Matrix

The phenotypic correlation analysis is being used to explore a linear relationship between various traits, which was visualized in the correlation matrix (**Figure 3**). In this analysis, DFF displayed a significant positive correlation with FL, while NDVI was correlated negatively. FW exhibited a strong negative correlation with TF and TF showed a negative correlation with YPP. FW and PH showed a moderate positive significant correlation with YPP (**Figure 3**).

3.7. Multivariate Analysis

Multivariate analysis is a tool to find patterns and relationships between several variables simultaneously. To understand the relationship among 130 eggplant genotypes with various morpho-physiological traits, principal component, biplot, and heatmap analysis were done which revealed different clusters of genotypes that performed better in different aspects. The genotypes by traits biplot were constructed from a two-way matrix of 10 morpho-physiological traits and 130 eggplant genotypes using the relative value of the trait (Figure 4). Again, biplot analysis showed the trait profiles of the genotypes, especially, those genotypes positioned far away from the origin and the results indicated a correlation between traits with genotypes. Again, traits on opposite sides of the origin are negatively correlated and traits near each other are positively correlated. Moreover, traits at 90° to each other are not correlated, concerning the origin. The principal component (PC) analysis identified a total of 10 principal components (PCs) for the morpho-physiological traits. Among them, the first two PC explained 82.26% of the entire morpho-physiological variations (Figure 4). This biplot revealed superior genotypes with higher levels of expression of favorable trait combinations. The total outcome proposed that TF, FD, SLA, FW, PH, and YPP could help to detect superior genotypes in elite germplasm.

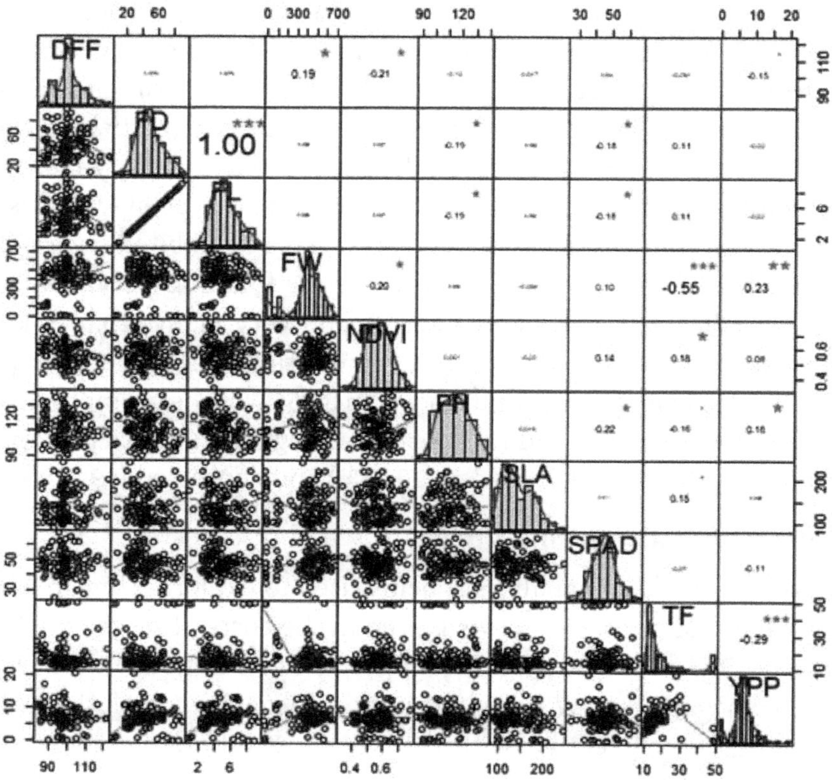

Figure 3. Correlation matrix, scatter plot and phenotypic frequency distribution of traits; * p <= 0.05; ** p <= 0.01; *** p > 0.001; DFF = Days to first flowering (day), FD = Fruit diameter (cm), FL = Fruit length (cm), FW = Fruit weight (g), YPP =Yield Plant-1 (kg), NDVI = Normalized difference vegetation index, PH = Plant height (cm), SLA = Single leaf area (cm^2), SPAD = Soil plant analyses development, TF = Total number of fruits.

3.8. Heatmap Analysis

The heatmap represented the overall performance of 10 observable traits among the 130 germplasm. A heatmap is a two-dimensional data visualization technique that uses color to show the scope of a phenomenon. Color variation by hue or intensity provides the reader with a visual representation of how the phenomenon is grouped or varies over space. It depicts the relative patterns of highly abundant features against a background of mostly low-abundance features (Figure 5).

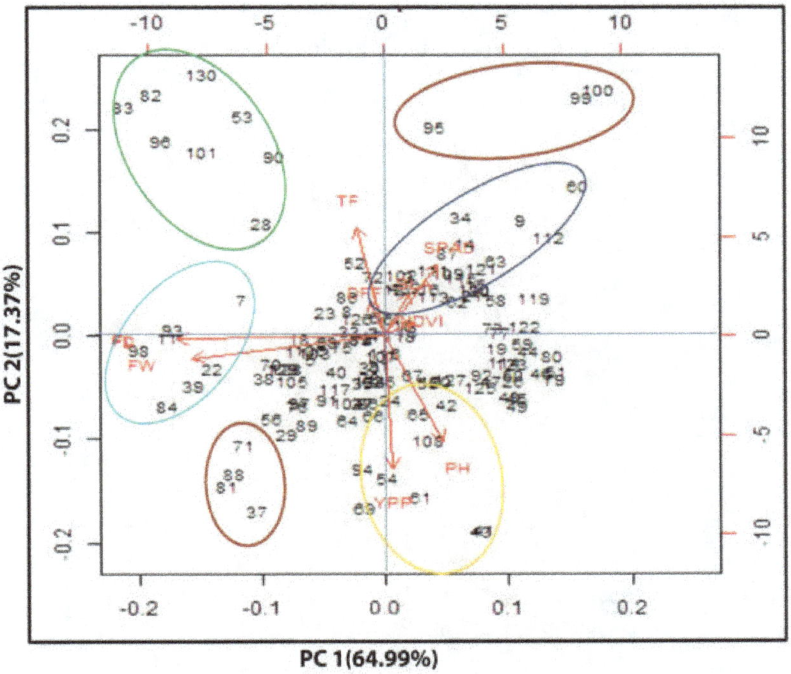

Figure 4. Genotypes by traits (G × T) biplot based on 130 germplasm and 10 quantitative traits of eggplant. DFF =Days to first flowering (day), FD = Fruit Diameter (cm), FL = Fruit Length (cm), FW = Fruit Weight (g), YPP =Yield Per Plant (kg), NDVI = Normalized Difference Vegetation Index, PH = Plant height (cm), SLA = Single Leaf Area (cm2), SPAD = Soil Plant Analyses Development, TF = Total Number of Fruits.

A heatmap analysis of characteristics was performed to demonstrate a chromatic examination of the genotypes. The heatmap analysis produced two dendrograms: one in the vertical direction, representing the germplasm, and one in the horizontal direction, representing the traits that caused the diffusion. The red diagonal depicts each accession's perfect connection to itself. The relationship measurements for pairs of germplasm are represented by the symmetric off-diagonal elements. Based on the morpho-physiological properties of the germplasm studied, six clusters emerged through hierarchical clustering (**Figure 5**). Group (a) included 43 germplasm from clusters v (30 germplasm) and vi (13 germplasm), while group (b) included the remaining 87 germplasm from four clusters. Out of a total of 87 germplasm, cluster-i received 33, cluster-ii received 13, cluster-iii received 30, and cluster-iv received 11 (**Figure 5**). Dendrogram two also revealed two significant groups: group (a) is associated with seven traits (FD, FL, FW, YPP, NDVI, SPAD, and TF), while group (b) is associated with three traits (DFF, PH, and SLA) (b). Surprisingly, the dendrogram two groups and sub-groups revealed the disparity effects of different eggplants.

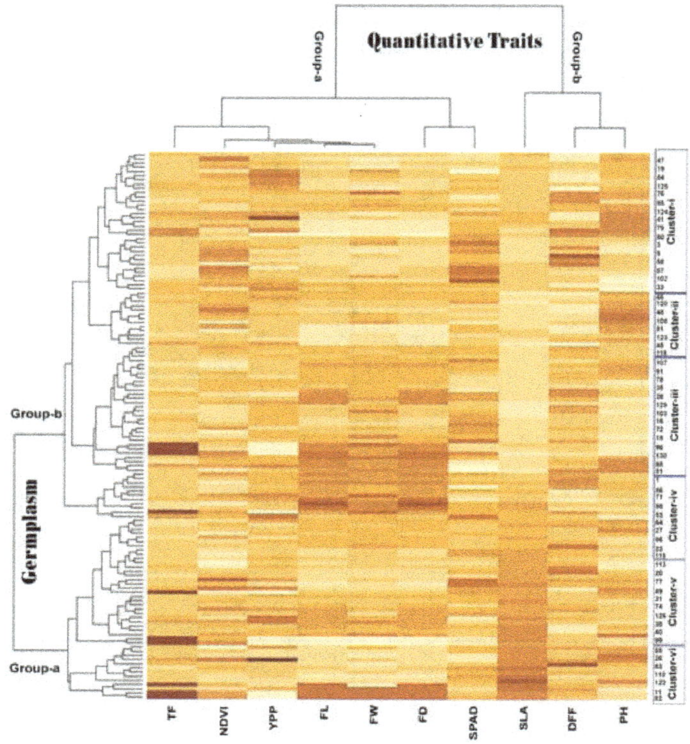

Figure 5. Heatmap showing the clustering pattern of 130 eggplant genotypes with 10 morpho-physiological traits. Heat map displaying the relationship matrix among Eggplant genotypes. The red diagonal represents a perfect relationship of each accession with itself. The symmetric off-diagonal elements represent the relationship measures for pairs of genotypes. The white of warmer colors on the diagonal show clusters of closely related genotypes. DFF = Days to first flowering (day), FD = Fruit diameter (cm), FL = Fruit length (cm), FW = Fruit weight (g), YPP = Yield plant^{-1} (kg), NDVI = Normalized difference vegetation index, PH = Plant height (cm), SLA = Single leaf area (cm2), SPAD = Soil plant analyses development, TF = Total number of fruits.

3.9. Multi-Trait Index Based on Factor Analysis and Genotype-Ideotype Distance (MGIDI)

The MGIDI index was intended to select the genotypes with respect to considering all measured traits. Based on the analysis, a highly significant genotypic effect was noted for 10 measured traits involving DFF (Day), PH (cm), FD, FL, FW, NDVI, SLA (cm2), SPAD, TF, and YPP (kg) (Table 1). The broad-sense heritability (h2) ranged from 71.07% (for DFF) to 98.25% (for SLA). All filtered traits are evaluated with high heritability values, which indicates that the selection gain of these traits is promising. Among selected traits, traits FD, FL, FW, TF, and YPP showed the highest genetic advanced mean. However, the genotypes selected using the MGIDI index were G80, G54, G66, G120, G46, G61, G65, G108, G4, G79, G42, G77, G47, G50, G51, G43,

G44, G48, and G49 (Figure 6). The strengths and weaknesses of all the genotypes are shown in Figure 7.

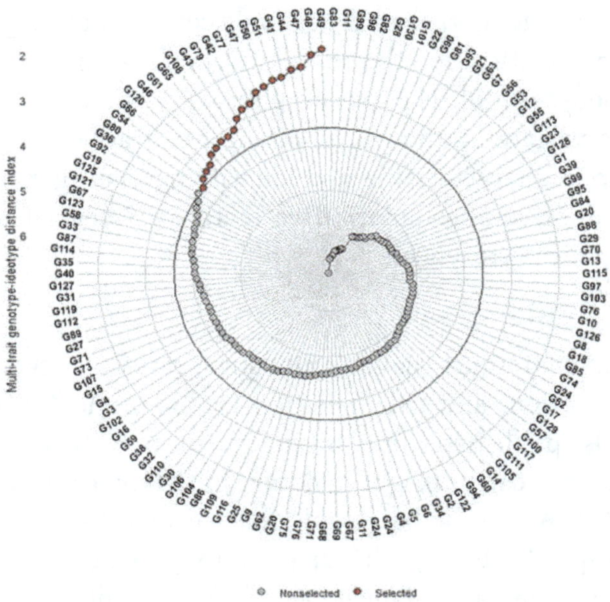

Figure 6. Germplasm ranking and selected germplasm from 130 local germplasm through multi-trait genotype-ideotypes distance index (MGIDI) considering 15% selection intensity.

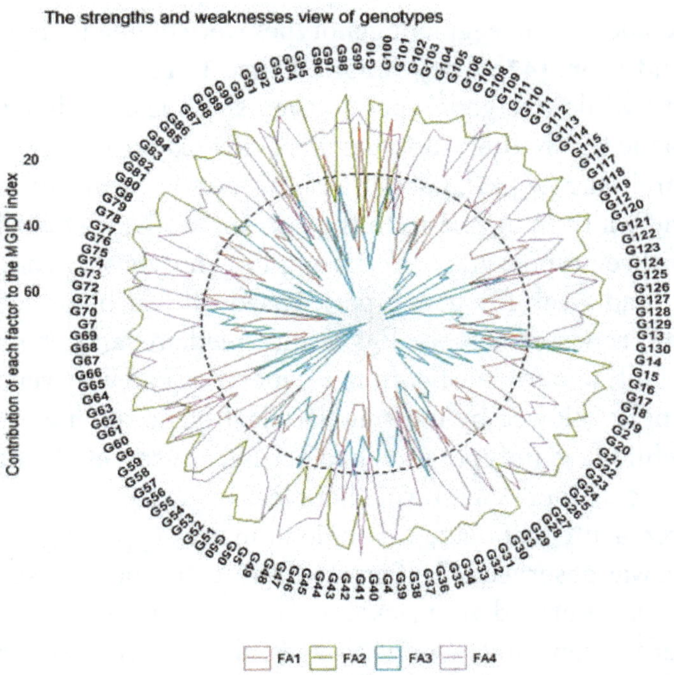

Figure 7. The strengths and weaknesses of all the genotypes.

4. DISCUSSION

In the present study, numerous qualitative and quantitative features were assessed in 130 brinjal genotypes in order to discover superior genotypes. Genetic diversity research is critical for the successful evaluation, preservation, and use of germplasm resources [40]. The breeding strategy is mostly determined by the degree of genetic variation, and morphological characteristics are viewed as a critical initial step in characterizing and identifying plant genetic resources [41]. Screening for qualitative and quantitative features is crucial for determining a plant's socioeconomic preferences.

4.1. Qualitative Traits

All qualitative traits are found to have significant variations except leaf hairs, corolla color, fruit cross-section, and fruit position (**Table 2** and **Table 3**). Some other previous reports also published a similar type of fruit curvature [**42,43,44**]. Eggplant is a herbaceous plant, mostly upright in nature [**45**]. Our results demonstrated that 17.69% upright, 49.23% intermediate, and 33.08% prostrate growth habit among all the studied genotypes (**Table 2**). Shekar et al. [**46**] clustered the eggplant plants into upright and intermediate. Islam et al. [**42**] saw 48% intermediate, 45% upright, and 7% prostrate growth habit at vegetative stage among the studied genotypes. They suggested that the plant growth habit is used for the identification of eggplant varieties.

As per leaf blade length, eggplant genotypes were divided into two groups, viz., intermediate and short [**47**]. They reported that 30 genotypes had intermediate leaf blade length while five genotypes showed short leaf blade length. Similarly, based on leaf blade width, the genotypes were divided into three types, viz., wide, intermediate, and narrow [**45**]. Osei et al. [**13**] found significant variation in leaf blade length and width in eggplant. In the current study, we distributed eggplant genotypes into three types, viz., very strong (5 genotypes), strong, intermediate (43 genotypes), and weak (59 genotypes), based on leaf blade lobing (**Table 2**). Sunseri et al. [**48**] reported 60% weak, 37% intermediate, and 3% strong leaf blade lobbing in eggplant. In our investigation, 6 genotypes had very acute, 28 had acute leaf blade tip angle followed by intermediate leaf blade tip angle, which includes 64 genotypes, while 32 genotypes each had obtuse. Sunseri et al. [**48**] observed 62% acute, 33% very acute, and 5% intermediate leaf blade tip angle in eggplant. Dash et al. [**49**] noticed acute leaf blade tip angle in many eggplant genotypes. A much higher variation was observed in leaf prickle (**Table 2**), where most of the genotype (72 genotypes) had intermediate prickle on the leaf. Sunseri et al. [**48**] noted that leaf prickles were absent in 25% of genotypes, whereas the remaining genotypes produced 40% very few, 25% few, and 5% both intermediate and many prickles. Tiwari et al. [**50**] noted prickles in stem, petiole, calyx including peduncle, and leaf including veins in eggplant. Many leaf hairs (98.46%) are found in most of the genotypes. Light violet (31.54%), pale violet (64.62%), and white (3.85%) colors were noted among the studied genotypes for corolla Color (Table 2).

In the present study, we described eight distinctive traits of eggplant fruits (**Table 3**). For traits related to the fruit calyx prickles, fruit color distribution, fruit color at ripening, and fruit flesh density much higher variation were observed in the cultivated eggplant (**Table 3**). Similar kinds of fruit calyx prickle distribution were also reported by [**45**]. It was noticed that 68 genotypes had no curvature on fruit; 37 genotypes showed slightly curved fruit and 25 genotypes had curved fruit. Sunseri et al. [**48,51**] reported sickle-shaped, snaked-shaped, curved, and U-shaped eggplant fruits along with fruits with no curvature. Eggplant genotypes were distributed into two types: rounded (39 genotypes) and depressed (91 genotypes), on the basis of fruit apex shape. Sunseri et al. [**48**] reported that 38, 34, and 28% of eggplant genotypes had depressed, rounded, and protruded types of fruit apex shape, respectively. Eggplant fruits traits variations are important in protection from UV irradiation, insect attack in plants as well as socio-economic value. More fruit color variation was noted by [**52**] and [**53**], which support our present study. Solaimana et al. [**51**] found uniform and stripped fruit color distribution. The variations in fruits color were also described by [**51,54,55**]. Tiwari et al. [**50**] divided the eggplant fruits into six color groups i.e., green (37.27%), purple (25.45%), milky white (13.62%), purple-black (12.72%), light purple or lilac grey (9.09%), and scarlet red (1.08%). Fruit flesh density is an important characteristic for the determination of fruit volume and weight. A wide range of variation was observed in fruit flesh density, the highest number of genotypes had very loose flesh density (74 genotypes).

4.2. Quantitative Traits

4.2.1. Genetic Components

The proposed index of 0–10% for low, 10–20% for moderate, and 20% for high variation was used to characterize the projected GCV and PCV values. In our experiment, closer PCV and GCV values were estimated in most of the traits which possibly were less influenced by the environment suggesting the reliability of selection based on these traits. The assessed 10 quantitative traits exhibited a wide range of variation and more or less similar results were observed by [48,53,56]. The selection procedure considers the differences between traits based on the degree of heredity. In order to understand the predicted selection benefits, assessing genetic progress may be an important method for improving crops. Several studies have shown that selection can effectively use available genetic variations with a certain degree of heritability to improve specific traits [26,57]. The concern of both heritability and genetic advance is more effective over the distinctively use of heritability. We found that the phenotype variance values of all traits were higher than the genotype variance, indicating that the environment regulates the expression of traits. The same kind of outcome also was gained from several studies for various traits in eggplant [51,58,59,60,61].

The evidence of the traits having essential potential in the assortment process due to low environmental impacts was calculated using strong GA with hBS for all yield-related traits except DFF. Heterosis breeding has the potential to improve traits with poor heritability and genetic advance [28]. Regarding the hBS and GA index [62], which were greater than 60% for high, 30–60% for moderate, and 0–30% for low, we discovered that all traits were highly heritable together with a high genetic advance mean value except DFF, implying that direct selection can be effective for eggplant crop improvement based on these traits with the effect of additive genes; a more or less similar result was obtained by many researchers [63,64]. Due to the strong influence of the environment on genetic effects, low to moderate heritability and genetic advanced values will inhibit the improvement of traits. Therefore, effective selection can only be achieved by selecting higher values of GCV, PCV, hBS, and GA, which means that the influence of additive genes is more stable than the influence of the environment. High GAM was also observed for all traits except DFF. This revealed that if the selection was carried out for the next generation for these features, a greater improvement in the population mean may be seen.

4.2.2. Correlation Matrix

The plant breeding correlation matrix is an outstanding method to assess the relationship between two or more variables. For higher genotype selection procedures, considering the correlation matrix can be a scaling measure [65]. In the correlation matrix, DFF displayed a strong positive correlation with FL, while NDVI was correlated negatively, FW and PH showed a moderate positive correlation with YPP (Figure 6). A moderate to positive significant association can be proposed; selection based on these traits can help increase the yield of this crop. More or less similar results were consistent with the study of [28,64,66,67].

4.3. Multivariate Statistical

Compared with univariate and bivariate statistical methods, multivariate statistical methods can analyze more than one relationship at the same time. There are many multivariate data analysis methods, each of which has a different purpose, such as Regression analysis, factor analysis, cluster analysis, analysis of variance, discriminant analysis, etc. [68]. Biplot analysis is usually employed to assess the component effects creating the genotypic variations.

The highest values indicate the highest influence of the trait on the total variation. Biplot analysis determines varietal stability in the multi-environmental trial [69]. It describes the relationship between different genotype traits. The association between morpho-physiological traits among the 130 genotypes was observed by the biplot analysis [70]. Again, the biplot analysis showed the trait profiles of the genotypes, especially, those genotypes positioned far away from the origin and the results indicated a correlation between traits with genotypes (Figure 4). An acute angle between two elements indicates a positive correlation, and an obtuse angle between two elements indicates a negative correlation. As a result, principal components (PC) analysis provides a good screening of available genotypes and aids in the selection of possible parents for crop breeding initiatives. In our data, the first two PC accounted for 82.36% of the overall variation (Figure 4). The yield potential of accessions was represented in PC 1; thus, the accessions contributing to this component are likely to undergo direct selection, or selected parents can be used in hybridization operations. These PC1 results are consistent with the results of the correlation analysis. The figure summarizes the information of the matrix in principal components, where the cosine of the angle between the vectors connecting the objects to the origin is proportional to the correlation coefficient between these objects. The heatmap shows the highest and lowest values of each genotype in different colors against all the traits comparing. The intensity of the color indicates the degree of high or low of the traits. Hierarchical clustering based on the morpho-physiological traits of the studied germplasm revealed six clusters (Figure 5). The heatmap analysis depicted the degree of correspondence among the morphological traits assessed in brinjal genotypes, and this result was consistently supported by [71,72].

4.4. Multi-Trait Index Based on Factor Analysis and Genotype-Ideotype Distance (MGIDI)

Experienced breeders frequently aim to merge multiple desirable features in order to create a new genotype that exhibits exceptional performance. When assessing various characteristics, it can be challenging to choose a certain genetic makeup from the idealized models. Various multivariate methods, including principal component analysis, factor analysis, cluster analysis, and other sampling techniques, are commonly employed to categorize measurable traits or choose test genotypes [73]. In this work, we employed a two-way heat map clustering pattern and PCA to establish a connection between test genotypes and measured qualities. Figures 6 and 7 illustrate this analysis. However, we were unable to identify specific genotypes. The MGIDI (multi-trait genotype-ideotypes distance index) is a novel method that simplifies the selection of genotypes with multiple features. It utilizes multiple trait information to aid in genotype selection. The eggplant genotypes were evaluated and graded according to data obtained from measuring numerous attributes (Figure 7). The MGIDI index identified the genotypes G80, G54, G66, G120, G46, G61, G65, G108, G4, G79, G42, G77, G47, G50, G51, G43, G44, G48, and G49 as very favorable eggplant genotypes. In addition to these genotypes, G80 was in close proximity to the cut-off point, suggesting that this genotype may possess desirable characteristics. Therefore, the researcher must carefully evaluate genotypes that are in close proximity to the designated threshold [27]. There is a forecasted substantial growth in the utilization of the MGIDI index in plant crop research. Moreover, this index was employed to identify the optimal strawberry genotype [74].

5. CONCISE SUMMARY

The present study definitively demonstrated that enhancing brinjal production and associated characteristics can be achieved by selecting and evaluating several genetic criteria. The majority of the qualitative parameters exhibited clear variances among the germplasm. The most significant variance was seen in the leaf prickles, fruit calyx prickles, fruit color at ripening, and fruit flesh density. The fruit yield per plant had the greatest quantitative variance, followed by the fruit weight. However, the current study has demonstrated that selecting based on the assessment of different genetic criteria such as GCV, PCV, hBS, and GA can lead to an enhanced eggplant production and associated characteristics. The recorded data and additional analyses (heatmap analysis, correlation matrix, PCA, MGIDI analysis) indicate that nearly all of the agronomic traits investigated in this study show significant variability. Hence, the germplasm G80, G54, G66, and G120 can be regarded as optimal progenitors for the forthcoming breeding program, taking into account both qualitative and quantitative traits. The current results demonstrate a significant genetic potential among the examined germplasm. In addition, the germplasm that shows promise in this study could potentially be utilized in future breeding operations to enhance the quality of eggplants.

AT A GLANCE

Eggplant is a ubiquitous and indispensable fruit vegetable that is available throughout the year. This study utilized 130 indigenous varieties of brinjal to identify a wide range of parents based on a selection index that considers numerous features. These parents will be used in future breeding programs. This study focused on analyzing the variation of 14 qualitative and 10 quantitative traits, as well as determining genetic parameters such as phenotypic and genotypic variance (PV and GV), genotypic and phenotypic coefficients of variation (GCV and PCV), broad-sense heritability (hBS), genetic advance, traits association, genotype by trait biplot (G × T), heatmap analysis, and multi-trait index based on factor analysis and genotype-ideotype distance (MGIDI). The examination of descriptive statistics and analysis of variance uncovered a significant amount of variability in morpho-physiological features. The estimated heritability (hBS) for all the examined qualities varied from 10.6% to 93%, suggesting that all the traits exhibited a significant degree of genetic inheritance. The genetic variations for most morpho-physiological variables ranged from low to large, suggesting a complex genetic architecture. The yield per plant showed a strong correlation with fruit diameter, number of fruits per plant, percentage of fruits infested by brinjal stem and fruit borer, and fruit weight. This suggests that selecting plants based on fruit number and weight alone could be enough to improve other features. The initial two principal components (PCs) accounted for approximately 81.27% of the overall variation observed in 38 brinjal morpho-physiological variables. The G × T biplot identified genotypes that shown greater performance through the combination of positive features. The mean genetic distance was 3.53, with a range of 0.25 to 20.01, demonstrating significant levels of heterogeneity within the germplasm. The heat map was utilized to determine the correlation matrix among all the brinjal genotypes. MGIDI is a suitable approach for selecting based on multiple trait information. After analyzing fourteen qualitative and ten quantitative features and evaluating numerous genetic criteria, it can be concluded that the germplasm G80, G54, G66, and G120 are the most suitable candidates to be used as parents in the future eggplant breeding program for development.

REFERENCES

1. Aubriot, X.; Daunay, M.-C. Eggplants and relatives: From exploring their diversity and phylogenetic relationships to conservation challenges. In *The Eggplant Genome*; Springer: Berlin/Heidelberg, Germany, 2019; pp. 91–134.
2. Minardi, A. Food Security in Developing Countries: The Case of "Production of Appropriate Food: Sufficient, Safe and Sustainable" Project. Ph.D Thesis, (Ph.D. in Agro-Food System, Academic Year 2017/2018). Catholic University of Sacred Heart, Via Emilia Parmense, Piacenza, Italy, 2019. Available online: **http://tesionline.unicatt.it/bitstream/10280/67851/1/tesiphd_completa_Minardi.pdf** (accessed on 15 July 2021).
3. Bhagirath, C.; Kadambini, G. *The Development and Regulation of Bt Brinjal in India (Eggplant/Aubergine)*; International Service for the Acquisition of Agri-biotech Applications: Ithaca, NY, USA, 2009.
4. BBS. *Statistical Yearbook of Bangladesh. Bangladesh Bureau of Statistics*; Ministry of Planning, Government of the Peoples' Republic of Bangladesh: Dhaka, Bangladesh, 2020; p. 139.
5. Prabakaran, R.K.-S.; Balakrishnan, S.; RameshKumar, S.; Arumugam, T.; Anandakumar, C. Genetic diversity, trait relationship and path analysis in eggplant landraces. *Electron. J. Plant* 2015, *6*, 831–837.
6. Van der Weerden, G.M.; Barendse, G.W. A web-based searchable database developed for the EGGNET project and applied to the Radboud University Solanaceae database. *Acta Hortic.* 2007, *745*, 503–506.
7. Boyaci, H.F.; Topcu, V.; Akin, T.; Yildirim, I.K.; Mehmet, O.; Aktas, A. Morphological and molecular characterization and relationships of Turkish local eggplant heirlooms. *Not. Bot. Horti Agrobot. Cluj Napoca* 2015, *43*, 100–107.

8. Mat Sulaiman, N.N.; Rafii, M.Y.; Duangjit, J.; Ramlee, S.I.; Phumichai, C.; Oladosu, Y.; Datta, D.R.; Musa, I. Genetic variability of eggplant germplasm evaluated under open field and glasshouse cropping conditions. *Agronomy* 2020, *10*, 436.

9. Muñoz-Falcón, J.; Prohens, J.; Vilanova, S.; Nuez, F. Diversity in commercial varieties and landraces of black eggplants and implications for broadening the breeders' gene pool. *Ann. Appl. Biol.* 2009, *154*, 453–465.

10. Gramazio, P.; Prohens, J.; Plazas, M.; Mangino, G.; Herraiz, F.J.; Vilanova, S. Development and genetic characterization of advanced backcross materials and an introgression line population of *Solanum incanum* in a *S. melongena* background. *Front. Plant Sci.* 2017, *8*, 1477.

11. Adeniji, O.; Aloyce, A. Farmer's knowledge of horticultural traits and participatory selection of African eggplant varieties (*Solanum aethiopicum*) in Tanzania. *Tropicultura* 2012, *30*, 185–191.

12. Datta, D.R.; Rafii, M.Y.; Misran, A.; Jusoh, M.; Yusuff, O.; Sulaiman, N.M.; Momodu, J. Genetic diversity, heritability and genetic advance of *Solanum melongena* L. from three secondary centers of diversity. *Bangladesh J. Plant Taxonom.* 2021, *28*, 155–169.

13. Osei, M.; Oluoch, M.; Osei, C.; Banful, B. Morphological characterisafion of African Eggplant (*Solanum* spp.) Germplasm in some African countries. *Agric. Innov. Sustain. Dev.* 2010. Available online: **https://library.faraafrica.org/wp-content/uploads/2019/11/Agricultural-Innovations-for-Sustainable-Development.pdf#page=171** (accessed on 15 July 2021).

14. Plazas, M.; Vilanova, S.; Gramazio, P.; Rodríguez-Burruezo, A.; Fita, A.; Herraiz, F.J.; Ranil, R.; Fonseka, R.; Niran, L.; Fonseka, H. Interspecific hybridization between eggplant and wild relatives from different genepools. *J. Am. Soc. Hortic. Sci.* 2016, *141*, 34–44.

15. Polignano, G.; Uggenti, P.; Bisignano, V.; Della Gatta, C. Genetic divergence analysis in eggplant (*Solanum melongena* L.) and allied species. *Genet. Resour. Crop. Evol.* 2010, *57*, 171–181.

16. Cericola, F.; Portis, E.; Toppino, L.; Barchi, L.; Acciarri, N.; Ciriaci, T.; Sala, T.; Rotino, G.L.; Lanteri, S. The population structure and diversity of eggplant from Asia and the Mediterranean Basin. *PloS ONE* 2013, *8*, e73702.

17. Syfert, M.M.; Castañeda-Álvarez, N.P.; Khoury, C.K.; Särkinen, T.; Sosa, C.C.; Achicanoy, H.A.; Bernau, V.; Prohens, J.; Daunay, M.C.; Knapp, S. Crop wild relatives of the brinjal eggplant (*Solanum melongena*): Poorly represented in genebanks and many species at risk of extinction. *Am. J. Bot.* 2016, *103*, 635–651.

18. Taher, D.; Solberg, S.Ø.; Prohens, J.; Chou, Y.-y.; Rakha, M.; Wu, T.-h. World vegetable center eggplant collection: Origin, composition, seed dissemination and utilization in breeding. *Front. Plant Sci.* 2017, *8*, 1484.

19. Collonnier, C.; Fock, I.; Kashyap, V.; Rotino, G.; Daunay, M.; Lian, Y.; Mariska, I.; Rajam, M.; Servaes, A.; Ducreux, G. Applications of biotechnology in eggplant. *Plant Cell Tissue Organ Cult.* 2001, *65*, 91–107.

20. Daunay, M.-C. Eggplant. In *Vegetables II*; Springer: Berlin/Heidelberg, Germany, 2008; pp. 163–220.

21. Mutegi, E.; Snow, A.A.; Rajkumar, M.; Pasquet, R.; Ponniah, H.; Daunay, M.C.; Davidar, P. Genetic diversity and population structure of wild/weedy eggplant (*Solanum insanum*, Solanaceae) in southern India: Implications for conservation. *Am. J. Bot.* 2015, *102*, 140–148.

22. Praneetha, K.J.D.S. Evaluation of brinjal (*Solanum melongena* L.) local types for yield and its quality characters. *IJCS* 2018, *6*, 292–297.

23. Rahman, M.; Islam, K.; Jahan, M.; Uddin, M. Efficacy of some botanicals in controlling brinjal shoot and fruit borer, Leucinodes orbonalis. *Progress. Agric.* 2009, *20*, 35–42.

24. Sharmin, D.; Meah, M.; Moniruzzaman, M. Inheritance of resistance to phomopsis blight and fruit rot in brinjal. *J. Agrofor. Environ.* 2010, 3, 135–140.

25. Rocha, J.R.D.A.S.D.C.; Machado, J.C.; Carneiro, P.C.S. Multitrait index based on factor analysis and ideotype-design: Proposal and application on elephant grass breeding for bioenergy. *Gcb Bioenergy* 2018, *10*, 52–60.

26. Sellami, M.H.; Lavini, A.; Pulvento, C. Phenotypic and Quality Traits of Chickpea Genotypes under Rainfed Conditions in South Italy. *Agronomy* 2021, *11*, 962.

27. Olivoto, T.; Nardino, M. MGIDI: Toward an effective multivariate selection in biological experiments. *Bioinformatics* 2021, *37*, 1383–1389.

28. Banerjee, S.; Verma, A.; Bisht, Y.S.; Maurya, P.; Jamir, I.; Mondal, S.; Bhattacharjee, T.; Chattopadhyay, A. Genetic variability, correlation coefficient and path coefficient analysis in brinjal germplasm. *Int. J. Chem. Stud.* 2018, *6*, 3069–3073.

29. Guide, F.F.R. *Bangladesh Agricultural Research Council (BARC)*; BARC: Dhaka, Bangladesh, 2012.

30. Withers, L.A.; Wheelans, S.K.; Williams, J.J.E. In vitro conservation of crop germplasm and the IBPGR databases. *Euphytica* 1990, *45*, 9–22.

31. Reynolds, M.; Balota, M.; Delgado, M.; Amani, I.; Fischer, R. Physiological and morphological traits associated with spring wheat yield under hot, irrigated conditions. *Funct. Plant Biol.* 1994, *21*, 717–730.

32. Radford, P. Growth analysis formulae-their use and abuse 1. *Crop. Sci.* 1967, *7*, 171–175.

33. Rocha, J.R.D.A.S.D.C.; Nunes, K.V.; Carneiro, A.L.N.; Marçal, T.D.S.; Salvador, F.V.; Carneiro, P.C.S.; Carneiro, J.E.S. Selection of superior inbred progenies toward the common bean ideotype. *Agron. J.* 2019, *111*, 1181–1189.

34. Pinherio, J.; Bates, D.; DebRoy, S.; Deepayan, S.; Team, R.C. Linear and Nonlinear Mixed Effects Models. R Package Version 3.1-145. 2007, Volume 3, pp. 1–89. Available online: **https://ftp.uni-bayreuth.de/math/statlib/R/CRAN/doc/packages/nlme.pdf** (accessed on 15 July 2021).

35. Aravind, J.; Mukesh, S.; Wankhede, D.; Kaur, V. Augmented RCBD: Analysis of Augmented Randomised Complete Block Designs. R Package Version 0.1. 2020, Volume 2. Available online: **https://aravind-j.github.io/augmentedRCBD/** (accessed on 15 July 2021).

36. Falconer, D.S. *Problems on Quantitative Genetics*; Longman: London, UK, 1983.

37. Burton, G.W. Quantitative inheritance in grasses. *6th Intl. Grassland Cong. Proc.* 1952, *1*, 227–238.

38. Olivoto, T.; Nardino, M. MGIDI: A novel multi-trait index for genotype selection in plant breeding. *Bioinformatics* 2020, 1–22.

39. Wickham, H.; Chang, W.; Wickham, M.H. Package 'ggplot2'. Create elegant data visualisations using the grammar of graphics. *R Package Version* 2016, *2*, 1–189.

40. Sarkar, M.; Kundagrami, S. Multivariate analysis in some genotypes of mungbean [*Vigna radiata* (L.) Wilczek] on the basis of agronomic traits of two consecutive growing cycles. *Legume Res. Int. J.* 2016, *39*, 523–527.

41. Carrillo-Perdomo, E.; Vidal, A.; Kreplak, J.; Duborjal, H.; Leveugle, M.; Duarte, J.; Desmetz, C.; Deulvot, C.; Raffiot, B.; Marget, P. Development of new genetic resources for faba bean (*Vicia faba* L.) breeding through the discovery of gene-based SNP markers and the construction of a high-density consensus map. *Sci. Rep.* 2020, *10*, 1–14.

42. Islam, M.; Chhanda, R.; Pervin, N.; Hossain, M.; Chowdhury, R. Characterization and genetic diversity of brinjal germplasm. *Bangladesh J. Agric. Res.* 2018, *43*, 499–512.

43. Begum, F.; Islam, A.A.; Rasul, M.G.; Mian, M.K.; Hossain, M.M. Morphological diversity of eggplant (Solanum melongena) in Bangladesh. *Emir. J. Food Agric.* 2013, *25*, 45–51.

44. Solaiman, A.H.M.; Nishizawa, T.; Khatun, M.; Ahmad, S. Morphological characterization and genetic diversity studies of promising brinjal genotypes for hybridization program in Bangladesh. *J. Adv. Agric.* 2014, 3, 218–228.

45. Parida, H.; Mandal, J.; Mohanta, S. A note on morphological characterization of brinjal (*Solanum melongena* L.) genotypes. *J. Crop. Weed* 2020, *16*, 250–255.

46. Shekar, K.C.; Ashok, P.; Sasikala, K. Characterization, character association, and path coefficient analyses in eggplant. *Int. J. Veg. Sci.* 2013, *19*, 45–57.

47. Kumar, G.; Meena, B.; Kar, R.; Tiwari, S.K.; Gangopadhyay, K.; Bisht, I.; Mahajan, R. Morphological diversity in brinjal (*Solanum melongena* L.) germplasm accessions. *Plant Genet. Resour.* 2008, *6*, 232–236.

48. Sunseri, F.; Polignano, G.B.; Alba, V.; Lotti, C.; Bisignano, V.; Mennella, G.; Drsquo, A.; Bacchi, M.; Riccardi, P.; Fiore, M.C.; et al. Genetic diversity and characterization of African eggplant germplasm collection. *Afr. J. Plant Sci.* 2010, 4, 231–241.

49. Dash, S.P.; Singh, J.; Sharma, D. Morphological characterization of brinjal (*Solanum melongena* L.) germplasm. *J. Pharmacogn. Phytochem.* 2019, *8*, 1574–1578.

50. Tiwari, S.K.; Bisht, I.S.; Kumar, G.; Karihaloo, J.L. Diversity in brinjal (*Solanum melongena* L.) landraces for morphological traits of evolutionary significance. *Veg. Sci.* 2016, *43*, 106–111.

51. Solaimana, A.H.M.; Nishizawa, T.; Khatun, M.; Ahmad, S. Physio-morphological characterization genetic variability and correlation studies in brinjal genotypes of Bangladesh. *Comput. Math. Biol.* 2015, *4*, 1–36.

52. Begum, M.N.S.; Shirazy, B.J.; Mahbub, M.M.; Siddikee, M.A. Performance of Brinjal (Solanum Melongena) Genotypes through Genetic Variability Analysis. *Am. J. Plant Biol.* 2017, *43*, 499–512.

53. Shinde, K.; Birajdar, U.; Bhalekar, M.; Patil, B. Genetic divergence in brinjal (*Solanum melongena* L.). *Veg. Sci.* 2012, *39*, 103–104.

54. Singh, S.; Khemariya, P.; Rai, A.; Rai, A.C.; Koley, T.K.; Singh, B. Carnauba wax-based edible coating enhances shelf-life and retain quality of eggplant (*Solanum melongena*) fruits. *LWT* 2016, *74*, 420–426.

55. Hassan, I.; Jatoi, S.A.; Arif, M.; Siddiqui, S.U. *Genetic Variability in Eggplant for Agro-Morphological Traits*; World Vegetable Center: Tainan, Taiwan, 2015.

56. Premabati Devi, C.; Munshi, A.D.; Behera, T.K.; Choudhary, H.; Saha, P. Characterisation of cultivated breeding lines of eggplant (*Solanum melongena* L.) and related wild Solanum species from India. *J. Hortic. Sci. Biotechnol.* 2016, *91*, 87–92.

57. Koundinya, A.; Das, A.; Layek, S.; Chowdhury, R.; Pandit, M.e. Genetic variability, characters association and path analysis for yield and fruit quality components in Brinjal. *J. Appl. Nat. Sci.* 2017, *9*, 1343–1349.

58. Kumar, S.R.; Arumugam, T.; An, C.; Premalakshmi, V. Genetic variability for quantitative and qualitative characters in Brinjal (*Solanum melongena* L.). *Afr. J. Agric. Res.* 2013, *8*, 4956–4959.

59. Sherly, J.; Shanthi, A. Variability, heritability and genetic advance in brinjal (*Solanum melongena* L.). *Res. Crop.* 2009, *10*, 105–108.

60. Singh, O.; Kumar, J. Variability, heritability and genetic advance in brinjal. *Indian J. Hortic.* 2005, *62*, 265–267.

61. Verma, P.; Kushwaha, M.; Panchbhaiya, A. Studies on variability, heritability and genetic advance for yield attributing traits in brinjal (*Solanum melongena* L.) for two different seasons. *Int. J. Curr. Microbiol. Appl. Sci.* 2018, *7*, 1543–1552.

62. Chattopadhyay, A.; Dutta, S.; Hazra, P. Characterization of genetic resources and identification of selection indices of brinjal (*Solanum melongena* L.) grown in eastern India. *Veg. Crop. Res. Bull.* 2011, *74*, 39.

63. Shende, R.; Desai, S.; Lachyan, T. Genetic variability and response to selection in brinjal (*Solanum melangena* L.). *Int. J. Curr. Res.* 2015, *7*, 21545–21547.

64. Calus, M.P.; Veerkamp, R.F. Accuracy of multi-trait genomic selection using different methods. *Genet. Sel. Evol.* 2011, *43*, 1–14.

65. Konyak, W.; Kanaujia, S.; Jha, A.; Chaturvedi, H.; Ananda, A. Genetic variability, correlation and path coefficient analysis of brinjal. *SAARC J. Agric.* 2020, *18*, 13–21.

66. Saha, S.; Haq, M.E.; Parveen, S.; Mahmud, F.; Chowdhury, S.R.; Harun-Ur-Rashid, M. Variability, correlation and path coefficient analysis: Principle tools to explore genotypes of brinjal (*Solanum melongena* L.). *Asian J. Biotechnol. Genet. Eng.* 2019, *2*, 1–9.

67. Dugard, P.; Todman, J.; Staines, H. *Approaching Multivariate Analysis: A Practical Introduction*; Routledge/ Taylor & Francis Group: Abingdon, UK, 2010.

68. Otieno, O.V.; Owuor, O.N. Multivariate genotype and genotype by environment interaction biplot analysis of sugarcane breeding data using R. *Int. J. Stat. Distrib. Appl.* 2019, 5, 22.

69. Yan, W.; Frégeau-Reid, J. Breeding line selection based on multiple traits. *Crop. Sci.* 2008, *48*, 417–423.

70. Khan, M.M.H.; Rafii, M.Y.; Ramlee, S.I.; Jusoh, M.; Al Mamun, M. Genetic analysis and selection of Bambara groundnut (*Vigna Subterranea* [L.] Verdc.) Landraces for high yield revealed by qualitative and quantitative traits. *Sci. Rep.* 2021, *11*, 1–21.

71. Meyer, R.S.; Bamshad, M.; Fuller, D.Q.; Litt, A. Comparing medicinal uses of eggplant and related Solanaceae in China, India, and the Philippines suggests the independent development of uses, cultural diffusion, and recent species substitutions. *Econ. Bot.* 2014, *68*, 137–152.

72. Virga, G.; Licata, M.; Consentino, B.B.; Tuttolomondo, T.; Sabatino, L.; Leto, C.; La Bella, S. Agro-morphological characterization of sicilian chili pepper accessions for ornamental purposes. *Plants* 2020, *9*, 1400.

73. Bhandari, H.; Bhanu, A.; Srivastava, K.; Singh, M.; Shreya, H.A. Assessment of genetic diversity in crop plants-an overview. *Adv. Plants Agric. Res.* 2017, 7, 00255.

74. Olivoto, T.; Diel, M.I.; Schmidt, D.; Lúcio, A.D.C. Multivariate analysis of strawberry experiments: Where are we now and where can we go? *BioRxiv* 2021, 1–10.

9 Chapter

ANALYSIS OF THE TRANSCRIPTOME SHOWS DISTINCT SETS OF GENES THAT RESPOND TO LOW NITROGEN LEVELS IN DIFFERENT PEPPER CULTIVARS

1. INTRODUCTION: AN OVERVIEW

Nitrogen (N) is a crucial macronutrient for plants, which can be taken up and incorporated by the roots in several forms, such as nitrate, ammonium, and amino acids. Plants mostly utilize nitrate and ammonium as their main sources of nutrients, with nitrate being the prevailing type [1]. Nitrogen metabolism can be categorized into three distinct processes: absorption, incorporation, and redistribution. During the absorption stage, Arabidopsis utilizes at least six transporters for nitrate uptake: NPF6.3/NRT1.1, NPF4.6/NRT1.2, NRT2.1, NRT2.2, NRT2.4, and NRT2.5 [2,3]. Nitrate is converted into ammonium by the action of nitrate and nitrite reductases during the assimilation process. Subsequently, glutamine synthetase (GS), glutamine antinotransferase (GOGAT), and asparagine synthetase (AS) are responsible for incorporating the ammonium into amino acids [4]. The process of N remobilization involves the breakdown of proteins and the transfer of amino acids. Multiple genes implicated in the ubiquitin-26S proteasome system [5] and responsible for encoding amino acid transporters [6,7] have also been found to be linked to these biological activities. These research have yielded crucial insights into the process of N metabolism. Nevertheless, our current understanding of the intricate N regulatory network is still lacking, and the underlying molecular mechanisms that control the genetic variance in N-use efficiency (NUE) among different crop cultivars are not yet fully understood.

Pepper, scientifically known as *Capsicum annuum* L., is a highly significant vegetable crop belonging to the Solanaceae family. It is cultivated globally for its use as a food source, medicinal properties, and as an attractive plant [8]. There is ample evidence to support the idea that there is a direct relationship between the rate at which pepper plants develop and the content of nitrogen in the soil, within a specific range. Nevertheless, improper N usage might result in unfavorable growth, productivity, and quality [9,10]. Recently, farmers in regions with fertile soil have been excessively using N fertilizer, leading to low nitrogen use efficiency (NUE) and significant environmental damage [11]. Conversely, in certain regions with low-quality soil, the requirement for a significant amount of nitrogen fertilizer poses a significant financial burden for pepper farmers [12]. Hence, it is imperative to enhance NUE in order to promote the advancement of sustainable pepper cultivation. Genetic diversity in nitrogen use efficiency (NUE) has been documented in various crops including rice [13], barley [14], wheat [15], rapeseed [16], maize [17], and cotton [18]. However, our understanding of the molecular basis of this genetic variation in NUE remains limited. A number of genes associated with enhanced nitrogen use efficiency (NUE) have been discovered in rice, such as DEP1 [19], OsNRT1.1B [20], OsNRT2.3b [21], ARE1 [22], OsNRT1.1A [23], and OsNPF6.1 [24]. Regrettably, there is a lack of research on the molecular control mechanism of nitrogen (N) metabolism in pepper, particularly in relation to variations in nitrogen use efficiency (NUE).

RNA-Seq, a cutting-edge technology for sequencing RNA, is highly efficient and may effectively uncover comprehensive changes in the entire genome caused by biotic or abiotic stressors. Additionally, it can offer valuable insights into the N metabolism network at a systemic level. RNA-Seq analysis has been used to study the transcriptome of a specific plant genotype's response to low nitrogen levels. This technique has been utilized to several plants including cucumber, maize, wheat, physic nut, and rice. Multiple potential genes implicated in low-N response were identified [25,27,28,29]. Moreover, researchers have also examined the potential regulatory functions of IncRNAs in response to nitrogen stress [26]. However, it is challenging to detect changes in nitrogen use efficiency (NUE) using a single genotype. Consequently, the use of comparative transcriptome analysis to study genotypes with varying levels of tolerance to low nitrogen has gained recognition as a valuable technique for comprehending nitrogen use efficiency (NUE) [30,31,32,33,34,35]. The N-stress tolerant sorghum genotypes have shown a significant presence of transcripts associated to high affinity nitrate transporters (NRT2.2, NRT2.3, NRT2.5, and NRT2.6) [31]. Additionally, the N-stress tolerant Tibetan wild barley genotype has been found to have an energy-saving assimilation pattern [32].

The objective of this work was to discover genes that respond to low nitrogen (N) levels and are expressed differently in genotypes that are resistant or susceptible to low N stress over a long period of time. The purpose is to gain more knowledge about the variance in nitrogen use efficiency (NUE). Initially, we assessed the low-nitrogen tolerance of 100 different pepper cultivars. Afterwards, we examined the overall alterations in gene expression across the entire genome of two different pepper cultivars that had varying levels of tolerance to low nitrogen (N) levels when subjected to low-N stress. Finally, the distinct low-N-responsive genes in the two cultivars were thoroughly examined by RNA-Seq and other bioinformatics techniques, uncovering probable novel candidate genes for enhancing the low-N tolerance of pepper cultivars.

2. MATERIALS AND METHODS

2.1. Screening of Pepper Cultivars with Different Low-N Tolerance

The low-N tolerance of 100 pepper cultivars was analyzed hydroponically in a greenhouse at the Chongqing Key Laboratory of Adversity Agriculture. Yamazaki nutrient solution for pepper was used as a sufficient-N solution containing 1.50 mM $Ca(NO_3)2{\cdot}4H_2O$, 6.00 mM KNO_3, 0.83 mM $NH_4H_2PO_4$, 0.75 mM $MgSO_4{\cdot}7H_2O$, 0.10 mM $FeSO_4{\cdot}7H_2O$, 0.10 mM Na2EDTA, 0.05 mM H_3BO_3, 0.0008 mM $ZnSO_4{\cdot}7H_2O$, 0.01 mM $MnSO_4{\cdot}4H_2O$, 0.0003 mM $CuSO_4{\cdot}5H_2O$, and 0.00002 mM $(NH_4)_6Mo_7O_24{\cdot}4H_2O$. Pepper seeds were soaked in warm water (55 °C) for 30 min, and then germinated in the dark on moist filter papers at 25–28 °C. Small buds were then transferred to foam boards floating in black plastic boxes filled with distilled water. The foam boards consisted of 15 uniform holes in which 15 plants grew. After seven days, the distilled water was changed to one-half-strength sufficient-N solution. A week later, the solution of half of the seedlings was changed to low-N solution, in which the concentrations of $Ca(NO_3)_2{\cdot}4H_2O$, KNO_3, and $NH_4H_2PO_4$ were %20 that of the sufficient-N solution, and the other nutrients were the same as the sufficient-N solution. Furthermore, K_2SO_4, KH_2PO_4, and $CaCl_2$ were added in moderation to the low-N solution to avoid K, P, and Ca deficiency. The rest of the seedlings grew under N-sufficient condition as controls. Each cultivar had three replicates (15 plants/ replicate) for each N condition (sufficient-N or low-N). The culture solution was refreshed every seven days at pH 6.4–6.2. All plants were grown in a greenhouse at 28 °C (day) and 25 °C (night) with a relative humidity of %80–70 and a photoperiod of 10/14 h, 300 μmol m-2 s-1. The leaf areas, shoot dry weights, root dry weights, and plant dry weights were evaluated 28 days later.

2.2. RNA Preparation and Sequencing

Seeds from the 750-1 (low-N-tolerant) and ZCFB (low-N-sensitive) pepper cultivars were germinated, and the seedlings were cultured under two different N conditions (sufficient-N and low-N) as described above. Four weeks later, the youngest fully expanded leaves and roots of the seedlings grown in sufficient-N solution and low-N solution were collected separately at 10:00–11:00 A.M., flash frozen in liquid N, and then stored at -80 °C. Samples were collected from 45 independent plants from three replicates (15 plants/replicates). Total RNA was extracted using an RNA

Plant Kit (Aidlab Biotech, Beijing, China). The quantity and quality of total RNA was examined on a NanoDrop 2000 spectrophotometer (Thermo Fisher Scientific, Madison, WI, USA) and an Agilent 2100 Bioanalyzer (Waldbronn, Germany), respectively. cDNA libraries were constructed and sequenced on a BGISEQ-500 platform at BGI (Shenzhen, China).

2.3. Transcriptome De Novo Assembly, Gene Functional Annotation, and Differentially Expressed Genes (DEGs) Analyses

The raw reads of transcriptome sequencing were filtered using SOAPnuke v.15.2 (BIG, Shenzhen, China), and a set of clean reads was obtained. After de novo assembling, the clean reads were mapped to the pepper reference genome [**36**]. Genes were functionally annotated based on the NCBI non-redundant (Nr) [**37**], Gene Ontology (GO) [**38**], and the Kyoto Encyclopedia of Genes and Genomes (KEGG) databases [**39**]. For gene expression analysis, the numbers of matched reads were calculated and then normalized to RPKM by RSEM v.1.2.12. Significant differential expression genes (DEGs) were identified as those with a fold-change ≥ 2.0 and an FDR ≤ 0.001. DEGs were clustered using GO-Term Finder software, and pathway enrichment analysis was performed based on terms from the KEGG database.

2.4. Quantitative Real-Time PCR (qRT-PCR) Analysis

To verify the reliability of RNA-Seq data, total RNA was extracted from the leaves and roots of biological triplicates from the two sequenced genotypes, and the template cDNA samples were prepared using an iScrip First Strand Synthesis System Kit (Bio-Rad Laboratories, Hercules, CA, USA). Primers for each PCR reaction were designed to have a melting temperature of 58–62 °C and to produce a PCR product between 100 and 200 bp (**Table S1**). The pepper actin gene (*AY572427*) was used as an internal control. qRT-PCR reactions were performed using a CFX96TMReal-Time System (Bio-Rad Laboratories, Hercules, CA, USA). The reaction conditions were as follows: 95 °C for 30 s, 40 cycles of 95 °C for 5 s, and 60 °C for 30 s. The 2-ΔΔCt method was used to calculate the expression levels of genes.

3. RESULTS

3.1. Screening for Pepper Cultivars with Contrasting Low-N Tolerance

To identify pepper cultivars with contrasting low-N tolerance, 100 pepper cultivars were screened based on various metrics. To avoid the effects of natural variation in individual embryos and/or endosperm size on initial growth from different genotypes, the relative value (low-N treated sample/control) of each character was used in the following analyses. Our results showed that peppers from these 100 cultivars had a wide range of relative values of leaf areas, shoots, roots, and plant dry weights, and the coefficient of variation (CV) of all of the relative values was above 15%, demonstrating the high variation in tolerance to low-N among cultivars (**Table 1** and **Table S2**).

Table 1. **Range of relative values of growth characteristics in pepper seedlings.**

Index	Mean	Range	CV (%)
Relative value of leaf area	0.63	0.38–0.95	18.9
Relative value of shoot dry weight	0.68	0.41–0.99	16.8
Relative value of root dry weight	1.08	0.66–1.61	15.6
Relative value of plant dry weight	0.73	0.44–1.03	15.4

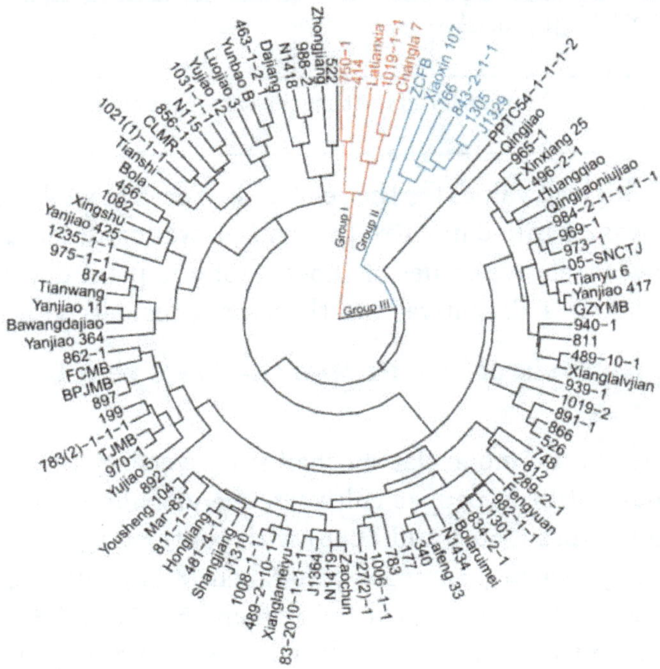

Figure 1. Cluster analysis of the low-nitrogen tolerance from 100 pepper cultivars.Cluster analysis was carried out based on the relative values of leaf area and shoot, root, and plant dry weights. Agglomerative hierarchical clustering was performed using the hclust algorithm in R, with dst = dist (dat1, method = "euclidean") and hclust (dst, method = "average"). Group I, low-N-tolerant group; Group II, low-N-sensitive group; the low-nitrogen tolerance of Group III was between that of Groups I and II.

Cluster analysis was then carried out based on the relative values of leaf areas, shoot dry weights, root dry weights, and plant dry weights. The 100 pepper cultivars could be grouped into three clusters, consisting of five (Group I), six (Group II), and 89 (Group III) cultivars (Figure 1). There were significant differences in each characteristic between the three groups (Table 2). The mean relative values of each characteristic in Group I were significantly higher than those of Group II, and the mean relative values of each characteristic in Group III were between those of

Groups I and II, indicating that Group I was a low-N-tolerant group, while Group II was a low-N-sensitive group.

Table 2. The low-nitrogen tolerance of each cluster group.

Cluster	Relative Value of Leaf Area		Relative Value of Shoot Dry Weight		Relative Value of Root Dry Weight		Relative Value of Plant Dry Weight	
	Mean	Range	Mean	Range	Mean	Range	Mean	Range
Group I	0.86a	0.75–0.95	0.90a	0.81–0.99	1.46a	1.31–1.61	0.97a	0.89–1.03
Group II	0.44c	0.38–0.48	0.47c	0.41–0.52	0.85c	0.66–1.03	0.52c	0.44–0.57
Group III	0.63b	0.41–0.91	0.68b	0.52–0.94	1.08b	0.67–1.44	0.73b	0.57–0.99

Note: Values followed by different letters within the same column are significantly different at the 0.05 level.

750-1 had the highest values for the relative leaf area as well as shoot and plant dry weights and was classified into Group I. Thus, it was considered low-N tolerant. ZCFB had the lowest relative values of shoot, root and plant dry weights and was classified into Group II. This cultivar was therefore considered low-N sensitive.

3.2. RNA-Seq Analysis of Two Pepper Cultivars with Contrasting Low-N Tolerances

To better understand the molecular mechanism governing the genetic variation of N-use efficiency (NUE) in pepper cultivars, the global gene expression profile of two pepper genotypes (750-1 and ZCFB) with contrasting low-N tolerance was next analyzed using RNA-Seq. Samples were taken 28 days after low-N treatment, and the following samples were used for sequencing: TCL1 (750-1-leaf control), TCR2 (750-1-root control), TTL3 (750-1-leaf low-N treatment), TTR4 (750-1-root low-N treatment), SCL5 (ZCFB-leaf control), SCR6 (ZCFB-root control), STL7 (ZCFB-leaf low-N treatment), and STR8 (ZCFB-root low-N treatment). Sequence data are available from the NCBI SRA database with accession number PRJNA700470. Over 65 million clean reads were generated from each library. A total of 91.75–93.55% of these clean reads could be well aligned to the pepper reference genome, at least 69% could be uniquely mapped to the reference genome (71.88% on average). The number of genes with at least one mapped transcript were 29,167, 29,817, 28,970, 29,981, 29,330, 29,814, 29,078, and 30,335 for libraries TCL1, TCR2, TTL3, TTR4, SCL5, SCR6, STL7, and STR8, respectively (**Table 3**).

3.3. Identification of DEGs and Validation of RNA-Seq by qRT-PCR

To identify low-N-responsive genes, comparisons were made between the low-N treated samples and controls (**Figure 2**). We found that 548/2073 differentially expressed genes (DEGs) (TCL1 vs. TTL3) were up/down-regulated in leaves, while 1385/1085 DEGs (TCR2 vs. TTR4) were up/down-regulated in the roots of 750-1. Additionally, 1239/2697 DEGs (TCL5 vs. TTL7) were up/down-regulated in leaves, and 2956/1262 DEGs (TCR6 vs. TTR8) were up/down-regulated in the roots of ZCFB.

Table 3. Summary of genome mapping.

Sample	Total Clean Reads	Total Mapping Ratio	Uniquely Mapping Ratio	Total Gene Number
TCL1	66,496,250	92.38%	70.18%	29,167
TCR2	66,186,556	92.03%	72.58%	29,817
TTL3	65,608,344	92.54%	69.65%	28,970
TTR4	65,388,728	92.26%	73.41%	29,981
SCL5	65,498,816	91.75%	69.00%	29,330
SCR6	65,052,664	91.53%	72.37%	29,814
STL7	66,211,026	93.55%	72.20%	29,078
STR8	66,062,774	93.09%	75.68%	30,335

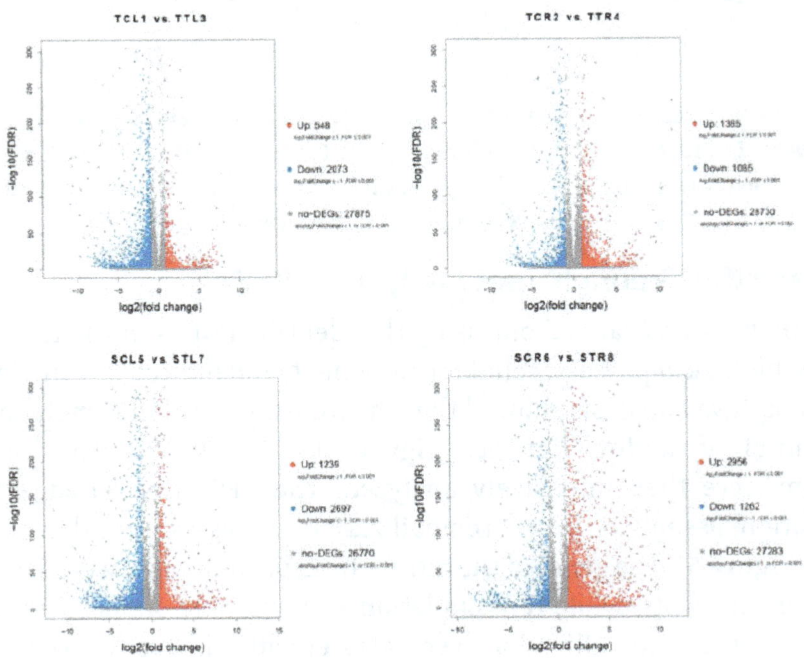

Figure 2. Volcano plot of DEGs. The X-axis represents the value of the difference multiple after log2 conversion, and the Y-axis indicates the significance value after log10 conversation. The three colors red, blue, and gray, represent up-regulated DEGs, down-regulated DEGs, and unchanged genes, respectively.

To validate our RNA-Seq data, the expression levels of eight genes (*LOC107841845*, *LOC107855963*, *LOC107858815*, *LOC107867964*, *LOC107870519*, *LOC107875600*, *LOC107875602*, and *LOC107875603*) in the leaves and roots of 750-1 and ZCFB were next confirmed by qRT-PCR. The expression trends of the eight genes were consistent with our results from RNA-Seq in the eight libraries (**Figure 3**).

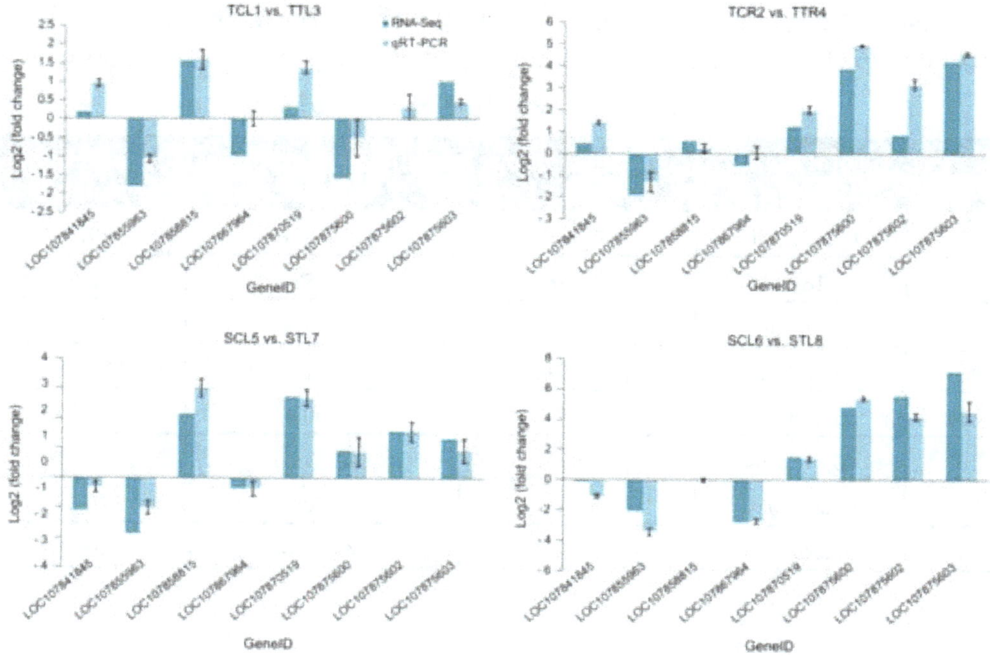

Figure 3. RNA-Seq results confirmed by qRT-PCR. A total of eight genes (*LOC107841845*, *LOC107855963*, *LOC107858815*, *LOC107867964*, *LOC107870519*, *LOC107875600*, *LOC107875602*, and *LOC107875603*) were examined by qRT-PCR. The pepper actin gene (AY572427) was used as an internal control.

3.4. GO and KEGG Enrichment Analysis of DEGs

Next, GO analysis was carried out using the identified DEGs in the three main GO categories: biological process, cellular component, and molecular function (**Figure** 4A). KEGG pathway analysis showed that the DEGs involved in N metabolism were enriched in all of the four libraries (**Figure** 4B). The DEGs involved in nitrogen metabolism were then intensively analyzed. The DEGs associated with nitrate transportation, assimilation, and remobilization processes were detected in each genotype (**Figure 5**). The genes of the nitrate transporter family were up-regulated, especially in the roots, while N assimilation systems were suppressed. The genes participating in N remobilization were also enhanced. Overall, the expression patterns of the DEGs involved in N metabolism were similar in both genotypes. Thus, further analysis is needed to find clues governing different NUEs between the two genotypes examined.

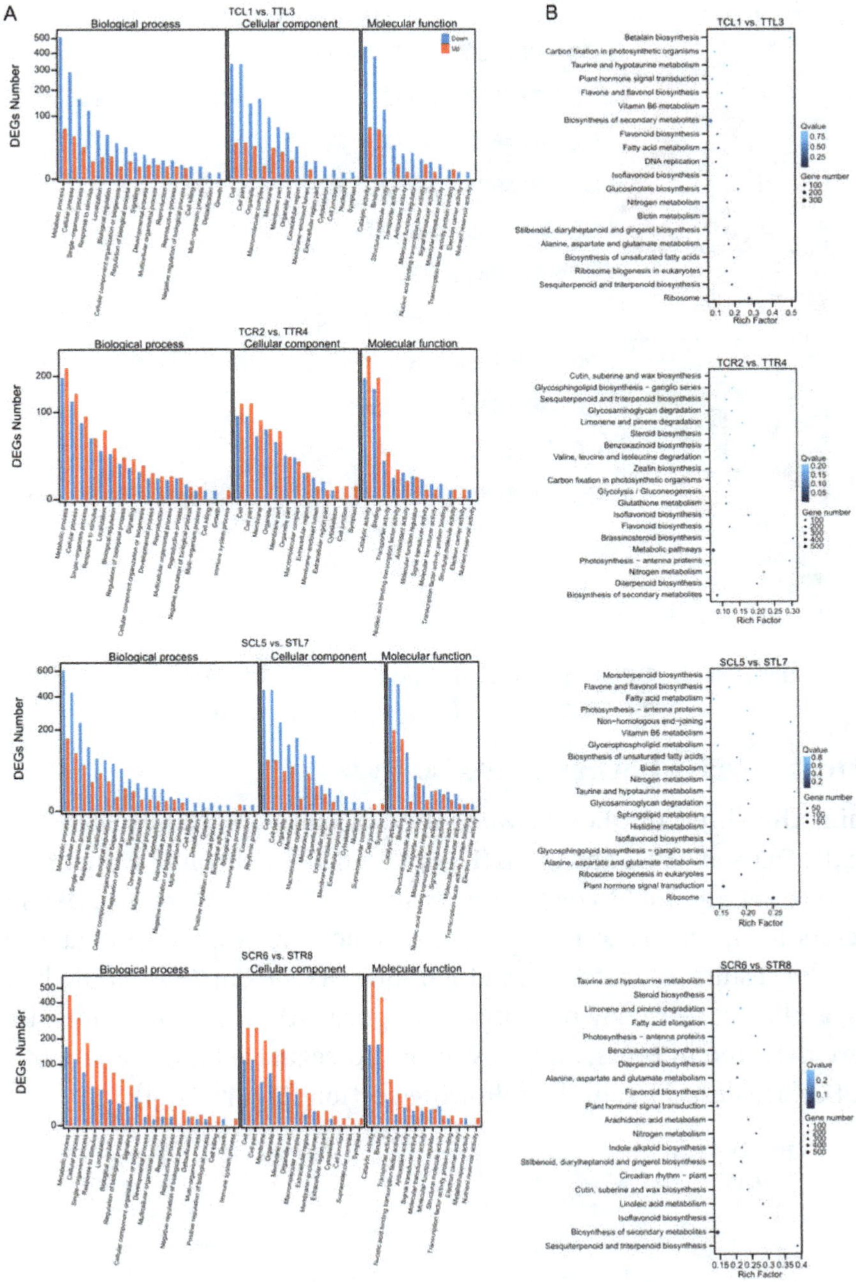

Figure 4. Function classifications of DEGs. (A): GO functional enrichment analysis of DEGs. The X-axis represents GO functional classification, and the Y-axis represents the number of up-regulated and down-regulated genes in each corresponding GO term. (B): Pathway enrichment analysis of DEGs. The X-axis represents enrichment factor values, and Y-axis represents pathway names.

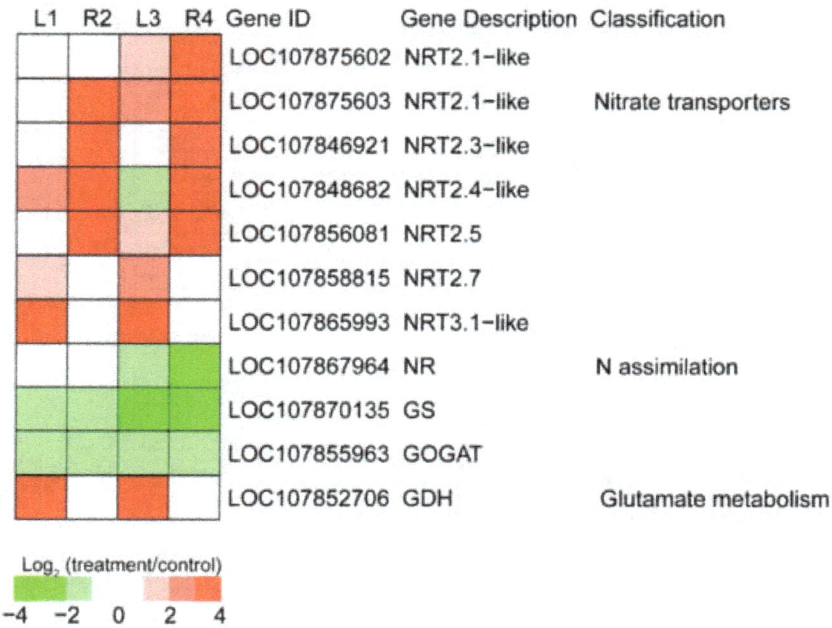

Figure 5. Heatmap of the DEGs involved in N metabolism. L1, TCL1 vs. TTL3; R2, TCR2 vs. TTR4; L3, TCL5 vs. TTL7; R4, TCR6 vs. TTR8.

3.5. Different DEGs between Two Cultivars

To examine the different DEGs between the two cultivars, comparisons were made between the DEGs of 750-1 and ZCFB (Figure 6A,B). The absolute values of log2 fold-changes were set as equal to or greater than five. Furthermore, the DEGs involved in N metabolism, which were analyzed previously, were excluded from these analyses. The results indicated that 110 unique DEGs from 750-1 could be detected in leaves, and most were down-regulated (Figure 6A), while 142 unique DEGs from ZCFB were detected in roots, and most were up-regulated (Figure 6B). Most of the unique DEGs are described in the following section (Figure 7A–J).

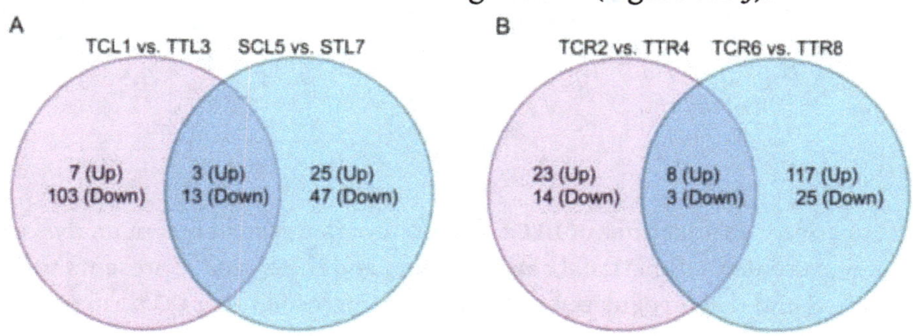

Figure 6. Venn diagram of unique DGEs in each genotype. (A): Unique DEGs in leaves. (B): Unique DEGs in roots. The absolute value of log2 fold change was equal to or greater than 5.

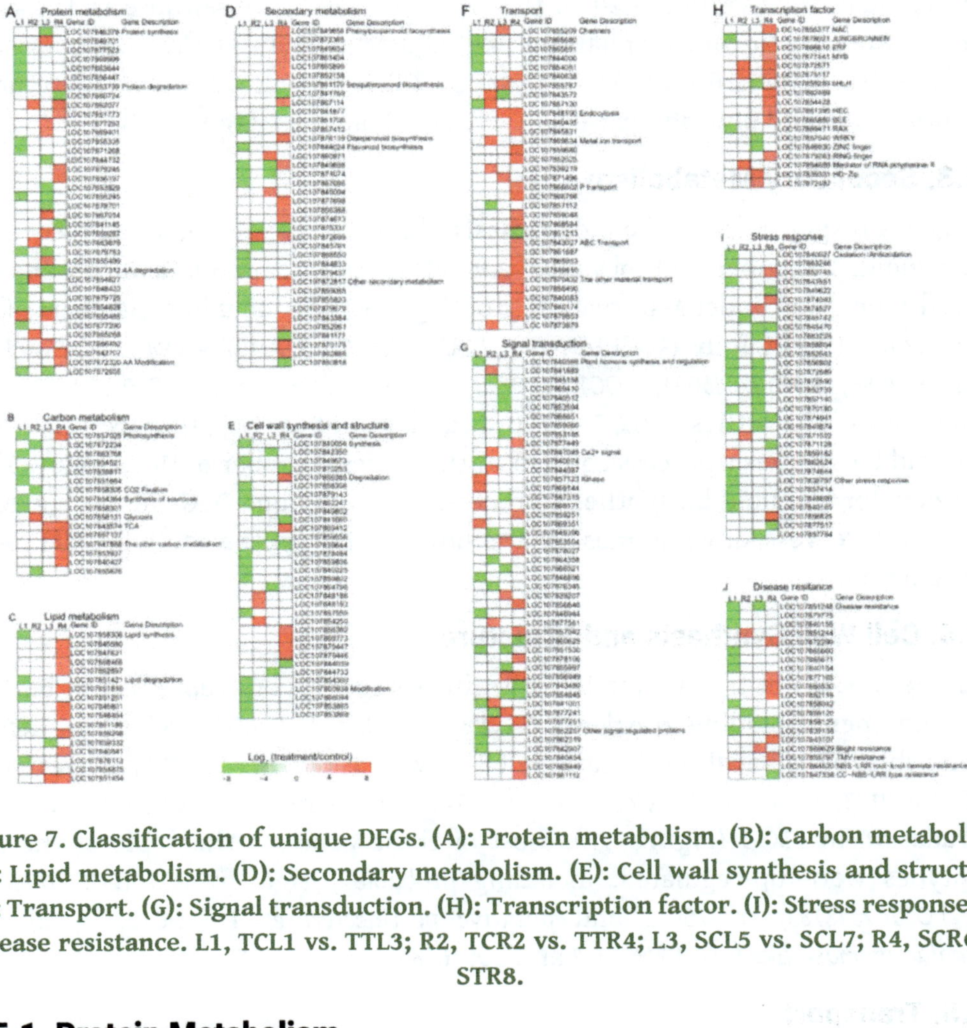

Figure 7. Classification of unique DEGs. (A): Protein metabolism. (B): Carbon metabolism. (C): Lipid metabolism. (D): Secondary metabolism. (E): Cell wall synthesis and structure. (F): Transport. (G): Signal transduction. (H): Transcription factor. (I): Stress response. (J): Disease resistance. L1, TCL1 vs. TTL3; R2, TCR2 vs. TTR4; L3, SCL5 vs. SCL7; R4, SCR6 vs. STR8.

3.5.1. Protein Metabolism

In 750-1, other than S-adenosylmethionine decarboxylase proenzyme (*LOC107855485*), all the unique DEGs required for protein metabolism were downregulated in the leaves (**Figure** 7A). In ZCFB, more than half of the genes associated with protein degradation were up-regulated, especially the members of ubiquitin-proteosome pathway (*LOC107853739*, *LOC107862077*, *LOC107851773*, *LOC107877293*, *LOC107869401*, *LOC10789245*, and *LOC107856197*).

3.5.2. Carbon and Lipid Metabolism

In the carbon metabolism process, genes associated with photosynthesis were down-regulated in the leaves of 750-1 (**Figure** 7B). Two genes (*LOC107843574* and *LOC107857137*) of the tricarboxylic acid cycle (TCA) were upregulated in the leaves and roots of ZCFB. In the lipid metabolism process, DEGs involved in lipid synthesis and degradation were inhibited in the leaves of 750-1, while those in the roots of ZCFB were significantly enhanced, including members of the 3-ketoacyl-CoA synthase (*LOC107845580*, *LOC107847631*, and *LOC107868466*), fatty acid desaturase (*LOC107851810*), and phospholipase (*LOC107859298*) families (**Figure** 7C).

3.5.3. Secondary Metabolism

The genes responsible for the biosynthesis of phenylpropanoid, sesquiterpenoid, diterpenoid, flavonoid, anthocyanin, and other secondary metabolites showed altered expression under N starvation in both genotypes, including members of the cytochrome P450 family (*LOC107844024*, *LOC107860971*, *LOC107849698*, *LOC107871074*, *LOC107867086*, *LOC107845094*, *LOC107877698*, *LOC107856388*, and *LOC107874613*) and anthocyanin biosynthesis (*LOC107845761* and *LOC107866550*) (**Figure** 7D). The flavonoid biosynthesis process consists of the largest number of DEGs followed by the phenylpropannoid biosynthesis process. All the unique DEGs in the leaves of 750-1 were repressed, while most of the unique DEGs in the roots of ZCFB were up-regulated.

3.5.4. Cell Wall Synthesis and Structure

In the leaves, all of the unique DEGs in this category were repressed in both of the genotypes, including *xyloglucan 6-xylosyltransferase 2* (*LOC107842350*), which is responsible for cell wall synthesis in ZCFB, as well as *Glucan edo-1,3-beta-glucosidase* (*LOC107879143*) and *beta-D-xylosidase* (*LOC107854300*), which functions in cell wall degradation in 750-1 (**Figure** 7E). Most of these genes in the roots of the two genotypes were up-regulated, including *probable pectate lyase 5* (*LOC107848186* and *LOC107848183*) in 750-1 and *putative pectinesterase 11* (*LOC107869412*) and *polygalacturonase-like* (*LOC107856362* and *LOC107869773*) in ZCFB.

3.5.5. Transport

In 750-1, all of the unique DEGs engaged in transport were down-regulated in the leaves, including genes associated with osmosis (*LOC107865680*) and stoma movement (*LOC107844000*), while in the roots, except for the gene encoding sodium-dependent phosphate transporter protein 1 (*LOC107857112*), all of the genes were up-regulated, including metal (*LOC107869634*) and sugar transport (*LOC107873879*) (**Figure** 7F). In ZCFB, most of the genes were up-regulated in both of the leaves and roots, including genes associated with channel (*LOC107855209*), endocytosis (*LOC107848190*), sodium/metabolite and urea transport (*LOC107871496* and *LOC107870432*) in the leaves, and gene encoding channel proteins (*LOC107840638*) and proteins involved in endocytosis (*LOC107848190*, *LOC107840435*, and *LOC107845831*) in the roots.

3.5.6. Signal Transduction and Transcription Factor

For hormone synthesis and regulation, all of the genes were inhibited in the leaves of both genotypes (**Figure** 7G), including those involved in ethylene synthesis (*LOC107840369*) and response (*LOC107865651*) in 750-1 and those associated with the auxin response (*LOC107840512* and *LOC107853594*) in ZCFB. In 750-1, the genes of the LRR receptor-like serine/threonine-protein kinase family (*LOC107869144*, *LOC107869171*, and*LOC107869351*) and those involved in Ca2+ signaling (*LOC107847049* and *LOC107844087*) were up-regulated in the leaves and roots, respectively. The genes of transcription factors (TFs) that showed changes in expression mainly fell into the *MYB*, *bHLH*, *NAC*, *WRKY*, *BEE*, and *ERF* families (**Figure** 7H).

3.5.7. Stress Response and Disease Resistance

In 750-1, most of the unique DEGs related to stress response and disease resistance were observable in the leaves, and except for *protein SRG1-like* (*LOC107859182*) and *NBS-LRR root-knot nematode resistance protein* (*LOC107864520*), all the DEGs of the two pathways were down-regulated (**Figure** 7I,J). Meanwhile, in ZCFB, DEGs involved in stress response and disease resistance were detected in both the leaves and roots, and most of the genes in the leaves were down-regulated, in contrast to in the roots.

4. DISCUSSION

In this study, a large number of low-N-responsive genes were identified from low-N-tolerant (750-1) and low-N-sensitive (ZCFB) pepper genotypes after long-term N deficiency, and the number of low-N-responsive genes in ZCFB was far greater than those in 750-1 (**Figure 2**). The expression profiles of the genes associated with nitrate transport, assimilation, and remobilization were similar between the 750-1 and ZCFB (**Figure 5**). However, many unique DEGs were detected between the two pepper genotypes after long-term N deficiency, implying different molecular mechanisms of resistance to N deficiency in these two genotypes.

4.1. Unique DEGs in the Low-N-Tolerant Genotype

In 750-1, the unique DEGs involved in primary metabolism, secondary metabolism, stress response, and disease resistances were down-regulated in the leaves after long-term N deficiency. However, the relative biomass of 750-1 was significantly higher than that of the other cultivars, indicating that the growth of 750-1 was repressed less than other cultivars under N starvation. The relative values of root dry weights and leaf areas of 750-1 were much higher than those of other cultivars, implying a larger absorption area of N and stronger photosynthesis in 750-1. Interestingly, genes of *somatic embryogenesis receptor kinase 2 (SERK2)-like* (*LOC107857123*) and *LRR receptor-like serine/threonine-protein kinase* (*LOC107869144* and *LOC107869171*) were up-regulated in the leaves of 750-1 (**Figure** 7G). Both SERK2 and LRR receptor-like serine/threonine-protein kinase belong to the receptor-like kinase (RLK) family. SERK2 is essential for male microsporogenesis in *Arabidopsis* [**40**]. It has been postulated that at an early low-N stress stage, processes such as absorption, transportation, and assimilation of N might be highly enhanced in 750-1 under low-N conditions, as well as primary metabolism, which could result in improved biomass accumulation in 750-1 relative to other cultivars. After long-term N limitation, 750-1 might slowly decrease vegetative growth and enter reproductive stage earlier to accelerate its life cycle and seed handing down.

4.2. Unique DEGs in the Low-N Sensitive Genotype

Lipids are a major subcellular component, the biosynthesis and composition of which are influenced by N [**41,42**]. As essential components of the membrane, lipids play an important role in endocytosis, in which plasma membrane lipids and associated proteins are internalized in vesicles that fuse with the endosomal system [**43**]. In our study, a number of DEGs involved in lipid metabolism were up-regulated in the roots of ZCFB (**Figure** 7C). Meanwhile, several genes associated with endocytosis were also enhanced (**Figure** 7F). In plants, endocytosis plays an important role not only in transporting membrane proteins, lipids, and extracellular molecules into the cell but also in nutrient delivery, toxin avoidance, and pathogen defense [**44,45**]. To our knowledge, few studies have detected the involvement of endocytosis in N starvation. Here, the detection of endocytosis-related genes being involved in N starvation could provide new insights into N metabolism in plants.

Secondary metabolites have no direct role in plant growth, but their roles in stress defense have garnered much interest in recent years [**46,47**]. In the present study, genes involved in the synthesis of secondary metabolites, such as phenylpropanoid and flavonoid, were significantly up-regulated after low-N stress in ZCFB, as well as those encoding ATP-binding cassette (ABC) transporters that are involved in the transport of plant secondary metabolites (**Figure** 7F) [**48,49**]. We also found that genes involved in the oxidation/antioxidation response (**Figure** 7H) and disease resistance (**Figure** 7I) were up-regulated in ZCFB, as well as the expression of genes belonging to the *NAC*, *ERF*, *MYB*, and *bHLH* families (**Figure** 7H), which play essential roles in stress response. Considering the low relative biomass of ZCFB (**Table S2**), enhanced secondary metabolism and stress response could suggest that there was a diversion of materials from primary metabolism to the defense system under N deficiency. Indeed, previous studies have shown that N limitation results in the reduction of plant growth; however, some secondary metabolites were also shown to accumulate [**25,50**]. Moreover, genes involved in the tricarboxylic acid cycle (TCA) were up-regulated in ZCFB (**Figure** 7B), implying that to survive N deficiency, ZCFB may require more energy to support enhanced biological processes, such as endocytosis, secondary metabolism, and transcriptional regulation.

4.3. P and N Crosstalk

Phosphorus (P) is acquired by plants primarily in the form of inorganic phosphate (Pi) by Pi transporters (PHT). The *Nitrogen limitation adaptation* (*NLA*) gene is involved in adaptive responses to low-N stress in *Arabidopsis* [**51**] and is an important regulator of PHTs under N deficiency [**52,53,54,55**]. Over-expression of the rice Pi transporter gene OsPT2 enhances tolerance to Pi deficiency, as well as N2 fixation and ammonium assimilation [**56,57**]. These studies indicate the intimate crosstalk between the N limitation and P deficiency response pathways. In the present study, *Sodium-dependent phosphate transport protein 1* (*LOC107857112*) was repressed in the roots of 750-1, and *PHT1;2* (*LOC107866602*) and *PHT1;9* (*LOC107866766*) were up-regulated in the roots of ZCFB under low-N stress (**Figure** 7F), providing new information regarding the low-N induced crosstalk between N and P.

5. CONCISE SUMMARY

Enhancing nitrogen use efficiency (NUE) is crucial for the advancement of sustainable pepper cultivation. Nevertheless, our understanding of the molecular mechanisms behind the genetic diversity in nitrogen-use efficiency (NUE) among different pepper cultivars is currently lacking. This work analyzed the transcriptome of two pepper genotypes, one with high tolerance to low-N (750-1) and the other with low tolerance (ZCFB), using RNA-Seq and various bioinformatics tools. The comparison was done after subjecting the plants to N famine for a period of 28 days. The findings demonstrated significant variations in the transcriptome reactions to low-N stress between 750-1 and ZCFB, particularly in genes unrelated to nitrogen metabolism. The distinct genes that respond specifically to low nitrogen levels in the two genotypes offer valuable insights into comprehending the variance in nitrogen use efficiency (NUE) across different genotypes.

AT A GLANCE

The molecular mechanisms responsible for the difference in nitrogen-use efficiency (NUE) in different genotypes of pepper (*Capsicum annuum* L.) are not well known. For this study, two genotypes (750-1, low-N tolerant; ZCFB, low-N sensitive) were chosen from a pool of 100 pepper cultivars based on their different levels of tolerance to low nitrogen (N). The selection was made by comparing their leaf areas, shoot dry weights, root dry weights, and overall plant dry weights during the seedling stage. Afterwards, the transcriptome of these two pepper genotypes was examined using RNA-Seq after subjecting them to 28 days of N deprivation. In the leaves of 750-1, we identified 2621 out of 2470 distinct expressed genes (DEGs). In the roots of ZCFB, we detected 3936 out of 4218 DEGs. The alterations in the expression of fundamental nitrogen metabolism genes exhibited a comparable pattern across the 750-1 and ZCFB samples. However, distinct differentially expressed genes (DEGs) that are not directly related to nitrogen (N) metabolism were found between the 750-1 and ZCFB cultivars. A total of 750-1,110 distinct differentially expressed genes (DEGs) were identified in the leaves. Among these, 103 genes were found to be down-regulated. These down-regulated genes are connected with various biological processes such as protein metabolism, photosynthesis, secondary metabolism, cell wall metabolism, stress response, and disease resistance. A total of 142 distinct differentially expressed genes (DEGs) were identified in the roots of ZCFB. Among these, 117 genes were found to be up-regulated, leading to the stimulation of various biological processes including protein degradation, synthesis of secondary metabolites, lipid metabolism, endocytosis, the tricarboxylic acid cycle (TCA), transcriptional regulation, stress response, and disease resistance. The results of our study not only enhance our understanding of the distinct regulatory mechanisms in low-N-tolerant and low-N-sensitive pepper cultivars, but also offer a wealth of potential candidate genes for enhancing the low-N tolerance of pepper cultivars.

REFERENCES

1. Liu, Q.; Chen, X.; Wu, K.; Fu, X. Nitrogen signaling and use efficiency in plants: What's new? *Curr. Opin. Plant Biol.* 2015, *27*, 192–198.

2. Krapp, A.; David, L.C.; Chardin, C.; Girin, T.; Marmagne, A.; Leprince, A.S.; Chaillou, S.; Ferrario-Méry, S.; Meyer, C.; Daniel-Vedele, F. Nitrate transport and signalling in Arabidopsis. *J. Exp. Bot.* 2014, *65*, 789–798.

3. Tegeder, M.; Masclaux-Daubresse, C. Source and sink mechanisms of nitrogen transport and use. *New Phytol.* 2018, *217*, 35–53.

4. Xu, G.; Fan, X.; Miller, A.J. Plant nitrogen assimilation and use efficiency. *Annu. Rev. Plant Biol.* 2012, *63*, 153–182.

5. Peng, M.; Bi, Y.M.; Zhu, T.; Rothstein, S.J. Genome-wide analysis of Arabidopsis responsive transcriptome to nitrogen limitation and its regulation by the ubiquitin ligase gene NLA. *Plant Mol. Biol.* 2007, *65*, 775–797.

6. Feng, L.; Yang, T.; Zhang, Z.; Li, F.; Chen, Q.; Sun, J.; Shi, C.; Deng, W.; Tao, M.; Tai, Y.; et al. Identification and characterization of cationic amino acid transporters (CATs) in tea plant (*Camellia sinensis*). *Plant Growth Regul.* 2018, *84*, 57–69.

7. Liu, S.; Wang, D.; Mei, Y.; Xia, T.; Xu, W.; Zhang, Y.; You, X.; Zhang, X.; Li, L.; Wang, N.N. Overexpression of *GmAAP6a* enhances tolerance to low nitrogen and improves seed nitrogen status by optimizing amino acid partitioning in soybean. *Plant Biotechnol. J.* 2020, *18*, 1749–1762.

8. Qin, C.; Yu, C.; Shen, Y.; Fang, X.; Chen, L.; Min, J.; Zhang, Z. Whole-genome sequencing of cultivated and wild peppers provides insights into *Capsicum* domestication and specialization. *Proc. Natl. Acad. Sci. USA* 2014, *111*, 5135–5140.

9. Ayodele, O.J.; Alabi, E.O.; Aluko, M. Nitrogen fertilizer effects on growth, yield and chemical composition of hot pepper (rodo). *Int. J. Agric. Sci.* 2015, *8*, 666–673.

10. Wahocho, N.A.; Zeshan Ahmed, S.; Jogi, Q.; Talpur, K.H.; Leghari, S.J. Growth and productivity of chilli (*Capsicum annuum* L.) under various nitrogen levels. *Sci. Int. Lahore* 2016, *28*, 1321–1326.

11. Zhu, J.H.; Li, X.L.; Christie, P.; Li, J.L. Environmental implications of low nitrogen use efficiency in excessively fertilized hot pepper (*Capsicum frutescens* L.) cropping systems. *Agric. Ecosyst. Environ.* 2005, *111*, 70–80.

12. Amare, T.; Nigussie, D.; Kebede, W.T. Performance of hot pepper (*Capsicum annuum*) varieties as influenced by nitrogen and phosphorus fertilizers at Bure, Upper Watershed of the Blue Nile in Northwestern Ethiopia. *Int. J. Agric. Sci.* 2013, 3, 599–608.

13. Sachiko, N.; Toriyama, K.; Yoshimichi, F. Genetic variations in dry matter production and physiological nitrogen use efficiency in rice (*Oryza sativa* L.) varieties. *Breed. Sci.* 2009, *59*, 269–276.

14. Anbessa, Y.; Juskiw, P.; Good, A.; Nyachiro, J.; Helm, J. Genetic variability in nitrogen use efficiency of spring barley. *Crop Sci.* 2009, *49*, 1259–1269.

15. Górny, A.G.; Banaszak, Z.; Ługowska, B.; Ratajczak, D. Inheritance of the efficiency of nitrogen uptake and utilization in winter wheat (*Triticum aestivum* L.) under diverse nutrition levels. *Euphytica* 2010, *177*, 191–206.

16. Bouchet, A.-S.; Laperche, A.; Bissuel-Belaygue, C.; Snowdon, R.; Nesi, N.; Stahl, A. Nitrogen use efficiency in rapeseed. *Agron. Sustain. Dev.* 2016, *36*, 38.

17. Rodrigues, M.C.; Rezende, W.M.; Silva, M.E.J.; Faria, S.V.; Zuffo, L.T.; Galvão, J.C.C.; DeLima, R.O. Genotypic variation and relationships among nitrogen-use efficiency and agronomic traits in tropical maize inbred lines. *Genet. Mol. Res.* 2017, *16*, gmr16039757.

18. Zhang, H.; Fu, X.; Wang, X.; Gui, H.; Dong, Q.; Pang, N.; Wang, Z.; Zhang, X.; Song, M. Identification and screening of nitrogen-efficient cotton genotypes under low and normal nitrogen environments at the seedling stage. *J. Cotton. Res.* 2018, *1*, 6.

19. Sun, H.; Qian, Q.; Wu, K.; Luo, J.; Wang, S.; Zhang, C.; Fu, X. Heterotrimeric G proteins regulate nitrogen-use efficiency in rice. *Nat. Genet.* 2014, *46*, 652–656.

20. Hu, B.; Wang, W.; Ou, S.; Tang, J.; Li, H.; Che, R.; Chu, C. Variation in *NRT1.1B* contributes to nitrate-use divergence between rice subspecies. *Nat. Genet.* 2015, *47*, 834–838.

21. Fan, X.; Tang, Z.; Tan, Y.; Zhang, Y.; Luo, B.; Yang, M.; Lian, X.; Shen, Q.; Miller, A.; Xu, G. Overexpression of a pH-sensitive nitrate transporter in rice increases crop yields. *Proc. Natl. Acad. Sci. USA* 2016, *113*, 7118–7123.

22. Wang, Q.; Nian, J.; Xie, X.; Yu, H.; Zhang, J.; Bai, J.; Zuo, J. Genetic variations in *ARE1* mediate grain yield by modulating nitrogen utilization in rice. *Nat. Commun.* 2018, *9*, 735.

23. Wang, W.; Hu, B.; Yuan, D.; Liu, Y.; Che, R.; Hu, Y.; Chu, C. Expression of the nitrate transporter gene *OsNRT1.1A/OsNPF6.3* confers high yield and early maturation in rice. *Plant Cell* 2018, *30*, 638–651.

24. Tang, W.; Ye, J.; Yao, X.; Zhao, P.; Xuan, W.; Tian, Y.; Wan, J. Genome-wide associated study identifies NAC42-activated nitrate transporter conferring high nitrogen use efficiency in rice. *Nat. Commun.* 2019, *10*, 5279.

25. Zhao, W.; Yang, X.; Yu, H.; Jiang, W.; Sun, N.; Liu, X.; Gu, X. RNA-Seq-based transcriptome profiling of early nitrogen deficiency response in cucumber seedlings provides new insight into the putative nitrogen regulatory network. *Plant Cell Physiol.* 2015, *56*, 455–467.

26. Lv, Y.; Liang, Z.; Ge, M.; Qi, W.; Zhang, T.; Lin, F.; Peng, Z.; Zhao, H. Genome-wide identification and functional prediction of nitrogen-responsive intergenic and intronic long non-coding RNAs in maize (*Zea mays* L.). *BMC Genom.* 2016, *17*, 350.

27. Curci, P.L.; Cigliano, R.A.; Zuluaga, D.L.; Janni, M.; Sanseverino, W.; Sonnante, G. Transcriptomic response of durum wheat to nitrogen starvation. *Sci. Rep.* 2017, *7*, 1176.

28. Kuang, Q.; Zhang, S.; Wu, P.; Chen, Y.; Li, M.; Jiang, H.; Wu, G. Global gene expression analysis of the response of physic nut (*Jatropha curcas* L.) to medium- and long-term nitrogen deficiency. *PLoS ONE* 2017, *12*, e0182700.

29. Xin, W.; Zhang, L.; Zhang, W.; Gao, J.; Yi, J.; Zhen, X.; Li, Z.; Zhao, Y.; Peng, C.; Zhao, C. An integrated analysis of the rice transcriptome and metabolome reveals differential regulation of carbon and nitrogen metabolism in response to nitrogen availability. *Int. J. Mol. Sci.* 2019, *20*, 2349.

30. Hao, Q.N.; Zhou, X.A.; Sha, A.H.; Wang, C.; Zhou, R.; Chen, S.L. Identification of genes associated with nitrogen use efficiency by genome-wide transcriptional analysis of two soybean genotypes. *BMC Genom.* 2011, *12*, 525.

31. Gelli, M.; Duo, Y.D.; Konda, A.R.; Zhang, C.; Holding, D.; Dweikat, I. Identification of differentially expressed genes between sorghum genotypes with contrasting nitrogen stress tolerance by genome-wide transcriptional profiling. *BMC Genom.* 2014, *15*, 179–194.

32. Quan, X.; Zeng, J.; Ye, L.; Chen, G.; Han, Z.; Shah, J.; Zhang, G. Transcriptome profiling analysis for two Tibetan wild barley genotypes in responses to low nitrogen. *BMC Plant Biol.* 2016, *16*, 30.

33. Li, W.; Xiang, F.; Zhong, M.; Zhou, L.; Liu, H.; Li, S.; Wang, X. Transcriptome and metabolite analysis identifies nitrogen utilization genes in tea plant (*Camellia sinensis*). *Sci. Rep.* 2017, *7*, 1693.

34. Sinha, S.K.; Sevanthi, V.A.M.; Chaudhary, S.; Tyagi, P.; Venkadesan, S.; Rani, M.; Mandal, P.K. Transcriptome analysis of two rice varieties contrasting for nitrogen use efficiency under chronic N starvation reveals differences in chloroplast and starch metabolism-related genes. *Genes* 2018, *9*, 206.

35. Iqbal, A.; Dong, Q.; Wang, X.R.; Gui, H.P.; Zhang, H.H.; Zhang, X.L.; Song, M.Z. Transcriptome analysis reveals differences in key genes and pathways regulating carbon and nitrogen metabolism in cotton genotypes under N starvation and resupply. *Int. J. Mol. Sci.* 2020, *21*, 1500.

36. Available online: **https://www.ncbi.nlm.nih.gov/genome/10896?genome_assembly_id=227510** (accessed on 3 August 2017).

37. Available online: **ftp://ftp.ncbi.nih.gov/blast/db** (accessed on 3 August 2017).

38. Available online: **http://www.geneontology.org/** (accessed on 3 August 2017).

39. Available online: **http://www.genome.jp/kegg/** (accessed on 3 August 2017).

40. Albrech, C.; Russinova, E.; Kemmerling, B.; Kwaaitaal, M.; de Vries, S.C. Arabidopsis somatic embryogenesis receptor kinase proteins serve brassinosteroid-dependent and -independent signaling pathways. *Plant Physiol.* 2008, *148*, 611–619.

41. Yang, Y.; Yu, X.; Song, L.; An, C. ABI4 activates *DGAT1* expression in Arabidopsis seedlings during nitrogen deficiency. *Plant Physiol.* 2011, *156*, 873–883.

42. Liu, M.Y.; Burgos, A.; Ma, L.; Zhang, Q.; Tang, D.; Ruan, J. Lipidomics analysis unravels the effect of nitrogen fertilization on lipid metabolism in tea plant (*Camellia sinensis* L.). *BMC Plant Biol.* 2017, *17*, 165.

43. D'Hondt, K.; Heese-Peck, A.; Riezman, H. Protein and lipid requirements for endocytosis. *Annu. Rev. Genet.* 2000, *34*, 255–295.

44. Chen, X.; Irani, N.G.; Friml, J. Clathrin-mediated endocytosis: The gateway into plant cells. *Curr. Opin. Plant Biol.* 2011, *14*, 674–682.

45. Fan, L.; Li, R.; Pan, J.; Ding, Z.; Lin, J. Endocytosis and its regulation in plants. *Trends Plant Sci.* 2015, *20*, 388–397.

46. Isah, T. Stress and defense responses in plant secondary metabolites production. *Biol. Res.* 2019, *52*, 39.

47. Khare, S.; Singh, N.B.; Singh, A.; Hussain, I.; Niharika, K.; Yadav, V.; Bano, C.; Yadav, R.K.; Amist, N. Plant secondary metabolites synthesis and their regulations under biotic and abiotic constraints. *J. Plant Biol.* 2020, *63*, 203–216.

48. Yazaki, K. ABC transporters involved in the transport of plant secondary metabolites. *FEBS Lett.* 2006, *580*, 1183–1191.

49. Lv, H.; Li, J.; Wu, Y.; Garyali, S.; Wang, Y. Transporter and its engineering for secondary metabolites. *Appl. Microbiol. Biotechnol.* 2016, *100*, 6119–6130.

50. Larbat, R.; Robin, C.; Lillo, C.; Drengstig, T.; Ruoff, P. Modeling the diversion of primary carbon flux into secondary metabolism under variable nitrate and light/dark conditions. *J. Theor. Biol.* 2016, *402*, 144–157.

51. Peng, M.; Hannam, C.; Gu, H.; Bi, Y.M.; Bothstein, S.J. A mutation in *NLA*, wich encodes a RING-type ubiquitin ligase, disrupts the adaptability of *Arabidopsis* to nitrogen limitation. *Plant J.* 2007, *50*, 320–337.

52. Kant, S.; Peng, M.; Rothstein, S.J. Genetic regulation by NLA and microRNA827 for maintaining nitrate-dependent phosphate homeostasis in *Arabidopsis*. *PLoS Genet.* 2011, *7*, e1002021.

53. Lin, W.Y.; Huang, T.K.; Chiou, T.J. Nitrogen limitation adaptation, a target of microRNA827, mediates degradation of plasma membrane-localized phosphate transporters to maintain phosphate homeostasis in *Arabidopsis*. *Plant Cell* 2013, *25*, 4061–4074.

54. Park, B.S.; Seo, J.S.; Chua, N.H. Nitrogen limitation adaptation recruits phosphate2 to target the phosphate transporter PT2 for degradation during the regulation of *Arabidopsis* phosphate homeostasis. *Plant Cell* 2014, *26*, 454–464.

55. Yue, W.; Ying, Y.; Wang, C.; Zhao, Y.; Dong, C.; Whelan, J.; Shou, H. OsNLA1, a RING-type ubiquitin ligase, maintains phosphate homeostasis in *Oryza sativa* via degradation of phosphate transporters. *Plant J.* 2017, *90*, 1040–1051.

56. Chen, G.H.; Yan, W.; Yang, S.P.; Wang, A.; Gai, J.Y.; Zhu, Y.L. Overexpression of rice phosphate transporter gene OsPT2 enhances tolerance to low phosphorus stress in soybean. *J. Agric. Sci. Technol.* 2015, *17*, 469–482.

57. Zhu, W.; Yang, L.; Yang, S.; Gai, J.; Zhu, Y. Overexpression of rice phosphate transporter gene OsPT2 enhances nitrogen fixation and ammonium assimilation in transgenic soybean under phosphorus deficiency. *J. Plant Biol.* 2016, *59*, 172–181.

10
Chapter

CAHSP18.1A, A SMALL HEAT SHOCK PROTEIN DERIVED FROM PEPPER (*CAPSICUM ANNUUM* L.), EXHIBITS A POSITIVE RESPONSE TO HEAT, DROUGHT, AND SALT TOLERANCE

1. INTRODUCTION: AN OVERVIEW

Plants have the ability to withstand significant biological and environmental pressures in their intricate and dynamic surroundings, such as drought, high salt levels, severe temperatures, and oxidation [1,2]. Plants have evolved many defensive methods to alleviate stress. Heat shock proteins (HSPs) play a crucial role in maintaining protein homeostasis and preventing or repairing protein misfolding in response to abiotic stressors. In addition, Heat Shock Proteins (HSPs) are molecular chaperones that have been conserved throughout evolution and are commonly found in different plant groups [3,4,5]. Plant heat shock proteins (HSPs) have essential functions in the folding, transportation, breakdown, and formation of proteins during both normal and stressful circumstances [6]. Plant cells respond to high temperatures by significantly elevating the levels of Heat Shock Proteins (HSPs) in order to safeguard against heat-induced harm and enhance the plant's ability to withstand heat, known as thermotolerance [7]. Furthermore, HSPs play a role in the regular growth and development of plants, including the formation of flowers, seeds, and fruits, as well as processes such as tuberization and nutrient absorption. Heat shock proteins (HSPs) are found in various cellular locations, including the cell membrane, cytoplasm, nucleus, and cell organelles such as mitochondria, chloroplasts, and endoplasmic reticulum [11,12].

HSPs are classified into distinct families based on their sequence homology and molecular weight. These families include HSP20s, HSP60s, HSP70s, HSP90s, and HSP100s [13,14]. HSP20s, one of the five conserved families, are also known as small heat shock proteins (sHSPs). HSP20s have molecular weights ranging from 15 to 42 kDa, as indicated by references [13,15,16]. Moreover, a notable feature of HSP20s is their capacity to interact with substrate proteins in the absence of ATP, and they also exhibit a robust affinity for denatured substrates [15,16,17,18]. Therefore, heat shock proteins (HSPs) possess a strong capacity to preserve the stability of exogenous proteins within cells, thereby preventing their aggregation. While there exist numerous varieties of substrate proteins, s HSPs possess a malleable N-terminus and α-crystallin domain (ACD) hydrophobic surface that can adjust to attach to these diverse protein substrates. Furthermore, s HSPs have the ability to interact with many substrates in diverse manners, enabling them to bind to a broader range of proteins and offer more intricate methods of action compared to other HSPs [19].

Korotaeva et al. [20] and Nieto-Sotelo et al. [21] shown that several heat shock proteins (HSPs) exhibit distinct patterns of expression across different species, and even within different genotypes of the same species. According to reports, the excessive expression of AtHSP17.6A has been found to enhance the resistance of Arabidopsis against penetration [22]. The introduction of AtHSP21 enhanced the thermotolerance of genetically modified Arabidopsis plants and prolonged the duration of their acquired heat resistance. Consequently, Arabidopsis exhibited increased resistance to heat stress upon subsequent exposure [23]. Additionally, certain investigations have indicated that the expression of HSP genes has a positive regulatory effect on the activities of protective enzymes. For instance, in the plant species Arabidopsis, the act of increasing the expression of a certain gene called AtHSP17.8 resulted in an increase in the activity of an enzyme called SOD, as observed in a study referenced as [24]. Furthermore, the upregulation of HSP16.9 in tobacco resulted in an augmentation of the enzymatic activity of POD, CAT, and SOD, as indicated by previous research [25].

Pepper, scientifically known as *Capsicum annuum* L., is a highly significant vegetable crop with both commercial and medical value on a global scale [26]. Pepper is often grown in tropical or subtropical areas with temperatures ranging from 15 to 34 °C [27]. Pepper growth and development can be hindered by salt, drought, and heat stress. These stressors can also cause significant harm to pepper pollination and seed formation, resulting in the shedding of flowers and fruits. As a consequence, the yield and quality of pepper fruits are reduced [28,29].Pepper HSP20s have a significant impact on how pepper plants respond to environmental stress. Guo et al. [30] discovered a total of 35 HSP20s in pepper. The nomenclature of all HSP20s was determined by their molecular weights. Furthermore, the promoter regions of these proteins were shown to contain stress-related cis-elements, including as heat shock elements (HSEs), TATA boxes, CCAAT motifs, and TC-rich repeats [26]. A significant number of CaHSP20 genes have no expression in several pepper tissues, including root, stem, leaf, and flower tissues. CaHSP22.4, CaHSP25.9, CaHSP16.4, CaHSP24.2, and CaHSP26 have been identified as functional proteins in recent years. CaHSP16.4 is found in both the cytoplasm and nucleus.

In Arabidopsis lines where CaHSP16.4 is overexpressed, there is a reported increase in heat stress tolerance [31]. According to Guo et al. [26], they discovered that when CaHSP22.4, a protein found in both the mitochondria and cytoplasm, was overexpressed, it enhanced the ability of Arabidopsis plants to withstand high temperatures. The expression of CaHSP22.4 rose when pepper plants were exposed to elevated temperatures. Furthermore, the CaHSP25.9 protein was shown to be present in both the cell membrane and cytoplasm. It plays a crucial role in enhancing the plant's ability to withstand high temperatures, high salt concentrations, and drought conditions. This discovery was made in pepper plants (*Capsicum annuum* L.) [32]. Pepper CaHSP24.2 is found in mitochondria, cytoplasm, and chloroplasts. It improves the ability of transgenic Arabidopsis plants to withstand high temperatures and controls the production of genes associated to heat stress [30]. In their study, He et al. [33] increased the expression of CaHSP26, resulting in improved resistance to heat stress in Arabidopsis. Remarkably, the ability to withstand high temperatures and high salt concentrations diminished in pepper plants when the CaHSP22.0 gene was suppressed [34]. These investigations indicate that sHSP20s may play a role in responding to heat stress [35] and help in developing pepper's ability to tolerate high temperatures [30].

Out of the 35 CaHSP20s that were studied, the expression level of CaHSP18.1a was shown to be higher in both the B6 and R9 lines when they were exposed to heat stress [30]. Furthermore, the analysis of the sequence revealed the presence of a Heat Shock Element (HSE) in CaHSP18.1a, along with the identification of other stress-related components [30]. Using the information provided, we examined the distribution within cells and the pattern of expression of CaHSP18.1a in several tissues of pepper. Additionally, we investigated how CaHSP18.1a reacts to salt, drought, and heat stresses. The technique of virus-induced gene silencing (VIGS) was initially employed to investigate the roles of CaHSP8.1a in the stress response of pepper plants. Furthermore, the overexpression (OE) of CaHSP18.1a in transgenic Arabidopsis thaliana demonstrated its positive regulatory function in the plant's ability to respond to heat, salt, and drought stress. The outcomes of our research serve as a foundation for conducting additional functional investigations on CaHSP18.1a in significant agricultural plants, as well as its involvement in stress resilience.

2. MATERIALS AND METHODS

2.1. Plant Materials and Growth Conditions

This study utilized the R9 pepper line, which is known for its ability to tolerate high temperatures, and the B6 pepper line, which is sensitive to temperature changes. The R9 line was obtained from the World/Asia Vegetable Research and Development

Center (PP0042-51), while the B6 line was picked by the Pepper Research Group at the College of Horticulture, Northwest A&F University in Yangling, China. Pepper seedlings were grown in a controlled environment with certain growth parameters before undergoing different treatments. These conditions included a daily cycle of 16 hours of light and 8 hours of darkness, as well as a relative humidity of 65%. The seedlings were kept under these conditions until they reached the stage of having 6-8 true leaves.

The temperature was changed throughout the course of the experiment. R9 peppers were grown under 25/20 °C day/night temperatures to enable analyzing gene expression [**30,36,37**]. However, the growing conditions for use of virus-induced gene silencing (VIGS) in the R9 pepper line were 22/18 °C day/night temperatures [**38**]. *Arabidopsis* ecotype Col0- variety seedlings were incubated at %65 relative humidity, 18/22 °C (day/night), and 16 h/8 h (light/dark) photoperiod conditions [**37,38**].

2.2. RNA Extraction and Real-Time Fluorescent Quantitative PCR qRT-PCR Analysis

Total RNA was extracted using the Trizol method [**28**]. Synthesis of cDNA was conducted with the PrimeScript™ kit (Takara, Dalian, China) according to the manufacturer's instructions. First, we downloaded the amino acid sequence of *CaHSP18.1a* from the Pepper Genomics Database (accessed date on 1 January 2020, **http://peppergenome.snu.ac.kr/**: Accession number: CA08g17060). Primer Premier 5.0 was used to design primers, and primer specificity was detected using NCBI Primer BLAST (**https://www.ncbi.nlm.nih.gov/tools/primer-blast/**, accessed on 5 January 2020) . The pepper ubiquitin binding gene *CaUbi3* (Accession number AY486137) was used as a reference gene [**39**]. qRT-PCR was performed using the iQ5.0 Bio-Rad iCycler thermal cycler (Bio-Rad, Hercules, CA, USA). The SYBR Green Super mix (Takara, Dalian, China) was used in the qRT-PCR reaction system following the manufacturer's instructions. The relative expression levels of the gene were analyzed using the 2-ΔΔCT method [**40**].

2.3. Subcellular Localization of CaHSP18.1a Protein

The ORF (open reading frame) of *CaHSP18.1a* without a termination codon was PCR-amplified using a specific primer pair. The resulting *CaHSP18.1a* fragment was cloned into the pVBG2307: GFP vector with *Xba*I and *Kpn*I restriction sites. The pVBG2307:*CaHSP18.1a*: GFP fusion protein transient expression vector and the control vector pVBG2307: GFP, after having been successfully constructed, were transformed into *Agrobacterium tumefaciens* strain GV3101, which was then injected into tobacco (*Nicotiana tabacum*) leaves to induce transient expression. After dark cultivation for approximately 36 h, epidermis samples of tobacco leaves were photographed under a fully automatic upright fluorescence microscope on the public platform of the College of Horticulture, Northwest A&F University, and the fluorescence patterns in the cells were observed; we specifically used the method described by Yu et al. [**41**].

2.4. Virus-Induced Gene Silencing of CaHSP18.1a

A 256-bp fragment of the *CaHSP18.1a* ORF was PCR-amplified using a specific primer pair. The underlined sequences are restriction enzyme cleavage sites (for *Xba*I and *Kpn*I). The resulting *CaHSP18.1a* fragment was inserted into TRV2:00 vectors, with the empty vector TRV2:00 and TRV2: *CaPDS* (phytoene desaturase gene) used as negative and positive controls. When R9 plants reached the two true leaves stage, we followed the method of Wang et al [**38**], which involved mixing the pTRV1 bacterial culture with an equal volume of the TRV2:00, TRV2: *CaPDS*, and TRV2: *CaHSP18.1a* cultures; this solution was injected into the leaves of R9 plants. After incubation in the dark at 18 °C for 2 days, plants were transferred to incubators under preset normal conditions. After 35 days, when most of the leaves of the TRV-2*CaPDS* pepper plants had become bleached, total RNA was extracted from the leaves of the silenced TRV2: *CaHSP18.1a* plants and the negative control TRV2:00 plants, and qRT-PCR was used to detect the *CaHSP18.1a* expression level, which was used to calculate silencing efficiency.

2.5. Generation of CaHSP18.1a-Overexpression Arabidopsis Lines

The entire coding regions of *CaHSP18.1a* were cloned into the pVBG2307 vector between the *Xba*I and *Kpn*I restriction sites to yield the final plasmid pVBG2307: *CaHSP18.1a* used for genetic transformation. The recombinant fusion vector was transformed into *Agrobacterium* strain GV3101 and transformed into *Arabidopsis thaliana* as described by Clough and Bent [**42**]. Transformed strains of pVBG2307 expression vector were screened with kanamycin and confirmed by PCR verification. We extracted DNA to detect the correctness of the target band. First, the fragment lengths of the bands were compared with the target band, obtaining, respectively, the OE1, OE2, OE3, OE4, and OE5 lines. Next, we performed real-time qRT-PCR quantitative analysis and detected the transcript from the inserted construct. *CaHSP18.1a* was thus determined to be expressed in large quantities in the OE3, OE2, and OE1 strains, but the wild-type gene was not detected. Both the target band and qRT-PCR results indicated that the *CaHSP18.1a* gene was successfully transferred into *Arabidopsis thaliana*, and the obtained T3-generation *Arabidopsis thaliana* could thus be used for further experiments.

2.6. Experimental Treatments and Sample Collection

The roots, stems, and leaves of R9 and B6 pepper seedlings (at the 4-to-6-leaf stage) grown under normal conditions were sampled in order to analyze expression of *CaHSP18.1a* in different tissues. For the thermotolerance treatment, R9 and B6 pepper seedlings (again, at the -4to-6-leaf stage) were grown at 42 °C for 24 h, and root, stem, and leaf samples were collected from stress-treated seedlings at ,0.5 ,0 3 ,1, and 6 h post-treatment. For the drought stress treatment, the roots of R9 seedlings were soaked in 300 mM mannitol, and root, stem, and leaf samples were collected from stress-treated seedlings at 12 ,6 ,3 ,0, and 24 h after treatment.

To analyze the function of *CaHSP18.1a* in response to pepper abiotic stress, silenced pepper seedlings and TRV2:00 pepper seedlings were grown at 42 °C for 24 h. For the drought and salt stresses, seedlings were treated with 300 mM mannitol and 300 mM NaCl for 24 h. Samples were collected and malondialdehyde (MDA) content, total chlorophyll content, and relative electrolyte leakage (REL) were determined. To identify the tolerance of *CaHSP18.1a*-overexpression in *Arabidopsis thaliana* in response to heat, salt, and drought stress, T-3generation *Arabidopsis thaliana* and wild-type lines were treated as described.

For heat stress, 2-week-old OE3 seedlings were treated at 42 °C for 24 h. For drought stress, water was withheld from 3-week-old transgenic *Arabidopsis* seedlings for 10 d. Samples were collected to measure the total chlorophyll contents, MDA content, REL, the activity levels of CAT, SOD, and ascorbic acid peroxidase (APX). For salt stress, the seeds of WT and transgenic lines were sown on MS medium with 100 ,0, and 150 mM NaCl, and the roots lengths were measured after 10 days of treatment. The germination rate was determined after 6 d. The -3week-old WT and transgenic plants were irrigated with 200 mM NaCl solution for 7 days, once every 2 days.

2.7. Measurement of Physiological Indicators

REL was estimated using the thiobarbituric acid reaction [**43**]. Total chlorophyll content was determined in the leaves according to methods previously described by Arkus et al [**44**]. Lipid peroxidation was determined by measuring the MDA content following the method of Campos et al [**45**]. POD and SOD activity levels were measured following the methods of Guo et al [**46**]. APX activity was measured using the methods of Nakano and Asada [**47**]. CAT activity was determined following AebiH [**48**].

2.8. Statistical Analyses

The experimental data were analyzed using Origin (Origin Lab, Northampton, MA, USA) and SPPS (SPSS Inc., Chicago, IL, USA). Significance tests for differences between control and stress treatments were assessed at a $p \leq 0.05$ level of significance. All experiments were performed and analyzed separately based on three biological replicates.

3. RESULTS

3.1. Expression of the CaHSP18.1a in Pepper Plants under Abiotic Stress

To confirm whether heat, drought, and salt have an effect on the expression of *CaHSP18.1a*, R9 and B6 pepper lines were used to analyze the expression of *CaHSP18.1a* under heat, drought, and salt stress. Under heat stress (**Figure 1**A,B), the expression levels of *CaHSP18.1a* were significantly upregulated (samples IV, V, and VI) at 3 h in the roots, stems, and leaves of R9 plants and peaked at 6 h (sample Point V) in B6 plants. However, during the 22 °C recovery stage, the recovery times of *CaHSP18.1a* in the R9 and B6 strains differed. In roots, the expression levels of *CaHSP18.1a* in R9 and B6 plants returned to baseline after a 6 h recovery at the normal temperature (sample VII point) (**Figure 1**B). In stems, the expression level of *CaHSP18.1a* returned to a normal level after 3 h under the 22 °C recovery conditions for both the R9 and B6 plants (point VI in samples) (**Figure 1**C).

In leaves, the expression level of *CaHSP18.1a* returned to normal levels at 24 h (sample VIII point) (**Figure 1**D). We also analyzed the expression pattern of *CaHSP18.1a* under salt and drought stresses (**Figure 1**E–H). After 6-h NaCl treatments at different concentrations, the expression of *CaHSP18.1a* in R9 and B6 leaves and roots was highest under the 150 mM NaCl, 100 mM NaCl treatments, respectively. The transcription of *CaHSP18.1a* was higher in R9 under different concentrations of NaCl treatment (**Figure 1**E, F). After 6-h treatments with different concentrations of mannitol, the expression of *CaHSP18.1a* in R9 and B6 leaves was the highest after the 150 mM mannitol treatment; the highest expression of *CaHSP18.1a* was observed in R9 and B6 roots subjected to the 50mM mannitol treatment (**Figure 1**G, H). In addition, the transcription of *CaHSP18.1a* was higher in R9 under different concentrations of mannitol. This analysis showed that the expression of *CaHSP18.1a* in pepper was induced by heat, salt, and drought. The response times of *CaHSP18.1a* in different organs of different pepper lines differed, and the response was more intense and more early in the R9 line, which suggests that *CaHSP18.1a* plays a substantial role in plant responses to heat stress.

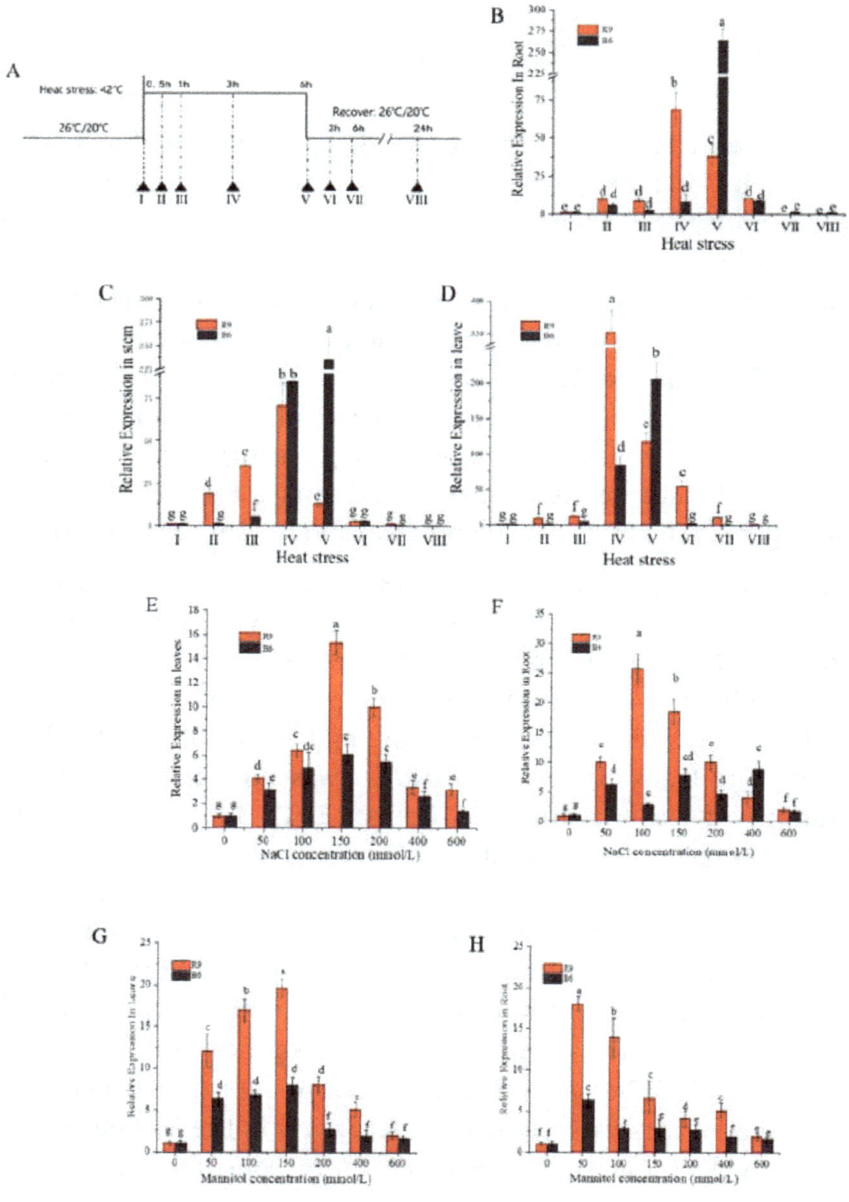

Figure 1. The expression characteristics of *CaHSP18.1a* in peppers in response to heat stress. (A) Time course of heat stress treatment and normal temperature recovery; the sampling time points are represented by triangles (pepper sample points I–VIII); (B–D) The expression levels of *CaHSP18.1a* in roots, stems, and leaves of R9 and B6 plants at each sampling time point; the expression levels of B6 and R9 plants were based on the reference level of their samples, and *CaUBI3-* was selected as the reference gene.(E–H) The expression levels of *CaHSP18.1a* following salt and drought treatment in R9 and B6 leaves and roots. The data presented are means with standard deviations of three biological replicates. Different letters denote statistical significance ($p \leq 0.05$).

3.2. Subcellular Localization of CaHSP18.1a Protein

To explore the subcellular localization of CaHSP18.1a, we constructed the pVBG2307:CaHSP18.1a: GFP fusion expression vector. Both pVBG2307: GFP and pVBG2307:CaHSP18.1a: GFP fusion plasmids were introduced into *Nicotiana tabacum* leaves, and fluorescence was confirmed in the transformed tobacco cells with a microscope (**Figure 2**). We found that the green fluorescence signal of pVBG2307: CaHSP18.1a: GFP was detected in the cell membrane (**Figure 2**A), while the fluorescence of the empty pVBG2307: GFP vector was distributed throughout the cell (**Figure 2**B), indicating that CaHSP18.1a is localized to the cell membrane.

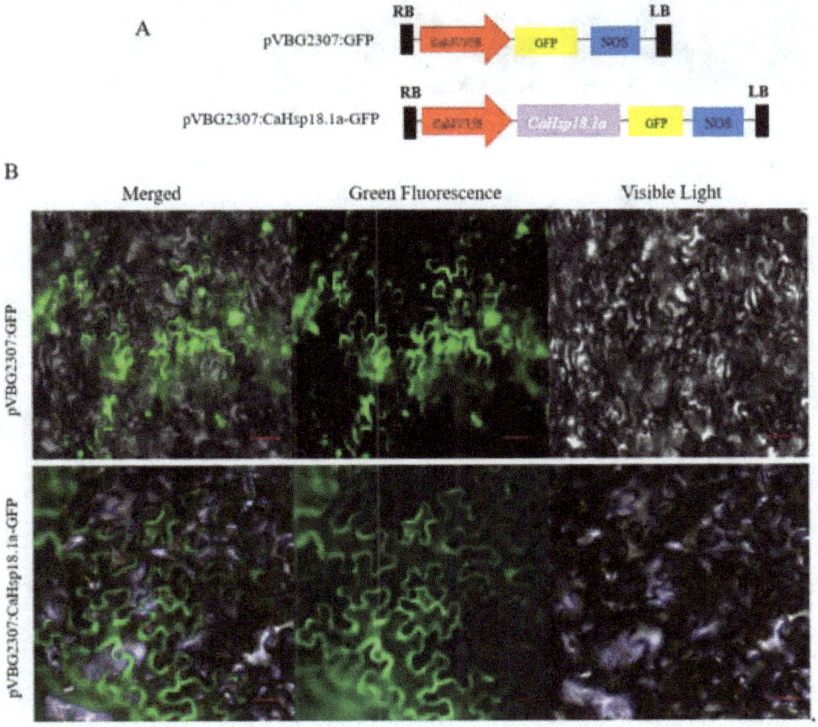

Figure 2. Transient expression of CaHSP18.1a in tobacco. (A) Schematic diagram of the CaHSP18.1a subcellular localization expression vector. (B) Subcellular localization of the CaHSP18.1a protein in tobacco leaves, with pVBG2307: GFP as control. Scale bar = 50 μm.

3.3. CaHSP18.1a-Silenced Plants Sensitive to Abiotic Stress

Confirming the VIGS procedure, after about 40–45 days, plants injected with the positive control TRV2: *CaPDS* showed a large area of typical white leaves, while under normal conditions there was no difference between *CaHSP18.1a*-silenced (TRV2:*CaHSP18.1a*) and negative control (TRV2:00) pepper plants. The silencing efficiency of *CaHSP18.1a*-silenced and TRV2:00 plants was assessed using q RT-PCR. The expression level of *CaHSP18.1a* in the silenced pepper plants decreased to less than %20 of that observed in the negative control plants. Thus, the silencing efficiency for *CaHSP18.1a*-silenced plants reached more than 80% . Therefore, control plants (TRV2:00) and silenced plants (TRV2:*CaHSP18.1a*) were used for the follow-up investigation.

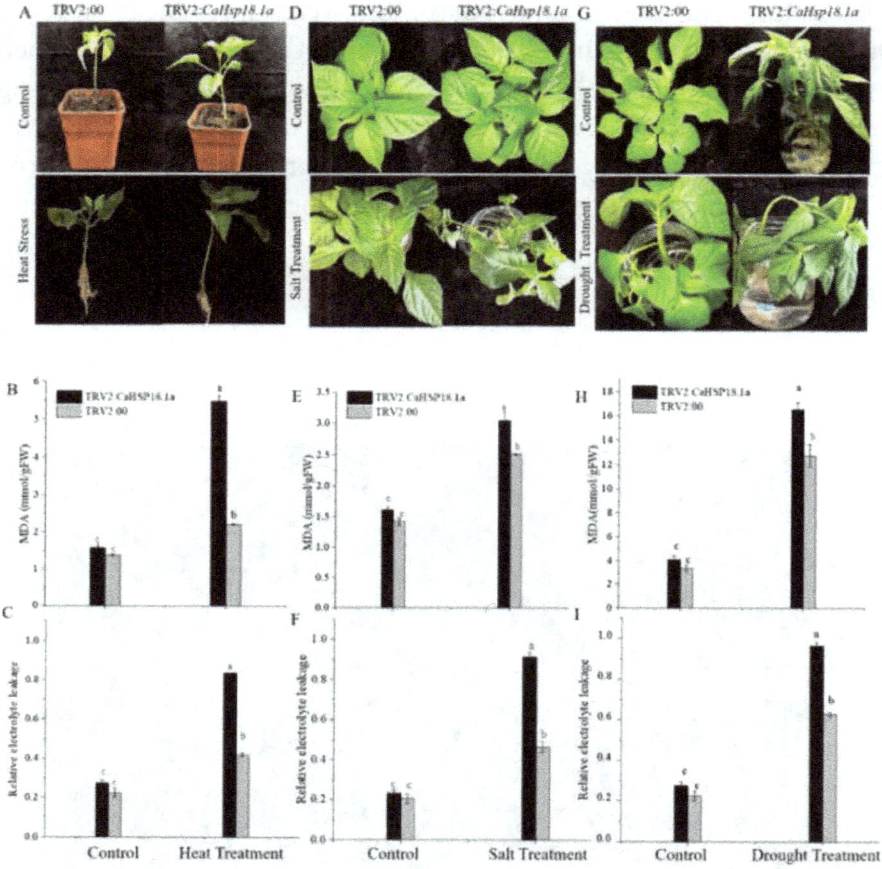

Figure 3. TRV2:*CaHSP18.1a* and TRV2:00 plant phenotypes under heat, drought, and salt treatments, respectively. (A–C) TRV2:*CaHSP18.1a* and TRV2:00 plant phenotypes, malonaldehyde (MDA) content, and relative electrolyte leakage (REL) under the 42 °C heat treatment for 24 h; (D–F) TRV2:*CaHSP18.1a* and TRV2:00 plant phenotypes, MDA content, and relative electrolyte leakage (REL) following salt stress by being soaked in 300 mM NaCl solution for 24 h; (G–I) TRV2:*CaHSP18.1a* and TRV2:00 plant phenotypes, MDA content, and REL following drought stress by being soaked in 300 mM mannitol solution for 24 h. Data are means with standard deviations of three biological replicates. Different letters denote statistical significance ($p \leq 0.05$).

HS (42 °C) was applied to *CaHSP18.1a*-silenced and control pepper plants for 3 h, and the silenced plants and the control group began to show different degrees of wilting. The heat-stress treatment (42 °C) induced significantly different symptoms after 24 h, such that the new growth of *CaHSP18.1a*-silenced plants was seriously wilted with curled leaves and shed lower leaves, while the leaves of the control plants were only slightly curled (**Figure** 3A). In addition, the MDA content and REL was lower in the control plants compared to *CaHSP18.1a*-silenced pepper plants (**Figure** 3B,C); however, the total chlorophyll content was higher in the control than in the *CaHSP18.1a*-silenced plants.

To study the salt-tolerance of silenced and control plants, we washed their roots and soaked them in 300 mM NaCl solution for 24 h. The leaves of silenced plants showed symptoms of wilting, shriveling, and serious yellowing, with lower leaves that had begun to absciss, while the leaves of the control plants only showed some yellowing and did not exhibit obvious wilting. The leaves of the control plants showed only yellowing and no apparent wilting (**Figure** 3D). The MDA content of both plants increased significantly, but that of silenced plants was higher than that of control plants (**Figure** 3E). Relative electrolyte leakage (REL) was higher in silenced plants compared to control plants (0.96 versus 0.65) (**Figure** 3F). To study the effects of *CaHSP18.1a*-silencing on drought tolerance, the silenced and control plants were soaked in 300 mM mannitol solution for 36 h. The *CaHSP18.1a*-silenced pepper showed severe loss of water and wilting, while control plants showed no obvious change (**Figure** 3G). Furthermore, the MDA content and REL both exhibited a similar increase in the silenced pepper plants (**Figure** 3H,I). This indicated that silencing of *CaHSP18.1a* reduced the drought tolerance of pepper plants.

3.4. Effect of CaHSP18.1a Overexpression on Transgenic Arabidopsis

3.4.1. Overexpression of CaHSP18.1a Enhances Plant Tolerance of Heat Stress

First, we transformed pVBG2307:*CaHSP18.1a* into *Agrobacterium tumefaciens* strain GV3101, which was used to transfect *Arabidopsis thaliana* using the dipping method; successful transformants were identified through resistance gene screening and molecular level detection until homozygous T3 lines were obtained. The wild-type (WT) line and five transgenic lines were cultured on Murashige and Skoog (MS) medium for 10 days, and the lengths of their roots were measured. The survival rate of WT plants was lower than those of the OE1, OE2, and OE3 seedlings. Transgenic plants and WT plants were cultured under normal growth conditions for 48 days, and the growth rates of the OE1, OE2, and OE3 lines exceeded those of WT plants. Thus, the OE3, OE2, and OE1 lines were selected for follow-up experiments.

The obtained transgenic lines and WT plants were heat treated (42 °C for 24 h) at the 3-week stage. After heat-stress treatment, the WT plants showed wilting symptoms. Notably, restorable wilt or indistinct-symptoms were observed among the *CaHSP18.1a*-OE seedlings (**Figure** 4A), indicating that *CaHSP18.1a* plays an active role in increasing the thermotolerance of transgenic *Arabidopsis*. In addition, REL and MDA content increased significantly in both the OE and WT lines after heat treatment, while the MDA content was notably lower in the transgenic lines relative to the WT plants (**Figure 4**B). In addition, the SOD and POD activities of *CaHSP18.1a*-OE seedlings were clearly higher than those of the WT plants (**Figure 4**C,D). However, the catalase (CAT) activity did not significantly differ between the WT and transgenic lines (**Figure 4**E). *CaHSP18.1a* played a role in transgenic *Arabidopsis*, probably by regulating the expression of endogenous genes.

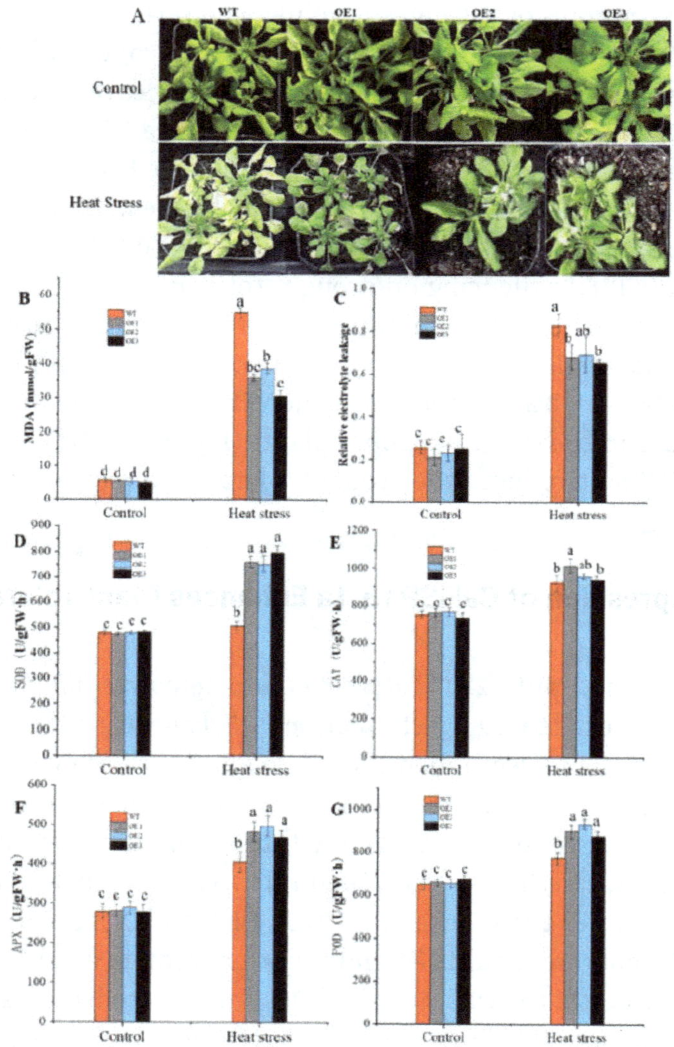

Figure 4. Heat resistance of transgenic *CaHSP18.1a*-OE *Arabidopsis* plants. (A) Phenotypes of 42 °C-treated wild-type (WT) and transgenic *Arabidopsis*; (B–C) Malonaldehyde (MDA) and relative electrolyte leakage (REL) of WT and transgenic *Arabidopsis*; (D–G) Superoxide dismutase (SOD), catalase (CAT), peroxidase (POD), and ascorbic acid peroxidase (APX) activity of WT and transgenic *Arabidopsis*. Data are means with standard deviations of three biological replicates. Different letters denote statistical significance ($p \leq 0.05$).

In the present study, among 18 stress-related genes, 12 were up-regulated in transgenic lines, while the other 6 did not change much (**Figure 5**). Among the up-regulated genes, *AtHSPC30, AtAPX3, AtCAT, AtHSP70*, and *AtRab1* were more prominently expressed in the transgenic OE3 line. Moreover, the expression of the 18 stress-related genes was markedly increased in both transgenic *Arabidopsis* and WT plants under heat stress. However, the expression of these genes in WT seedlings was lower than that in transgenic seedlings (**Figure 5**).

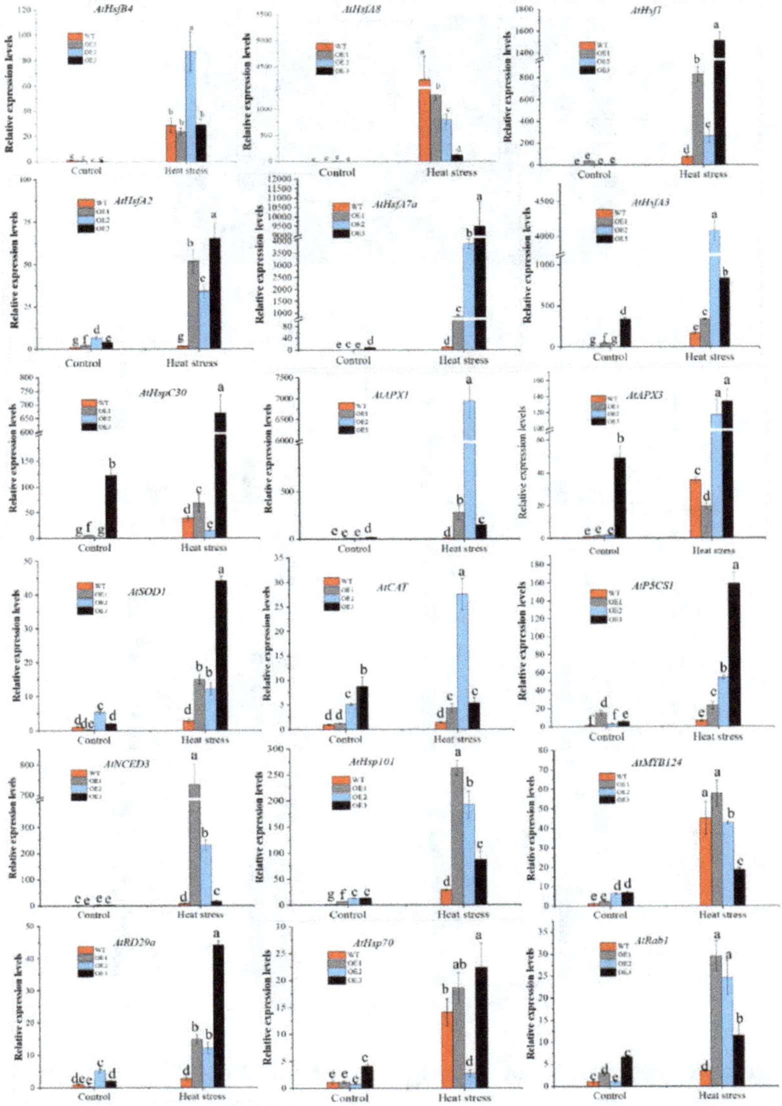

Figure 5. Expression patterns of stress-response genes in wild-type (WT), OE1, OE2, and OE3 lines before and after 42 °C heat treatment for 24 h. Data are means with standard deviations of three biological replicates. Different letters denote statistical significance ($p \leq 0.05$).

3.4.2. Overexpression of CaHSP18.1a Enhances Plant Tolerance to Drought Stress

CaHSP18.1a is a molecular chaperone, though its response to drought stress is still unclear. To further study its function under drought and salt stress, *CaHSP18.1a* transgenic *Arabidopsis thaliana* and WT seedlings were drought treated (**Figure 6**A).

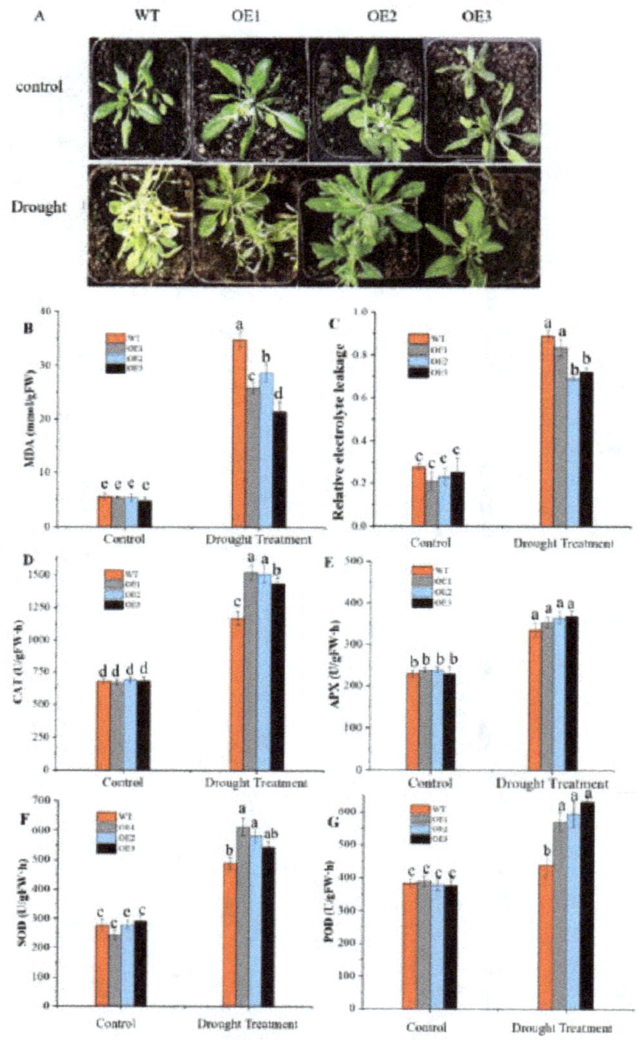

Figure 6. Drought resistance of transgenic *CaHSP18.1a*-OE *Arabidopsis* plants. (A) Phenotypes of wild-type (WT) and *CaHSP18.1a*-OE *Arabidopsis*; (B,C) Malonaldehyde (MDA) content and relative electrolyte leakage (REL) of WT and *CaHSP18.1a*-OE *Arabidopsis*; (D–G) Superoxide dismutase (SOD), catalase (CAT), peroxidase (POD), and ascorbic acid peroxidase (APX) activity of WT and *CaHSP18.1a*-OE *Arabidopsis* without watering for 10 days. Data are means with standard deviations of three biological replicates. Different letters denote statistical significance ($p \leq 0.05$).

After water control treatment was conducted for 10 d on 3-week-old plants with consistent growth, WT plants showed severe wilting; the leaves turned yellow, while overexpression plants grew better than WT plants (**Figure 6**A). These results indicated that *CaHSP18.1a* increases drought tolerance of transgenic *Arabidopsis*. In addition, the MDA content and REL were increased in both the WT and OE lines, whereas the MDA content of transgenic seedlings was obviously lower than that of WT plants (**Figure 6**B,C). Thus, the degree of damage in OE plants was lower than that in WT plants. The SOD, CAT, and peroxidase (POD) activity showed an upward trend in both the WT and OE lines, but the activity level of SOD, CAT, and POD in transgenic seedlings was notably higher than that in WT plants (**Figure 6**D,F,G). While ascorbic acid peroxidase (APX) activity increased, there was, however, no visible difference between *CaHSP18.1a*-OE and WT lines (**Figure 6**E). The expression levels of the 18 stress-related genes were induced to varying degrees by drought stress. However, the expression of *AtHsfA2*, *AtHSPC30*, and *AtAPX1* exhibited almost no change in WT seedlings, and all of them were strongly increased in *CaHSP18.1a*-OE lines after drought stress. In addition, the expression levels of other genes were higher in transgenic lines compared to the WT plants after drought stress (**Figure 7**). Thus, the 18 stress-related genes examined may be involved at different levels in the response of *CaHSP18.1a*-OE lines to drought stress.

3.4.3. Overexpression of CaHSP18.1a Enhances Plant Tolerance to Salt Stress

To study the role of *CaHSP18.1a* in salt stress, transgenic *Arabidopsis thaliana* was subjected to salt stress. First, the germination rate of transgenic seeds under salt stress was observed. OE2, OE1, and WT seeds exhibited normal germination on MS plates without NaCl; the germination rate and seedling growth of transgenic lines were almost unaffected by 100 mM NaCl MS plates. However, the seed germination rate differed when the NaCl concentration was increased to 150 mM, and the germination rate of WT seeds was slightly lower than that of transgenic seeds. The salt tolerance of *CaHSP18.1a* transgenic *Arabidopsis* seeds increased. After the germinated seedlings were moved to MS plates with NaCl concentrations of 100 ,0, and 150 mM, compared with the untreated seedlings, the growth of OE2, OE3, OE1, and WT seedlings after 6 d under 100 and 150 mM NaCl treatments was worse ; root elongation was significantly inhibited, and the root length of transgenic lines seedlings was greater than that of WT plants (**Figure 8**A).

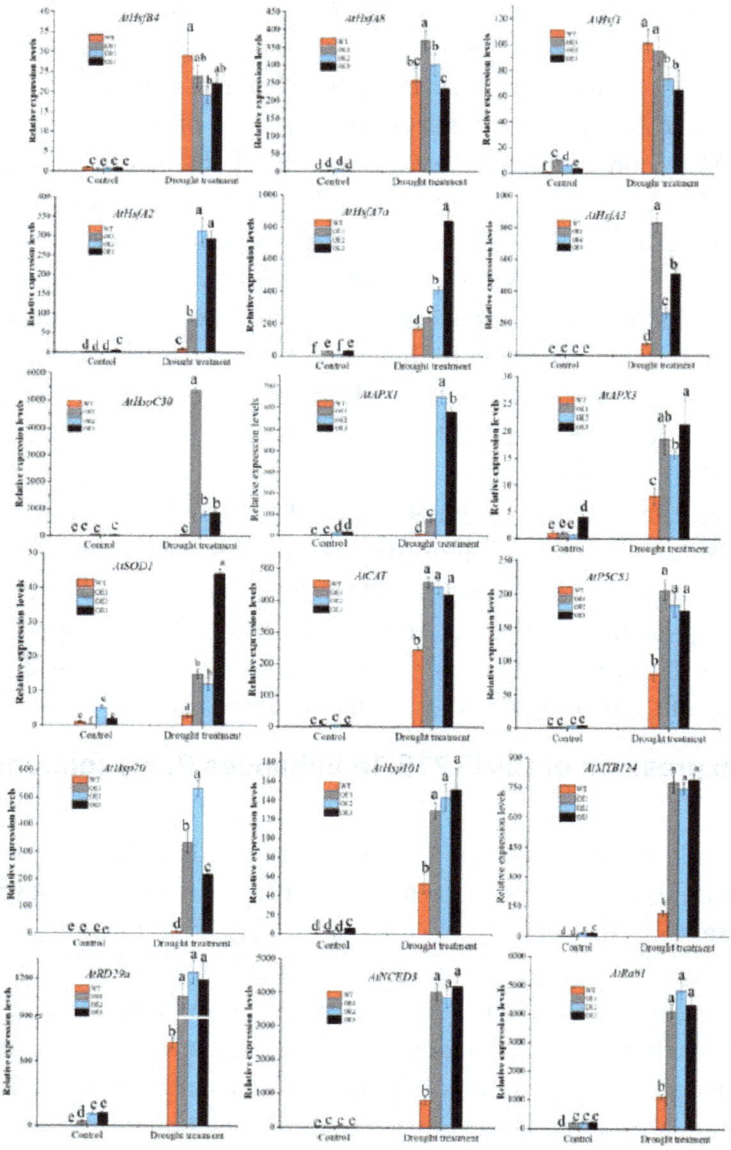

Figure 7. Expression pattern of stress-response genes in wild-type (WT), OE1, OE2, and OE3 lines before and after 10 days of drought treatment. Data are means with standard deviations of three biological replicates. Different letters denote statistical significance ($p \le 0.05$).

To explore the effect of *CaHSP18.1a* on the salt tolerance of *Arabidopsis thaliana*, 3-week-old WT and transgenic plants were irrigated with 200 mM NaCl solution for 7 days. As shown in **Figure 8**B, WT plants showed dehydration and wilting with weak growth. However, the transgenic *Arabidopsis thaliana* showed no significant difference from the control plants except for a slight yellowing phenotype in their leaves.

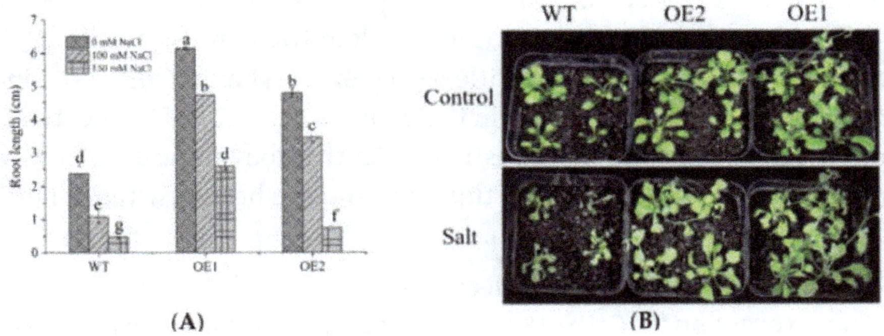

Figure 8. Overexpression of *CaHSP18.1a* enhanced tolerance to salt stress. (A) Root length of wild-type (WT) and *CaHSP18.1a*-OE *Arabidopsis* lines grown for 10 d on Murashige and Skoog (MS) medium containing 100 ,0, and 150 mM NaCl. (B) Seedling growth for WT and transgenic plants exposed to NaCl. Three-week-old seedlings were watered with 200 mM NaCl once every 1–2 days. The images were taken after 7 days.

4. DISCUSSION

Plants are inevitably subjected to various extreme environmental conditions, such as heat, drought, oxidation and salt damage [**36**]. Under such adverse conditions, sHsP20s make a valuable protective contribution [**19**]. Plant HSPs are linked to heat tolerance and have been confirmed in many species [**31,49**]. Guo et al. [**30**] identified sHsP20s in pepper, and showed that *CaHSP18.1a* was induced in different tissues of pepper plants under heat stress, but the function of *CaHSP18.1a* under heat, salt, and drought stress has not been further studied. In this study, we identified that *CaHSP18.1a* is positively involved in plant tolerance to heat and salt, drought stress.

CaHSP18.1a was responsive to heat stress in both R9 and B6 plants and strongly induced [**30**]. In accordance with the results, we also found that the expression level of *CaHSP18.1a* was strongly induced in both the R9 and B6 lines after heat stress treatment (**Figure** 1B–D). However, under heat stress, the expression of *CaHSP18.1a* in R9 plants was higher than that in B6 plants (**Figure** 1B–D). This may be because R9 is a thermo-tolerant cultivar, it has better thermo-tolerance and adaptability than B6 under heat stress. The heat tolerance of plants is related to the dynamic expression patterns of heat stress-related genes [**50**]. Under heat stress, other HSP20s or HSPs in R9 are also strongly and rapidly induced in the early stage of heat stress (0.5–1 h). It had also been reported that the expression level of *CaHSP25.8* and *CaHSP30.1* in R9 was higher than B6, but with the extension of heat stress treatment time, the expression level of these two genes in B6 were higher than R**30**] 9]. However, these results also showed that the expression of *CaHSP18.1a* was lower at V in R9 than in B6. The expression of pepper HSP20s is regulated by many transcription factors, such as HSFs [**51**]. Under heat stress, it is because HSFs that regulate the expression of *CaHSP18.1a* in pepper variety R9 and B6 are different, or the expression of the same HSFs that regulate the expression of *CaHSP18.1a* is different in R9 and B6, causing the differential expression of *CaHSP18.1a* in B6 and R**51**] 9,**52**]. Therefore, the difference in the expression of *CaHSP18.1a* between heat-resistant and heat-sensitive varieties is due to the above reasons. However, the relationship between the function of this gene and the heat-resistance mechanism of pepper still needs further research.

In addition, *CaHSP18.1a* was induced under salt and drought stress (**Figure 1**E–H). The expression of *CaHSP18.1a* in R9 leaves and roots was highest under the 150 mM, 100 mM NaCl treatments, respectively (**Figure 1**E,F). The expression of *CaHSP18.1a* in R9 leaves was the highest after the 150 mM mannitol treatment; the highest expression of *CaHSP18.1a* was observed in R9 roots subjected to the 50 mM mannitol treatment (**Figure 1**G,H). However, the expression of the *CaHSP18.1a* decreased at higher NaCl and mannitol concentrations. The response pattern of *CaHSP18.1a* that rapidly and sharply responded to salt and drought stress in a short time, and then had slight variations, was similar to quite a few HSP20s such as *TaHSP***53**] *23.9*] and *ClHSP22.8* [**54**]. Thus, *CaHSP18.1a* may play a role in pepper which rapidly adapts to drought and salt stress.

VIGS technology is an important method used to study gene function under adverse environments [**55**]. In the R9 line, silencing of *CaHSP16.4* reduces heat tolerance and drought resistance of pepper plants [**31**]; *CaHSP22.0*-silenced peppers showed more sensitivity to salt and heat stress, which was mainly reflected in decreased antioxidant enzyme activity, increased leaf conductivity, and increased superoxide anion and MDA contents [**34**]. MDA content and REL are products of cell membrane lipid peroxidation, which damages the integrity of plasma membranes under salt or heat stress [**35**] and may sensitize plants to subsequent stress [**56**]. MDA content, total chlorophyll content, and REL are widely used to determine the degree to which plants have been damaged by abiotic stress [**4**]. It has also been reported that proline content, MDA content, and POD and SOD activity of pepper were significantly related to the variation in heat tolerance and temperature stress time, which can be used as an index for heat resistance identification [**57**].

In this study, after treatments with high temperature, salt, and drought stress, the content of MDA and REL in peppers that had been silenced for *CaHSP18.1a* was higher than that of the controls, indicating that the damage to cell membranes increased in *CaHSP18.1a*-silenced plants (**Figure 3**A–C). It was also found that the *CaHSP18.1a*-silenced plants had lower total chlorophyll content when exposed to heat stress (**Figure 3**B and **Supplementary Figure S1C**). These results demonstrated that silencing of *CaHSP18.1a* reduced pepper stress tolerance (**Figure 3**). In contrast, overexpression of *CaHSP18.1a* in *Arabidopsis thaliana* transgenic lines was associated with minimal injury symptoms, increased REL, and decreased MDA content compared with WT plants (**Figure 4** and **Figure 6**). These results showed that *CaHSP18.1a* increases plant tolerance to heat, salt, and drought stresses.

HSP20s are widely distributed in plants, and their location may be related to their function, as exemplified by *AtHsP21* being localized to chloroplasts [**23**]. *CaHSP18.1a* was predicted to have cytoplasm localization [**30**]. Subcellular localization of *CaHSP18.1a* is shown in **Figure 2**, which confirmed that it is localized to the cell membrane.

Studies have shown that HSP20 is a molecular chaperone that can also participate in antioxidant mechanisms of plants [**6,58**]. HSP20scan cooperate with the plant's antioxidant scavenging system to protect plants from secondary damage [**59,60**]. For example, overexpression of *AtHSP17.6* can increase CAT enzyme activity and further regulate abiotic stress responses [**61**]. Furthermore, the over-expression of *ZmHSP16.9* in tobacco can increase the activities of POD, CAT and SOD, and enhance oxidative stress tolerance [**62**]. In this study, SOD, CAT, POD, and APX enzyme activities in overexpression of *CaHSP18.1a* plants were significantly enhanced under heat and drought stress. This is similar to the results obtained with *Arabidopsis* transformed with *CaHSP25.9*; that is, by increasing the activities of ROS-scavenging related antioxidant enzymes, the heat, salt, and drought tolerance of plants can be increased [**31,63**]. It has been reported that plants have built defense mechanisms that scavenge excess reactive oxygen species (ROS) throughout their long evolutionary histories [**64,65,66**], such as ROS-scavenging non-enzymatic antioxidants (e.g., ascorbic acid (AsA), glutathione, and proline) [**67**] and antioxidant enzymes (e.g., peroxidase (POD), catalase (CAT), superoxide dismutase (SOD), and glutathione peroxidase (GPX)) that prevent secondary oxidative stress caused by abiotic stress [**68,69,70**]. Moreover, the expression levels of *AtSOD1, AtAPX1, AtAPX3*, and *AtCAT1* were also increased by heat and drought stress, and were higher in the *CaHSP18.1a* transgenic *Arabidopsis* than in the WT. This indicated that *CaHSP18.1a* may improve stress resistance through the ROS-scavenging system, but the specific mechanism needs further study.

Many stress-related genes are involved in plant responses to heat, salt, and drought stresses. It has been reported that *AtHsfA2* is a heat shock transcription factor that enables prolonged acquired thermo-tolerance, and it can enhance tolerance to salt and osmotic stresses [**71,72,73**]. s HSPs are downstream target genes of HsfA2 [**74**]. Burke [**75,76**] have also shown that *AtHSA32* and *AtHSP101* expression can be induced by high temperature and participate in the acquired thermo-tolerance of plants. *AtMYB44* can be induced by salt, drought, and other stresses to participate in the abscisic acid (ABA) signaling pathway; Refs. [**77,78**] found that the ABA signaling response gene *AtDREB2A* can be induced by low temperature stress. The drought responsive gene *AtRD29A* was up-regulated under heat, salt, and drought stresses [**79**], while the molecular chaperone HSP70 participates in drought and heat stress responses [**28**]. *NCED3* is related to biological metabolism and also participates in defense responses to drought stress [**80**]. Notably, HSP20s can regulate many of these stress-related genes [**31,36**]. For example, *CaHSP16.4* and *OsMSR4-* can increase the expression of these genes in transgenic seedlings, thereby enhancing stress resistance [**31,81**]. In this study, we assessed the expression levels of 13 stress-related genes in WT and transgenic plants. *CaHSP18.1a* enhanced heat tolerance in transgenic *Arabidopsis*, which may be closely related to its regulation of the expression of many heat-stress-related genes in *Arabidopsis*. Overexpressed genotypes compared with wild type under normal conditions also showed higher values of stress-related genes expression, higher expression of *AtP5CS*, higher expression of *AtNCED*, higher expression of *AtMYB*, and higher expression of *AtRD29, AtHsfA2, AtRab1* and *AtHSP30*. These results showed that *CaHSP18.1a* may play an important role in regulation of these genes. Under heat stress, the expressions levels of *AtHsfB4, AtHSFA8, AtHSFA2, AtHSFA7a, AtHSPC30, AtHSFA3, AtHSP70,* and *AtHSP101* in transgenic plants were significantly higher than WT plants (**Figure 5**). Specifically, the levels of AtHSPC30 and AtHSP70 transcripts were almost three times greater in transgenic seedlings compared to WT plants. Additionally, the transcript level of AtHsfA2 was similarly increased in transgenic plants. Under drought stress, the CaHSP18.1a transgenic Arabidopsis plants exhibited greater expression levels of AtHSP70, AtHSP101, AtDREB2A, AtMYB124, AtNCED3, AtRD29A, and AtRAB1 compared to the WT plants (Figure 7). Feng and Huang [31,32] have observed similar findings. CaHSP18.1a is likely to react to heat and drought stress by utilizing its intricate regulatory network.

5. CONCISE SUMMARY

This work initially examined the expression of CaHSP18.1a in R9 and B6 pepper lines and established that CaHSP18.1a was expressed in response to abiotic stressors such as high temperature, drought, and high salinity. Suppression of CaHSP18.1a reduced the ability of pepper plants to withstand high temperatures, lack of water, and high salt levels by affecting various molecular and physiological processes. Overexpression investigations of CaHSP18.1a in transgenic Arabidopsis provided more evidence that CaHSP18.1a has a beneficial role in the plant's ability to respond to heat, drought, and salt stresses. The levels of expression of other genes associated to stress were also assessed, and it was shown that several of them were considerably influenced by the overexpression of CaHSP18.1a. Our findings provide additional evidence that the CaHSP18.1a protein is specifically located in the cellular membrane. Overall, these findings indicate that CaHSP18.1a most likely functions as a promoter of the response to non-living environmental factors in pepper plants.

AT A GLANCE

Pepper is a thermophilic crop, which means it thrives at high temperatures. It is a shallow-rooted plant that is very susceptible to abiotic stresses including heat, salt, and drought. Pepper crops experience significant impediments to their growth and development due to unfavorable stressors, leading to reductions in both the quantity and quality of the harvest. Small heat shock proteins (s HSPs) have a vital function in safeguarding plant cells from different types of stress. A prior investigation conducted in our laboratory demonstrated that the expression of CaHSP18.1a was significantly upregulated in response to heat stress. However, the precise role and method by which CaHSP18.1a responds to abiotic stimuli remain unclear. This study initially examined the expression of CaHSP18.1a in the thermo-sensitive B6 line and thermo-tolerant R9 line. It was found that the transcription of CaHSP18.1a was significantly increased in response to heat stress, salt, and drought stress in both R9 and B6. Furthermore, the response was more pronounced and occurred earlier in the R9 line. CaHSP18.1a suppression in the R9 line resulted in reduced tolerance to high temperatures, water scarcity, and salinity. CaHSP18.1a suppression led to notable elevations in relative electrolyte leakage (REL) and malonaldehyde (MDA) levels, accompanied by a decrease in total chlorophyll content after heat, salt, and drought stressors. Transgenic Arabidopsis plants were used to conduct overexpression analysis of CaHSP18.1a, which provided further confirmation that CaHSP18.1a plays a beneficial role in enhancing resistance to heat, drought, and salt stressors. The transgenic Arabidopsis exhibited elevated levels of chlorophyll content and increased activities of superoxide dismutase, catalase, and ascorbate peroxidase compared to the wild type (WT). Nevertheless, the transgenic Arabidopsis exhibited reduced relative conductivity and MDA concentration in comparison to the wild type (WT). Additionally, we demonstrated that the CaHSP18.1a protein is specifically located in the cellular membrane. The data suggest that CaHSP18.1a may function as a stimulator of responses to abiotic stressors.

REFERENCES

1. Ahuja, I.; de Vos, R.C.H.; Bones, A.M.; Hall, R.D. Plant molecular stress responses face climate change. *Trends Plant Sci.* 2010, *15*, 664–674.
2. Wang, W.; Vinocur, B.; Altman, A. Plant responses to drought, salinity and extreme temperatures: Towards genetic engineering for stress tolerance. *Planta* 2003, *218*, 1–14.
3. McLoughlin, F.; Basha, E.; Fowler, M.E.; Kim, M.; Bordowitz, J.; Katiyar-Agarwal, S.; Vierling, E. Class I and II Small Heat Shock Proteins Together with HSP101 Protect Protein Translation Factors during Heat Stress. *Plant Physiol.* 2016, *172*, 1221–1236.

4. Muthusamy, S.K.; Dalal, M.; Chinnusamy, V.; Bansal, K.C. Genome-wide identification and analysis of biotic and abiotic stress regulation of small heat shock protein (HSP20) family genes in bread wheat. *J. Plant Physiol.* 2017, *211*, 100–113.

5. Xiang, J.; Chen, X.; Hui, W.; Xiang, Y.; Yan, M.; Wang, J. Overexpressing heat-shock protein OsHSP50.2 improves drought tolerance in rice. *Plant Cell Rep.* 2018, *37*, 1585–1595.

6. Wang, W.; Vinocur, B.; Shoseyov, O.; Altman, A. Role of plant heat-shock proteins and molecular chaperones in the abiotic stress response. *Trends Plant Sci.* 2004, *9*, 244–252.

7. Zhang, L.; Liu, J. Research Progress of Heat Shock Protein. *Guangdong Seric.* 2006, *40*, 39–42.

8. Beck, E.H.; Fettig, S.; Knake, C.; Hartig, K.; Bhattarai, T. Specific and unspecific responses of plants to cold and drought stress. *J. Biosci.* 2007, *32*, 501–510.

9. Agrawal, L.; Narula, K.; Basu, S.; Shekhar, S.; Ghosh, S.; Datta, A.; Chakraborty, S. Comparative Proteomics Reveals a Role for Seed Storage Protein AmA1 in Cellular Growth, Development, and Nutrient Accumulation. *J. Proteome Res.* 2013, *12*, 4904–4930.

10. Shekhar, S.; Mishra, D.; Gayali, S.; Buragohain, A.K.; Chakraborty, S.; Chakraborty, N. Comparison of proteomic and metabolomic profiles of two contrasting ecotypes of sweetpotato (*Ipomoea batata* L.). *J. Proteom.* 2016, *143*, 306–317.

11. Waters, E.R.; Lee, G.J.; Vierling, E. Evolution, structure and function of the small heat shock proteins in plants. *J. Exp. Bot.* 1996, *47*, 325–338.

12. Boston, R.S.; Viitanen, P.V.; Vierling, E. Molecular chaperones and protein folding in plants. *Plant Mol. Biol.* 1996, *32*, 191–222.

13. Hu, W.; Hu, G.; Han, B. Genome-wide survey and expression profiling of heat shock proteins and heat shock factors revealed overlapped and stress specific response under abiotic stresses in rice. *Plant Sci.* 2009, *176*, 583–590.

14. Li, J.; Zhang, J.; Jia, H.; Li, Y.; Xu, X.; Wang, L.; Lu, M. The *Populus trichocarpa* PtHSP17.8 involved in heat and salt stress tolerances. *Plant Cell Rep.* 2016, *35*, 1587–1599.

15. Eyles, S.J.; Gierasch, L.M. Nature's molecular sponges: Small heat shock proteins grow into their chaperone roles. *Proc. Natl. Acad. Sci. USA* 2010, *107*, 2727–2728.

16. Martin, H. A first line of stress defense: Small heat shock proteins and their function in protein homeostasis. *J. Mol. Biol.* 2015, *427*, 1537–1548.

17. Tyedmers, J.; Mogk, A.; Bukau, B. Cellular strategies for controlling protein aggregation. *Nat. Rev. Mol. Cell Biol.* 2010, *11*, 777–788.

18. Waters, E.R. The evolution, function, structure, and expression of the plant sHSPs. *J. Exp. Bot.* 2013, *64*, 391–403.

19. Lambert, W.; Koeck, P.J.B.; Ahrman, E.; Purhonen, P.; Cheng, K.; Elmlund, D.; Hebert, H.; Emanuelsson, C. Subunit arrangement in the dodecameric chloroplast small heat shock protein HSP21. *Protein Sci.* 2011, *20*, 291–301.

20. Korotaeva, N.E.; Antipina, A.I.; Grabelnykh, O.I.; Varakina, N.N.; Borovskii, G.B.; Voinikov, V.K. Mitochondrial Low-Molecular-Weight Heat-Shock Proteins and the Tolerance of Cereal Mitochondria to Hyperthermia. *Russ. J. Plant Physiol.* 2001, *48*, 798–803.

21. Nieto-Sotelo, J.; Martínez, L.M.; Ponce, G.; Cassab, G.I.; Alagón, A.; Meeley, R.B.; Ribaut, J.-M.; Yang, R. Maize HSP101 Plays Important Roles in Both Induced and Basal Thermotolerance and Primary Root Growth. *Plant Cell Online* 2002, *14*, 1621–1633.

22. Sun, W.; Bernard, C.; Cotte, B.V.D.; Montagu, M.V.; Verbruggen, N. At-HSP17.6A, encoding a small heat-shock protein in Arabidopsis, can enhance osmotolerance upon overexpression. *Plant J.* 2010, *27*, 407–415.

23. Sedaghatmehr, M.; MuellerRoeber, B.; Balazadeh, S. The plastid metalloprotease FtsH6 and small heat shock protein HSP21 jointly regulate thermomemory in Arabidopsis. *Nat. Commun.* 2016, 7, 12439.

24. Kim, D.H.; Xu, Z.; Hwang, I. AtHSP17.8 overexpression in transgenic lettuce gives rise to dehydration and salt stress resistance phenotypes through modulation of ABA-mediated signaling. *Plant Cell Rep.* 2013, *32*, 1953–1963.

25. Nicky, D.; Xu, J.; Peters, J.L.; Sunghun, P.; Ivo, R. Multi-Level Interactions Between Heat Shock Factors, Heat Shock Proteins, and the Redox System Regulate Acclimation to Heat. *Front. Plant Sci.* 2015, *6*, 999.

26. Guo, M.; Liu, J.H.; Ma, X.; Zhai, Y.F.; Gong, Z.H.; Lu, M.H. Genome-wide analysis of the HSP70 family genes in pepper (*Capsicum annuum* L.) and functional identification of CaHSP2-70 involvement in heat stress. *Plant Sci.* 2016, *252*, 246–256.

27. Zou, X. Cultivation Seasons and Cropping patterns of Capsicum in China. *J. China Capsicum* 2002, 3, 1672–4542.

28. Guo, M.; Zhai, Y.-F.; Lu, J.-P.; Chai, L.; Chai, W.-G.; Gong, Z.-H.; Lu, M.-H. Characterization of CaHSP70-1, a Pepper Heat-Shock Protein Gene in Response to Heat Stress and Some Regulation Exogenous Substances in *Capsicum annuum* L. *Int. J. Mol. Sci.* 2014, *15*, 19741–19759.

29. Pagamas, P.; Nawata, E. Sensitive stages of fruit and seed development of chili pepper (*Capsicum annuum* L. var. Shishito) exposed to high-temperature stress. *Sci. Hortic.* 2008, *117*, 21–25.

30. Guo, M.; Liu, J.H.; Lu, J.P.; Zhai, Y.F.; Wang, H.; Gong, Z.H.; Wang, S.B.; Lu, M.H. Genome-wide analysis of the CaHSP20 gene family in pepper: Comprehensive sequence and expression profile analysis under heat stress. *Front. Plant Sci.* 2015, *6*, 806.

31. Huang, L.; Cheng, G.; Khan, A.; Wei, A.; Gong, Z. CaHSP16.4, a small heat shock protein gene in pepper, is involved in heat and drought tolerance. *Protoplasma* 2019, *256*, 39–51.

32. Feng, X.; Zhang, H.; Ali, M.; Gai, W.; Gong, Z. A small heat shock protein CaHSP25.9 positively regulates heat, salt, and drought stress tolerance in pepper (*Capsicum annuum* L.). *Plant Physiol. Biochem.* 2019, *142*, 151–162.

33. He, L.; Gao, N.; Shan, Z.; Guo, S. Pepper Chloroplast Small Heat Shock Proteins CaHSP26 Enhanced the Ability of Anti-high Temperature Stress in Arabidopsis. *North. Hortic.* 2011, *6*, 140–143.

34. Sun, J.; Cheng, G.; Huang, L.; Liu, S.; Ali, M.; Khan, A.; Yu, Q.; Yang, S.; Luo, D.; Gong, Z. Modified expression of a heat shock protein gene, CaHSP22.0, results in high sensitivity to heat and salt stress in pepper (*Capsicum annuum* L.). *Sci. Hortic.* 2019, *249*, 364–373.

35. Yu, J.; Cheng, Y.; Feng, K.; Ruan, M.; Ye, Q.; Wang, R.; Li, Z.; Zhou, G.; Yao, Z.; Yang, Y.; et al. Genome-Wide Identification and Expression Profiling of Tomato HSP20 Gene Family in Response to Biotic and Abiotic Stresses. *Front. Plant Sci.* 2016, 7, 1215.

36. Wang, M.; Zou, Z.; Li, Q.; Sun, K.; Chen, X.; Li, X. The CsHSP17.2 molecular chaperone is essential for thermotolerance in Camellia sinensis. *Sci. Rep.* 2017, 7, 1237.

37. Zhai, Y.; Wang, H.; Liang, M.; Lu, M. Both silencing- and over-expression of pepper CaATG8c gene compromise plant tolerance to heat and salt stress. *Environ. Exp. Bot.* 2017, *141*, 10–18.

38. Wang, J.E.; Liu, K.K.; Li, D.W.; Zhang, Y.L.; Zhao, Q.; He, Y.M.; Gong, Z.H. A novel peroxidase CaPOD gene of pepper is involved in defense responses to Phytophthora capsicum infection as well as abiotic stress toleranc. *Int. J. Mol. Sci.* 2013, *14*, 3158–3177.

39. Wan, H.; Yuan, W.; Ruan, M.; Ye, Q.; Wang, R.; Li, Z.; Zhou, G.; Yao, Z.; Zhao, J.; Liu, S.; et al. Identification of reference genes for reverse transcription quantitative real-time PCR normalization in pepper (*Capsicum annuum* L.). *Biochem. Biophys. Res. Commun.* 2011, *416*, 24–30.

40. Livaka, K.J.; Schmittgen, T.D. Analysis of relative gene expression data using real-time quantitative PCR and the 2-ΔΔCT Method. *J. Methods* 2001, *25*, 402–408.

41. Yu, C.; Zhan, Y.; Feng, X.; Huang, Z.A.; Sun, C. Identification and Expression Profiling of the Auxin Response Factors in *Capsicum annuum* L. under Abiotic Stress and Hormone Treatments. *Int. J. Mol. Sci.* 2017, *18*, 2719.

42. Clough, S.J.; Bent, A. Floral dip: A simplified method for Agrobacterium -mediated transformation of Arabidopsis thaliana. *Plant J.* 1998, *16*, 735–743.

43. Yin, Y.X.; Guo, W.L.; Zhang, Y.L.; Ji, J.J.; Xiao, H.J.; Yan, F.; Zhao, Y.Y.; Zhu, W.C.; Chen, R.G.; Chai, W.G. Cloning and characterisation of a pepper aquaporin, CaAQP, which reduces chilling stress in transgenic tobacco plants. *Plant Cell Tissue Organ Cult.* 2014, *118*, 431–444.

44. Arkus, K.A.J.; Cahoon, E.B.; Jez, J.M. Mechanistic analysis of wheat chlorophyllase. *Arch. Biochem. Biophys.* 2005, *438*, 146–155.

45. Campos, P.S.; nia Quartin, V.; chicho Ramalho, J.; Nunes, M.A. Electrolyte leakage and lipid degradation account for cold sensitivity in leaves ofCoffea sp. plants. *J. Plant Physiol.* 2003, *160*, 283–292.

46. Guo, W.L.; Chen, R.G.; Gong, Z.H.; Yin, Y.X.; Ahmed, S.S.; He, Y.M. Exogenous abscisic acid increases antioxidant enzymes and related gene expression in pepper (*Capsicum annuum*) leaves subjected to chilling stress. *Genet. Mol. Res. GMR* 2012, *11*, 4063–4080.

47. Nakano, Y.; Asada, K. Hydrogen Peroxide is Scavenged by Ascorbate-specific Peroxidase in Spinach Chloroplasts. *Plant Cell Physiol.* 1981, *22*, 867–880.

48. Aebi, H. [13] Catalase in vitro. *Methods Enzymol.* 1984, *105*, 121–126.

49. Jiang, C.; Xu, J.; Zhang, H.A.; Zhang, X.; Shi, J.; Li, M.I.; Ming, F. A cytosolic class I small heat shock protein, RcHSP17.8, of Rosa chinensis confers resistance to a variety of stresses to Escherichia coli, yeast and Arabidopsis thaliana. *Plant Cell Environ.* 2009, *32*, 1046–1059.

50. Lopes-Caitar, V.S.; de Carvalho, M.C.; Darben, L.M.; Kuwahara, M.K.; Nepomuceno, A.L.; Dias, W.P.; Abdelnoor, R.V.; Marcelino-Guimarães, F.C. Genome-wide analysis of the HSP20 gene family in soybean: Comprehensive sequence, genomic organization and expression profile analysis under abiotic and biotic stresses. *BMC Genom.* 2013, *14*, 1471–2164.

51. Wang, X.; Huang, W.; Yang, Z.; Liu, J.; Huang, B. Transcriptional regulation of heat shock proteins and ascorbate peroxidase by CtHsfA2b from African bermudagrass conferring heat tolerance in Arabidopsis. *Sci. Rep.* 2016, *6*, 28021.

52. Schramm, F.; Ganguli, A.; Kiehlmann, E.; Englich, G.; Walch, D.; von Koskull-Döring, P. The Heat Stress Transcription Factor HsfA2 Serves as a Regulatory Amplifier of a Subset of Genes in the Heat Stress Response in Arabidopsis. *Plant Mol. Biol.* 2006, *60*, 759–772.

53. Wang, J.; Gao, X.; Dong, J.; Tian, X.; Wang, J.; Palta, J.A.; Xu, S.; Fang, Y.; Wang, Z. Over-Expression of the Heat-Responsive Wheat Gene TaHSP23.9 in Transgenic Arabidopsis Conferred Tolerance to Heat and Salt Stress. *Front Plant Sci.* 2020, *11*, 243.

54. He, Y.; Yao, Y.; Li, L.; Li, Y.; Gao, J.; Fan, M. A heat-shock 20 protein isolated from watermelon (ClHSP22.8) negatively regulates the response of Arabidopsis to salt stress via multiple signaling pathways. *PeerJ.* 2021, *9*, e10524.

55. Yao, B.; Yue, S.; Sun, L.; Ma, Z.; Su, J.; Liu, X. Development of Virus Induced Gene Silencing and Application of VIGS in Plant Abiotic Stresses. *Chin. Agric. Sci. Bull.* 2016, *32*, 131–136.

56. Ammar, J.B.; Lanoiselle, J.-L.; Lebovka, N.I.; Hecke, E.V.; Vorobiev, E.J.J. Impact of a Pulsed Electric Field on Damage of Plant Tissues: Effects of Cell Size and Tissue Electrical Conductivity. *J. Food Sci.* 2011, *76*, E90–E97.

57. Ma, B.; Lu, M.; Gong, Z.H. Responses of growth and physiology of pepper (*Capsicum annuum* L.) seedlings to high temperature stress. *J. Northwest A F Univ.* 2013, *41*, 112–118.

58. Shang, H.; Cao, S.; Yang, Z.; Cai, Y.; Zheng, Y. Effect of Exogenous γ-Aminobutyric Acid Treatment on Proline Accumulation and Chilling Injury in Peach Fruit after Long-Term Cold Storage. *J. Agric. Food Chem.* 2011, *59*, 1264–1268.

59. Rossel, J.B.; Wilson, I.W.; Pogson, B.J. Global Changes in Gene Expression in Response to High Light in Arabidopsis. *Plant Physiol.* 2002, *130*, 1109–1120.

60. Aghdam, M.S.; Sevillano, L.; Flores, F.B.; Bodbodak, S. Heat shock proteins as biochemical markers for postharvest chilling stress in fruits and vegetables. *Sci. Hortic.* 2013, *160*, 54–64.

61. Li, G.; Li, J.; Hao, R.; Guo, Y. Activation of catalase activity by a peroxisome-localized small heat shock protein HSP17.6CII. *J. Genet. Genom.* 2017, *44*, 395–404.

62. Sun, L.; Liu, Y.; Kong, X.; Zhang, D.; Pan, J.; Zhou, Y.; Wang, L.; Li, D.; Yang, X. ZmHSP16.9, a cytosolic class I small heat shock protein in maize (*Zea mays*), confers heat tolerance in transgenic tobacco. *Plant Cell Rep.* 2012, *31*, 1473–1484.

63. Meifang, L.; Xiujian, W.; Shangjing, G. Obtain of Transgenic Sweet Pepper Plants with Gene from Endoplasmic Reticulum Small Heat Shock Protein CaHSP22.5. *Mol. Plant Breed.* 2018, *7*, 2212–2218.

64. Zhao, J.; Liang, J.; Wu, X.; Liu, X.; Li, H.; Zhu, S. Expression Profiling of Rice ROS Scavenging System Related Genes under Salt or Low Temperature Stress. *Acta Bot. Boreali-Occident. Sin.* 2015, *35*, 872–883.

65. Kim, J.M.; Woo, D.H.; Kim, S.H.; Lee, S.Y.; Park, H.Y.; Seok, H.Y.; Chung, W.S.; Moon, Y.H. Arabidopsis MKKK20 is involved in osmotic stress response via regulation of MPK6 activity. *Plant Cell Rep.* 2012, *31*, 217–224.

66. Tuteja, N.; Banu, M.S.A.; Huda, K.M.K.; Gill, S.S.; Jain, P.; Xuan, H.P.; Tuteja, R. Pea p68, a DEAD-Box Helicase, Provides Salinity Stress Tolerance in Transgenic Tobacco by Reducing Oxidative Stress and Improving Photosynthesis Machinery. *PLoS ONE* 2014, *9*, e98287.

67. Ron, M.; Sandy, V.; Martin, G.; Frank, V.B. Reactive oxygen gene network of plants. *Trends Plant Sci.* 2004, *9*, 490–498.

68. Cao, S.; Yang, Z.; Cai, Y.; Zheng, Y. Fatty acid composition and antioxidant system in relation to susceptibility of loquat fruit to chilling injury. *Food Chem.* 2011, *127*, 1777–1783.

69. Foyer, C.H.; Noctor, G. Redox homeostasis and antioxidant signaling: A metabolic interface between stress perception and physiological responses. *Plant Cell Online* 2005, *17*, 1866–1875.

70. Moller, I.M. Plant Mitochondria and Oxidative Stress: Electron Transport, Nadph Turnover, and Metabolism of Reactive Oxygen Species. *Annu. Rev. Plant Physiol. Plant Mol. Biol.* 2001, *52*, 561–591.

71. Charng, Y.; Liu, H.; Liu, N.; Chi, W.; Wang, C.; Chang, S.; Wang, T. A Heat-Inducible Transcription Factor, HsfA2, Is Required for Extension of Acquired Thermotolerance in Arabidopsis. *Plant Physiol.* 2007, *143*, 251–262.

72. Heerklotz, D.; Döring, P.; Bonzelius, F.; Winkelhaus, S.; Nover, L. The balance of nuclear import and export determines the intracellular distribution and function of tomato heat stress transcription factor HsfA2. *Mol. Cell. Biol.* 2001, *21*, 1759–1768.

73. Ogawa, D.; Kazuo, Y.; Nishiuchi, T. High-level overexpression of the Arabidopsis HsfA2 gene confers not only increased themotolerance but also salt/osmotic stress tolerance and enhanced callus growth. *J. Exp. Bot.* 2007, *58*, 3373–3383.

74. Pv, K.D.; Scharf, K.D.; Nover, L. The diversity of plant heat stress transcription factors. *Trends Plant Sci.* 2007, *12*, 452–457.

75. Burke, J.J.; Chen, J. Enhancement of reproductive heat tolerance in plants. *PLoS ONE* 2015, *10*, e0122933.

76. Charng, Y.Y.; Liu, H.C.; Liu, N.Y.; Hsu, F.C.; Ko, S.S. Arabidopsis Hsa32, a novel heat shock protein, is essential for acquired thermotolerance during long recovery after acclimation. *Plant Physiol.* 2006, *4*, 1297–1305.

77. Nguyen, N.H.; Nguyen, C.T.T.; Jung, C.; Cheong, J.-J. AtMYB44 suppresses transcription of the late embryogenesis abundant protein gene AtLEA4-5. *Biochem. Biophys. Res. Commun.* 2019, *511*, 931–934.

78. Nakashima, K.; Shinwari, Z.K.; Sakuma, Y.; Seki, M.; Miura, S.; Shinozaki, K.; Yamaguchi-Shinozaki, K. Organization and expression of two Arabidopsis DREB2 genes encoding DRE-binding proteins involved in dehydration- and high-salinity-responsive gene expression. *Plant. Mol. Biol.* 2000, *42*, 657–665.

79. Verbruggen, N.; Hermans, C. Proline accumulation in plants: A review. *Amino Acids* 2008, *35*, 753–759.

80. Rymaszewski, W.; Vile, D.; Bediee, A.; Dauzat, M.; Luchaire, N.; Kamrowska, D.; Granier, C.; Hennig, J. Stress-Related Gene Expression Reflects Morphophysiological Responses to Water Deficit. *Plant. Physiol.* 2017, *174*, 1913–1930.

81. Yin, X.; Huang, L.; Zhang, X.; Wang, M.; Xu, G.; Xia, X. Expression of rice geneOsMSR4confers decreased ABA sensitivity and improved drought tolerance in *Arabidopsis thaliana*. *Plant. Growth Regul.* 2015, *75*, 549–556.

11 Chapter

ANALYSIS OF SIMPLE SEQUENCE REPEATS IN CUCURBITA SPECIES: A COMPREHENSIVE STUDY OF GENOME-WIDE CHARACTERIZATION, COMPARATIVE ANALYSIS, AND GENETIC DIVERSITY

1. INTRODUCTION: AN OVERVIEW

The Cucurbita genus, which has a chromosome count of 2n = 2x = 40, is a part of the Cucurbitaceae family and consists of over 13 species [1]. The majority of Cucurbita species are found in the wild, whereas just three species, namely Cucurbita maxima, Cucurbita moschata, and Cucurbita pepo, have been successfully domesticated and are extensively grown as significant food crops worldwide [2]. Currently, Asia boasts the largest pumpkin cultivation area, with China serving as the primary pumpkin producer. In 2012, the planting area of pumpkin was approximately 3.8 × 104 Hm2 in China, and the total output reached 10 × 7.06 tons (**http://www.fao.org/faostat/zh/#data/QC/visualize**, 2020). Because of its extensive cultivation and domestication history, Cucurbita species exhibit a wider range of fruit shapes, sizes, and colors compared to other Cucurbitaceae species [3]. In addition, Cucurbita species possess robust roots and demonstrate excellent adaptation to many biotic and abiotic stressors, including cold, viruses, and salt. As a result, they are extensively utilized as rootstocks in grafting [4,5]. Despite being a widely grown crop worldwide, there is a dearth of comprehensive genetic study on Cucurbita, and only a limited number of studies have been carried out to enhance the cultivation and breeding of this plant genus.

Simple sequence repeats (SSRs) are commonly employed in genetic mapping, genetic diversity analysis, and genome-wide association studies because of their high prevalence, multiple alleles, co-dominance, and cost-effectiveness [6,7].

The complete sequencing of the genomes of Cucumis sativus, *Cucumis melon*, and Citrullus lanatus, which belong to the Cucurbitaceae family, has been accomplished [8,9,10]. Additionally, genome-wide SSR markers have been identified and created for these crops, leading to significant advancements in their utilization for gene and quantitative trait locus (QTL) mapping, as well as comparative genomics [11,12,13]. Comparative mapping utilizing 199 SSR markers derived from cucumber [14] indicated a tentative syntenic connection between melon (2n = 2x = 24) and cucumber (2n = 2x = 14) chromosomes. Subsequently, Yang et al. (2014) created a more concentrated map of Cucumis hystrix, which included a total of 416 SSR markers. Out of these markers, 151 were obtained from cucumber and 50 were obtained from melon. The chromosome-level syntenic relationships were effectively established across the three Cucumis species through the use of common markers. This was further validated by fluorescence in situ hybridization (FISH) [15]. Cucumber and melon shared a total of 91 syntenic blocks, while cucumber and Cucumis hystrix had 53 syntenic blocks in common. In addition, the use of genome-wide SSR markers derived from melon and watermelon has allowed for a more precise characterization of chromosomal syntenic relationships. The intricate mosaic patterns of chromosome synteny among melon, watermelon, and cucumber have been thoroughly established using cross-species SSR markers [12,15]. Nevertheless, the connections and structural changes in chromosomes between different species of Cucurbita and other crops in the Cucurbitaceae family are now fragmented and incomplete.

Previous research have produced amplified fragment length polymorphism (AFLP), random amplified polymorphic DNA (RAPD), and SSR markers based on the conserved sequences found among species or genera [16,17,18,19,20]. Nevertheless, these limited indicators are inadequate for conducting studies on genetic diversity, genetic mapping, and comparative genomics. Esteras et al. (2012) created the initial genetic map of pumpkin by utilizing 304 single nucleotide polymorphisms (SNPs) and 11 SSR markers. They discovered that the linkage groups of pumpkin were partially homoeologous to the chromosomes of cucumber. The utility of these expressed sequence tag (EST)-SNP markers is now constrained by the restricted quantity of markers, the expensive enzymes required, and the complex operational method [21]. Comprehensive application and comparative analysis are still not possible due to the limited coverage of the entire genome and the absence of polymorphic markers. Due to advancements in high-throughput sequencing technology, there has been a surge in research on Cucurbita, leading to the availability of whole-genome sequences for three significant cucurbit crops. According to the data from these SNPs, the analysis of the arrangement of genes in the entire genome showed that both the *C. maxima* and *C. moschata* genomes experienced a whole-genome duplication event. Additionally, it was observed that pairs of regions in the *C. maxima* (or *C. moschata*) genomes that are similar to each other are found in chromosomes corresponding to the two sub-genomes [22]. Montero et al. discovered that a whole genome duplication (WGD) event had occurred in the majority of the *C. pepo* genome's covered areas [23]. In addition, several transcriptomes of Cucurbita species have been made accessible, and EST-SSRs were derived from these transcriptomes [24,25,26,27,28]. Currently, there is a scarcity of SSR markers developed for Cucurbita species. The comprehensive assembly of the whole-genome sequences of *C. moschata*, *C. maxima*, and *C. pepo* will significantly facilitate the extensive development of SSR markers. This will enable the creation of high-resolution maps, gene mapping, and genome-wide association studies (GWAS). This study involved a comprehensive analysis of SSR motifs in three Cucurbita species at a genome-wide level. We examined the distribution and frequency of various types of repeats, identified SSR markers that can be transferred across different species using in silico PCR analysis, and investigated the chromosome synteny between *C. pepo* and other Cucurbitaceae crops. Furthermore, a total of 66 fundamental SSR markers were discovered within the genomes of Cucurbita. These markers were then employed to assess the genetic variability and population structure of 61 *C. pepo* germplasms. The findings of our work will be valuable for conducting research on the population structure, genetic diversity, molecular-assisted selection, and map-based cloning in Cucurbita species.

2. METHODS

2.1. Plant Materials

All the pumpkin accessions utilized in this investigation were acquired from the National Crop Germplasm Resource Platform of China (namely, the platform dedicated to vegetable germplasm resources) in the year 2018. Four accessions originated from Russia, one from America, and 56 accessions were collected from 17 provinces in China.

2.2. Genome SSR Identification and Development in Cucurbita Genomes

The genome information of watermelon, melon, cucumber, and pumpkin was downloaded from **http://cucurbitgenomics.org/** (2020). To develop a set of higher polymorphic SSR primers for the future study, the criteria used for microsatellite identification in this study was from 2 to 8 bp, and mononucleotides were not considered due to the difficulty in distinguishing bona fide microsatellites from sequencing or assembly error. The microsatellite identification tool (MISA) was used to identify and analyze SSR markers including perfect and compound microsatellites. The specific screening details were as follows: repeats with a minimum length of 18 bp (for di- and tetra-nucleotides), 20 bp (for penta-nucleotides), 24 bp (for hexa-nucleotides), 21 bp (for hepta-nucleotides), and 24 bp (for octa-nucleotides). The oligonucleotide primers for these SSRs were designed according to the flanking genomic sequence using Primer3 software (v.1.1.4). Primers were designed to generate amplicons of 100–300 bp in length with the following minimum, optimum, and maximum values for Primer3 parameters: primer length (bp): 18–20–24; Tm (°C): 50–55–60. Other parameters used the default program values.

2.3. In Silico PCR and Synteny Analysis of Cross-Species SSR Markers

Using the SSR markers from pumpkin (*C. pepo* MU-CU16-) genome as a reference, we comparatively analyzed the genome SSR information of cucumber (Gy14), melon (DH92), watermelon (97103), *C. moschata* cv. Rifu, and *C. maxima* cv. Rimu. This was performed with a custom Perl script that used the NCBI BLASTN program as a search engine with an expected value of 10 and filtering. We allowed up to five nucleotide mismatches at the 5'-end of the primer, no mismatches at the 3'-end, and a minimum of %90 overall match homology. To establish the syntenic relationships of chromosomes between *C. pepo* with *C. sativus*, *C. lanatus*, *C. melo*, *C. maxima*, and *C. moschata*, we discarded these SSR markers with multiply physical locations in the same genome, only retaining the SSR markers in the genomes which had a single in silico PCR product. In addition, these shared SSR markers located on the unanchored scaffolds of the chromosome were further filtered. The SSR marker-based syntenic relationships were finally visualized with visualization blocks in Circos software v.0.55 [**29**].

2.4. Genomic DNA Extraction, PCR Amplification, and Electrophoresis Detection

Genomic DNA of all the materials was extracted using 1 g of young leaf sample with the cetyl trimethyl ammonium bromide (CTAB) method [**30**]. The extracted DNA was dissolved in 1× Tris-EDTA (TE) buffer (Solarbio, Cat: T1121). The concentration and purity were detected by the Nanodrop-2000 nucleic acid analyzer. The extracted DNA was diluted to 30 ng/μL as working solution and kept at 4 °C.

Each PCR reaction contained 1 μL of template DNA, 0.5 μM each of forward and reverse primers, 5 μL mastermix (GenStar, Cat: A012-105), and 3 μL ddH_2O. The amplification was carried out as follows: An initial denaturing step at 95 °C for 5 min, 94 °C for 30 s, followed by 6 cycles of 68–58 °C for 45 s. Each cycle was reduced by 2 °C, each annealing time was 1 min, and 72 °C for 1 min; 30 cycles of 94 °C for 30 s, 50 °C for 30 s, and 72 °C for 1 min. In the last cycle, primer extension was performed at 72 °C for 10 min.

The PCR products were analyzed by 9% polyacrylamide gel electrophoresis, and a 100 bp DNA ladder was used as the reference marker. After electrophoresis, silver staining was performed to display the PCR products, and photos were taken for preservation.

2.5. Calculation of Clustering

The heterozygosity (He), observer gene number (Na), effective alleles (Ne), observed heterozygosity (Ho), and the Shannon–Weaver index (I) were calculated using Pop-gen software v.1.32 (Canada, University of Alberta). Polymorphic information content (PIC) of SSR markers was computed using EXCEL (China, WPS of JINSHAN). When the PIC of an SSR marker was below 0.25, it was considered as a low polymorphic marker, and a marker was considered highly polymorphic if its PIC was above 0.5.

These amplification bands of each SSR primer were separated using polyacrylamide gel-electrophoresis. The band patterns were visualized with silver staining, and gel images were taken with a digital camera. In the same location, the presence of a band was marked as "1", the absence of a band was marked as "0", and a missing band was marked as "-1". In this study we used Genalex-6 software [**31**] to conduct the matrix calculation of SSR marker data which had been assigned a value, then transformed it into a triangle matrix, saved it as a mega-file, finally, imported the mega-file into the Mega-6.0 software (USA, Tamura, K team), and selected the unweighted pair group method with arithmetic (UPGMA) algorithm in the "phylogeny" dropdown menu to draw the cluster diagram [**32**].

The software Structure v.2.3 (USA, UChicago; Britain, Oxon) was used to analyze the population structure [**33,34**]. An admixture model and correlated allele frequencies were used to estimate the number of the populations. For each of the K-values (ranging from 1 to 5), ten independent runs were performed with a burn-in period of 100,000 followed by 500,000 Markov chain Monte Carlo runs. The optimal K-values depends on the peak of K = mean (|Ln"P(D)|)/(sdLnP(D)). Based on the structure results, the most probable K-value was analyzed using Structure Harvester (**http://taylor0.biology.ucla.edu/struct_harvest/**, 2020).

3. RESULTS

3.1. The Frequency and Distribution of Different SSR Types in Cucurbita Genomes

A total of 103,056 microsatellite sequences were identified in the *Cucurbita* genome, including 34,375 SSR loci in the 269.9 Mb draft genome sequence of *C. moschata* cv. Rifu, 30,577 SSR loci in the 271.4 Mb draft genome sequence of *C. maxima* cv. Rimu, and 38,104 SSR loci in the 263 Mb draft genome sequence of *C. pepo* MU-CU16-

(**Table S2**). *Cucurbita pepo* had the largest number of markers with the smallest reference genome size, indicating the highest average density of markers (145 SSR/Mb). To obtain more information, we used *C. pepo* with a higher marker density as the control for the following comparative genomic analysis.

Here, we analyzed repeat types ranging from di-nucleotide to octa-nucleotide. Among all of these nucleotide motifs, di-nucleotide motifs (41.0%) were the most common type, accounting for 41.78%, 39.90%, and 41.01% of the total SSR loci discovered in *C. moschata*, *C. maxima*, and *C. pepo*, respectively, followed by tri-nucleotide motifs (16.97%, 19.19%, and 17.88%, respectively), whereas octa-nucleotide motifs (3.78%, 3.76%, and 3.38%, respectively) were the least represented repeat type in the three *Cucurbita* genomes. In general, the frequency of the total SSR loci decreased with the increase in motif length, except for hepta-nucleotide SSRs.

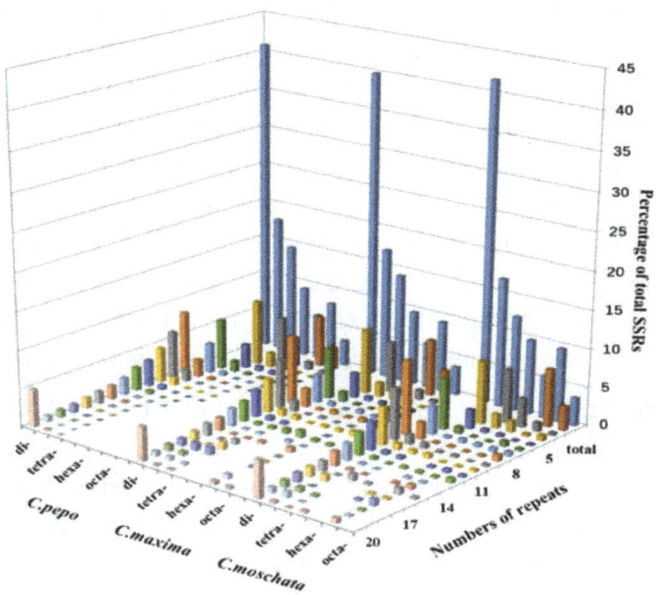

Figure 1. Distribution of SSR motif repeat numbers and relative frequency in Cucurbita genome. The vertical axis shows the abundance of microsatellites that have different motif repeat numbers (from 3 to >20) with different colors.

We further examined the distribution of SSR motifs with regard to their repeat numbers (**Figure 1**). For all the repeat types, with an increase in the repeat number, the SSR frequency decreased sharply, and this change was more obvious in the longer SSR motifs (**Figure 1**). Consequently, the mean repeat numbers in the di-nucleotides were the highest of all of the repeat types. The analysis of individual SSR types revealed that some specific motifs were more prevalent than others in each class. For example, the AT motif was the most frequent di-nucleotide type in all three genomes, accounting for 31.61% (in *C. moschata*), 28.81% (in *C. maxima*), and 30.45% (in *C. pepo*) of the total di-nucleotide loci. Similarly, the AAT, AAAT, AAAAT, AAAAAT, AAAAAAT, and AAAAAAAT motifs (AATAATAT motif in *C. maxima*) were the most frequent types in each class. These results indicated that AnT-rich motifs were the most abundant in all SSR motifs in the *C. moschata*, *C. maxima*, and *C. pepo* genomes.

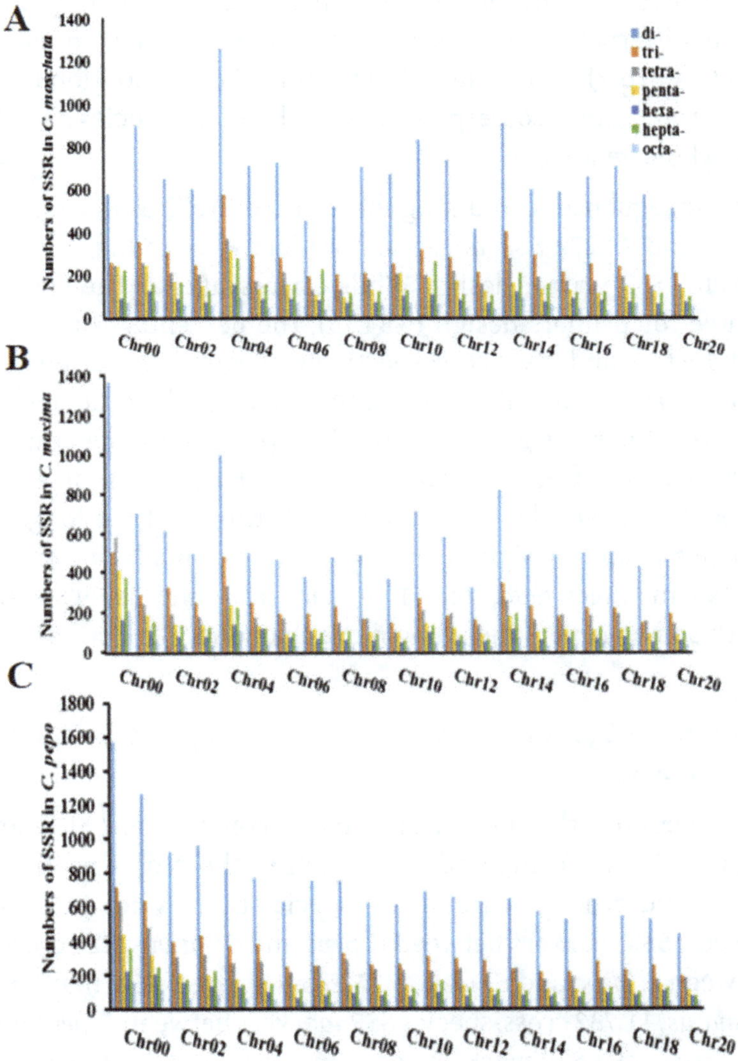

Figure 2. (A) The distribution of SSR repeat types on each chromosome in *C. moschata*. (B) The distribution of SSR repeat types on each chromosome in *C. maxima*. (C) The distribution of SSR repeat types on each chromosome in *C. pepo*. The vertical axis shows the number of microsatellites from di-nucleotide to octo-nucleotide which are discriminated by different colors. The horizontal axis shows different chromosomes of C. ssp, and LG00 means all the chromosome unanchored scaffolds.

We also investigated the SSR density in each chromosome of the three *Cucurbita* species and found that the density of microsatellite loci was not correlated with the chromosome size. For example, in the *C. moschata* genome, the SSR density of the longest chromosome (Chr04) had a medium density of SSRs, while Chr02, which is much shorter than Chr04, had the highest SSR density. A similar trend was also observed in the other two genomes, indicating that the distribution of SSRs was uneven in the *Cucurbita* chromosomes. To better understand the distributions of different SSR motifs, we further checked their frequencies on each chromosome (**Figure 2**). Our results showed that the distribution of different SSR types on the chromosomes corresponded with their frequencies and SSR density in the *Cucurbita* whole genomes.

The genomic sequences containing these microsatellites were screened for PCR primer design, and 94,272 SSR microsatellite loci were found to contain suitable flanking sites for SSR primer design. While *C. moschata* had the lowest proportion of SSRs suitable for primers design (%84.75), the percentages in *C. maxima* and *C. pepo* reached %94.53 and %95.09, respectively. Though the di-nucleotide repeat types were the most frequent in all three genomes, they did not exhibit good performance in primer design. Interestingly, the hexata-nucleotide repeat types had the highest ratio of SSRs suitable for primer design in all three genomes, followed by penta-nucleotide repeat types, indicating that the longer motifs were more suitable for primer design in *Cucurbita* species. Finally, a total of 91,248 SSR primers (28,194 in *C. moschata*, 28,061 in *C. maxima*, and 34,993 in *C. pepo*) were designed, with some primers including more than one SSR locus as the compound SSR.

3.2. Chromosome Synteny Relationships of *C. pepo* with Other Cucurbitaceae Species

In order to understand the universality and correlation of SSR markers among Cucurbitaceae crops, we compared and analyzed the cross-species SSR markers between *C. pepo* and other Cucurbitaceae species by in silico PCR. We identified 391 cross-species SSR markers between *C. pepo* and *C. sativus*, 425 cross-species SSR markers between *C. pepo* and *C. melo*, 717 cross-species SSR markers between *C. pepo* and *C. lanatus*, 11,732 cross-species SSR markers between *C. pepo* and *C. maxima*, and 15,274 cross-species SSR markers between *C. pepo* and *C. moschata*. The collinear blocks to inversion blocks ratio was 26:26 between the *C. pepo* and *C. sativus* genomes, 25:36 between the *C. pepo* and *C. melo* genomes, 51:38 between the *C. pepo* and *C. lanatus* genomes, 154:158 between the *C. pepo* and *C. maxima* genomes, and 153:152 between the *C. pepo* and *C. moschata* genomes. Interestingly, the ratio of collinear blocks to inversion blocks was nearly 1:1 among the three *Cucurbita* species. Each *C. pepo* chromosome shared 36–3 SSR markers with *C. sativus*, *C. lanatus*, or *C. melo*.

However, most of the *C. pepo* chromosome shared a larger number of SSR markers (1,436-3) with *C. maxima* or *C. moschata*. The *C. pepo* syntenic block, CpeCma7, had the largest number of shared SSR markers (i.e., 296) between *C. pepo* chromosome Cpe1 and *C. maxima* chromosome Cma4.

The physical positions of those common shared markers were compared. The main syntenic relationships between *C. pepo* and other Cucurbitaceae species are listed in **Table 1**, and the syntenic relationships visualized for *C. pepo* with *C. lanatus*, *C. melo*, and *C. sativus* are shown in **Figure 3**.

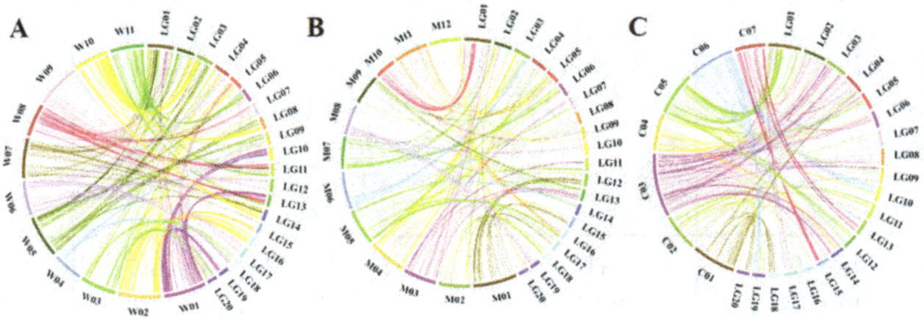

Figure 3. Syntenic relationships of *C. pepo* with (A) C. lanatus, (B) C. melo, and (C) C. sativus. Chromosome synteny between *C. pepo* and C. sativus was based on 391 cross-species markers; synteny between *C. pepo* and C. melo was based on 425 cross-species markers; synteny between *C. pepo* and C. lanatus was based on 717 cross-species markers. W1–W11 represent C. lanatus' eleven chromosomes, M01–M12 represent C. melo's twelve chromosomes, C01–C07 represent C. sativus's seven chromosomes, and LG01–LG20 represent *C. pepo*'s twenty chromosomes. Syntenic blocks are connected by the same color lines from *C. pepo* chromosomes.

The main syntenic relationships among the chromosomes revealed complex mosaic patterns. In **Figure 3**, each *C. pepo* chromosome was syntenic to more than two chromosomes in other Cucurbitaceae species. The *C. pepo* chromosomes Cpe9 and Cpe16 had the simplest syntenic pattern with watermelon, and each of them was mainly syntenic to one watermelon chromosome (**Table 1**). Cpe9 was syntenic to watermelon chromosome W5, and 14 commonly shared SSR markers were found between Cpe9 and W5. From the markers CpeSSR15544 to CpeSSR16107, there were three blocks belonging to watermelon chromosome W5, and each block contained at least four SSR markers. According to the continuous physical positions of these markers on both of the reference genomes, the syntenic blocks CpeWM37 and CpeWM38 showed an inversion pattern, and the syntenic block CpeWM39 showed a collinear pattern between *C. pepo* and *C. lanatus*. Similar comparisons were carried out between *C. pepo* and *C. sativus* or *C. pepo* and *C. melo* using the cross-species SSR markers. The *C. pepo* chromosomes Cpe7, Cpe8, Cpe11, and Cpe20 had the simplest syntenic pattern with *C. sativus*, and each of them was only syntenic to one cucumber chromosome. Meanwhile, the simplest syntenic patterns between

C. pepo and *C. melo* were mainly found on chromosomes Cpe15, Cpe18, Cpe19, and Cpe20. The most complicated syntenic pattern was found on *C. pepo* chromosome Cpe1, which corresponded to five chromosomes of *C. moschata*, four chromosomes of *C. maxima*, seven chromosomes of *C. lanatus*, three chromosomes of *C. sativus*, and five chromosomes of *C. melo*.

Table 1. The main syntenic relationships of *C. pepo* with other Cucurbitaceae species.

C. pepo	*C. moschata*	*C. maxima*	*C. lanatus*	*C. sativus*	*C. melo*
Cpe1	Cmo3(4), Cmo4(1,436), Cmo9(5), Cmo10(3), Cmo17(24)	Cma3(5), Cma4(1,103), Cma9(9), Cma17(19)	Cla1(5), Cla5(21), Cla6(4), Cla7(13), Cla8(5), Cla10(3), Cla11(14)	Csa3(6), Csa5(28), Csa6(3)	Cme3(4), Cme6(4), Cme7(3), Cme9(5), Cme10(16)
Cpe2	Cmo1(913), Cmo10(3), Cmo18(12)	Cma1(674), Cma18(8)	Cla5(8), Cla7(8), Cla10(13), Cla11(7)	Csa3(7), Csa4(6), Csa7(3)	Cme1(3), Cme4(10), Cme7(9)
Cpe3	Cmo4(3), Cmo14(1,080)	Cma4(4), Cma14(859)	Cla5(24), Cla7(8), Cla10(36)	Csa3(34), Csa4(4)	Cme4(25), Cme6(17), Cme7(6)
Cpe4	Cmo10(3), Cmo11(822)	Cma11(640)	Cla2(6), Cla3(5), Cla6(12), Cla10(3)	Csa1(15), Csa3(3)	Cme2(11)
Cpe5	Cmo2(904), Cmo10(8)	Cma2(692), Cma10(8)	Cla1(8), Cla2(20), Cla9(5)	Csa3(5), Csa5(8), Csa6(8)	Cme4(3), Cme5(4), Cme9(6), Cme11(10)
Cpe6	Cmo9(551)	Cma9(396)	Cla5(12), Cla8(4), Cla9(5), Cla11(9)	Csa3(6), Csa4(3)	Cme4(6), Cme7(9)
Cpe7	Cmo5(3), Cmo12(587), Cmo14(20)	Cma5(4), Cma12(452)	Cla2(3), Cla8(6)	Csa2(5)	Cme3(5), Cme5(3)
Cpe8	Cmo6(785)	Cma6(431)	Cla5(13), Cla10(16)	Csa3(6)	Cme4(10), Cme6(6)
Cpe9	Cmo18(544), Cmo19(6)	Cma2(3), Cma18(440)	Cla5(14)	Csa1(4), Csa3(5), Csa5(5)	Cme6(4), Cme10(3), Cme12(4)
Cpe10	Cmo3(659), Cmo18(5)	Cma3(547), Cma18(6)	Cla1(23), Cla4(13)	Csa4(3), Csa6(19)	Cme8(20)

C. pepo	C. moschata	C. maxima	C. lanatus	C. sativus	C. melo
Cpe11	Cmo5(707), Cmo10(3)	Cma5(574)	Cla2(4), Cla8(17), Cla11(14)	Csa2(14)	Cme3(11), Cme5(9)
Cpe12	Cmo17(665)	Cma17(516)	Cla6(12), Cla9(15)	Csa6(9), Csa7(14)	Cme1(17), Cme11(6)
Cpe13	Cmo8(9), Cmo15(649)	Cma4(3), Cma8(7), Cma15(468)	Cla1(21), Cla8(11), Cla11(8)	Csa2(3), Csa5(13), Csa6(6)	Cme3(8), Cme9(12)
Cpe14	Cmo16(565)	Cma16(409)	Cla5(4), Cla7(17), Cla10(14)	Csa3(15), Csa4(8)	Cme6(13), Cme7(5)
Cpe15	Cmo19(493)	Cma19(356)	Cla2(16), Cla7(4), Cla9(11)	Csa3(4), Csa7(11)	Cme1(12)
Cpe16	Cmo20(526)	Cma20(394)	Cla2(23)	Csa2(4), Csa6(4)	Cme5(5), Cme11(10)
Cpe17	Cmo4(3), Cmo8(634), Cmo9(7), Cmo14(3), Cmo17(3)	Cma8(472), Cma14(3), Cma17(4)	Cla6(11), Cla9(17)	Csa6(3), Csa7(3)	Cme1(8), Cme11(4)
Cpe18	Cmo10(500), Cmo14(3)	Cma10(354), Cma14(3)	Cla3(8), Cla6(22)	Csa1(16)	Cme2(17)
Cpe19	Cmo7(658)	Cma7(462)	Cla1(20), Cla4(7)	Csa4(4), Csa6(16)	Cme8(23)
Cpe20	Cmo13(468)	Cma13(330)	Cla1(4), Cla3(12), Cla4(4)	Csa1(11)	Cme12(8)

The syntenic relationships among different *Cucurbita* species were simple and clear. For instance, each of the 20 chromosomes in *C. pepo* was mainly syntenic with one chromosome in *C. moschata* or *C. maxima* (**Figure 4**), implying that the chromosomes in the *Cucurbita* genomes were highly conserved during evolution. Our results also showed that there were three main relationship patterns among the *C. pepo*, *C. maxima*, or *C. moschata* genomes, including (1) the eleven linear relationship chromosomes between *C. pepo* and *C. maxima* or *C. moschata* such as Cpe2–Cmo1–Cma1. Most of the cross-markers in the corresponding chromosomes showed collinear patterns. (2) There were eight inverted relationship chromosomes between *C. pepo* and *C. maxima* or *C. moschata*. For example, the chromosome Cpe1 of *C. pepo* was inverted to the chromosome Cmo4 of *C. moschata* and Cma4 of *C. maxima*. (3) There was a mosaic pattern between *C. pepo* and *C. maxima* or *C. moschata*, for example, Cpe4–Cmo11–Cma11.

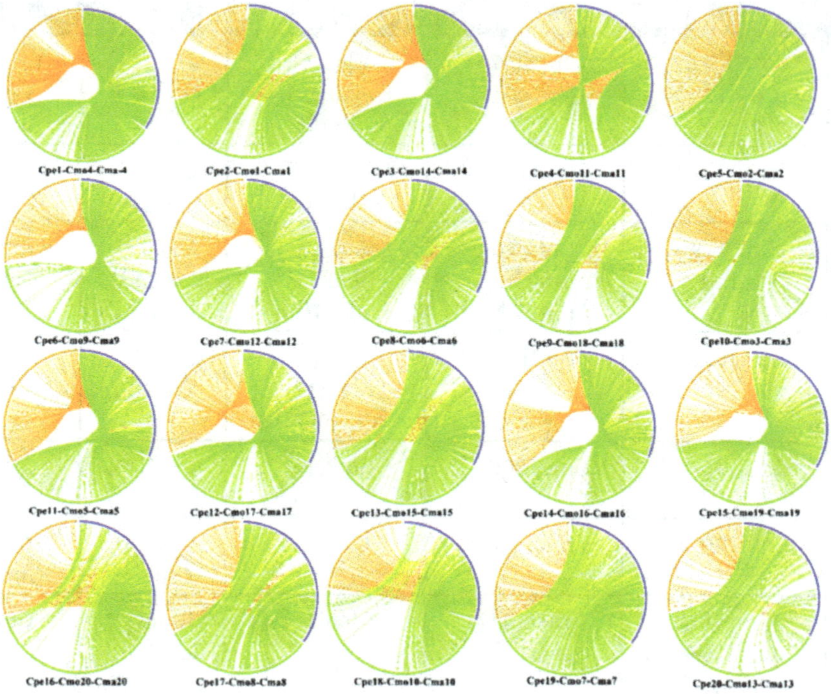

Figure 4. Chromosome synteny of *C. pepo* (blue) with *C. moschata* (green) and *C. maxima* (yellow). The physical positions of chromosomes of each crop in the figure are arranged clockwise. Chromosome synteny between *C. pepo* and *C. moschata* was based on 14,276 cross-species markers; synteny between *C. pepo* and *C. maxima* was based on 10,655 cross-species markers. Cpe1–Cpe20 represent *C. pepo*'s twenty chromosomes, Cmo1–Cmo20 represent *C. moschata*'s chromosomes, and Cma1–Cma20 represent *C. maxima* chromosomes. The syntenic relationship between *C. pepo* and *C. moschata* are connected with the green color lines, and the syntenic relationship between *C. pepo* and *C. maxima* are connected with the yellow color lines.

3.3. The Genetic Diversity and Population Structure Analysis of the *C. pepo* Germplasm

In our preliminary study, approximately 400 SSR markers were screened using 61 accessions of *C. pepo* germplasm. Finally, a total of 66 core SSR markers were selected based on the allelic number, the genomic coverage, and the efficiency of PCR amplification. These markers exhibited clear band spectrums and were evenly distributed on the chromosomes. In this study, 276 alleles were detected by the 66 SSR markers in the 61 *C. pepo* accessions with an average of 4.18 loci per SSR marker. The number of Na ranged from two to nine. The highest number of Na was nine, which was detected by SSR010246, SSR026560, SSR026918, SSR027656, and SSR026980, followed by SSR011546, SSR003315, and SSR026797 with eight alleles. The number of Ne varied from 1.03 to 6.07 with an average of 2.31. The SI ranged from 0.083 to 1.96 with an average of 0.83. The PIC value ranged from 0.03 to 0.83 with an average of 0.43.

We further used a model-based approach for population structure analysis of the 61 *C. pepo* accessions. According to the results of the structural operation, when K = 2, ΔK showed a significant peak value, indicating that the 61 accessions used in this study could be obviously divided into two groups, named group I and group II. The five *C. pepo* subsp. *ovifer* accessions (31 ,30 ,29 ,2, and 45) were clustered into group I (%8.20), and all of them were wild materials. Most of the *C. pepo* subsp. *pepo* accessions were clustered into group II (%91.80), which were all cultivated materials (**Figure** 5A). This indicated that the SSR markers we used could clearly distinguish the cultivated materials from the wild materials. The backgrounds of the cultivated accessions were narrow, except for accession 45 in group I, which should have a complex genetic background, similar to accession 14 and 16 in group II. The UPGMA analysis revealed that the 61 *C. pepo* accessions were divided into two clusters (**Figure** 5B), which was consistent with their population structure. The five *C. pepo* subsp. *ovifer* accessions were clustered together at the base of the phylogenetic tree, which further supported our population structure analysis.

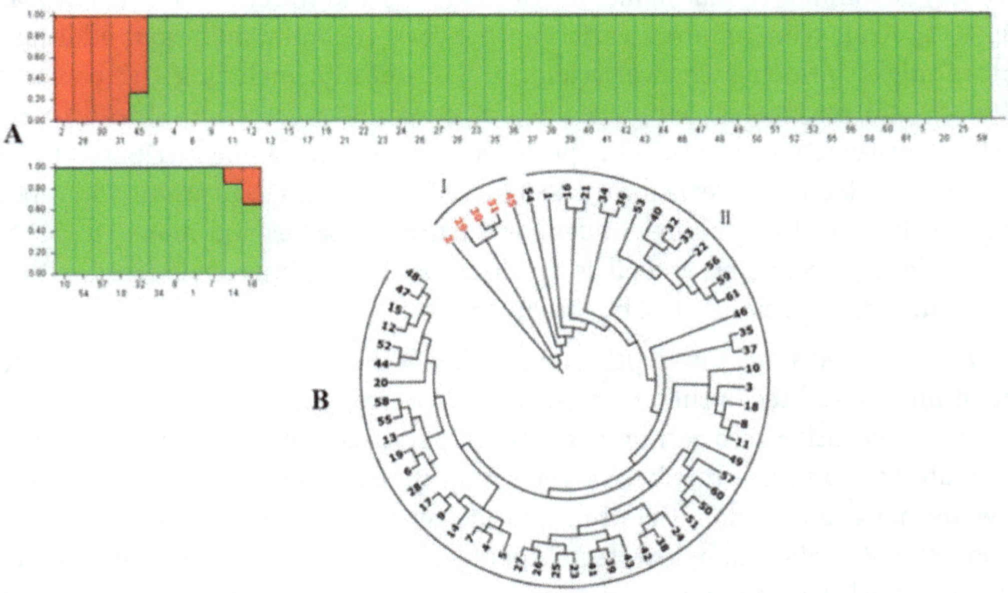

Figure 5. The genetic diversity of the 61 accessions based on SSR markers. (A) Population structure of 61 accessions in *C. pepo* by the model-based analysis. The scale of the *y*-axis represents the percentage of genetic components, and the *x*-axis represents the different materials. (B) Phylogenetic tree of 61 accessions in the *C. pepo* by UPGMA analysis; Group I (red makers) and Group II correspond to the structure analysis.

4. DISCUSSION

4.1. Frequency, Distribution, and Characterization of Microsatellites in Three Cucurbita Genomes

Sequencing technology has facilitated the identification and utilization of genomic SSR loci in other plant species, including cotton [35,36], foxtail millet [37], cucumber [11], watermelon [13], tobacco [38], and melon [12]. The species Cucurbita moschata, *C. maxima*, and *C. pepo* are widely grown around the world, and their graft genomes were published several years ago. Nevertheless, there is a scarcity of data regarding the creation of genome-wide SSR markers in Cucurbita species, which has significantly hindered genetic study in this area. The current work focused on identifying and characterizing genome-wide microsatellites in three Cucurbita species. The genomes of *C. moschata, C. maxima*, and *C. pepo* contained 34,375, 30,577, and 38,104 SSR loci, respectively. Cucurbita pepo had the shortest genome size and the highest number of microsatellites, suggesting that there was no direct relationship between genome size and the number of microsatellites. The density of SSR markers in the three Cucurbita species was around 113-145 SSR per megabase (SSR/Mb), which is lower than the density seen in cucumber (552 SSR/Mb) but similar to that in melon (109 SSR/Mb) and watermelon (111 SSR/Mb) [11,12,13]. Aside from inherent variations between genomes, numerous additional factors might influence the variations in SSR density, including the choice of software and parameters employed for microsatellite detection. The disparity in SSR density between Cucurbita species and cucumber can be attributed to the contrasting selection criteria for the SSR loci. This includes the variation in repeat types, such as di- to octa-nucleotides for Cucurbita species and mono- to penta-nucleotides for cucumber, as well as the discrepancy in minimum lengths, with 18 bp for Cucurbita species and 12 bp for cucumber.

We conducted a more in-depth examination of the distribution and occurrence rate of microsatellites in the three Cucurbita species, as depicted in Figure 1 and Figure 2. A negative association was generally noticed between the frequency of microsatellites and the number of repetition units. In line with prior research on watermelon and melon, the di-nucleotide repeats were shown to be the most prevalent SSRs, followed by tri-, tetra-, penta-, hepta-, hexa-, and octo-nucleotide repeats [12,13]. This phenomenon exhibits variation among different species. For instance, C. sativus had the highest density of tetra-nucleotide repeats at 164.2 SSR/Mb, followed by Populus trichocarpa at 144.9 SSR/Mb, Medicago truncatula at 102.8 SSR/Mb, and Vitis vinifera at 171.3 SSR/Mb. On the other hand, Arabidopsis thaliana had the highest density of tri-nucleotide repeats at 146.6 SSR/Mb, followed by Glycine max at 103.1 SSR/Mb, and Oryza sativa at 220.1 SSR/Mb [11]. Several studies have found that di-nucleotide motifs with a large number of repeats are more common and diverse than those with shorter repeat units [39]. Di-nucleotide repeats occur less frequently in coding areas compared to non-coding regions [40,41]. Additionally, it has been reported that the exon region contains a greater number of triplet SSRs compared to other types of repeats. These triplet SSR motifs may be associated with the higher occurrence of specific amino acids [42,43]. These SSRs within the coding sequence have the potential to impact various genetic functions, such as gene regulation, development, and evolution. Nevertheless, the function of genes that possess simple sequence repeats (SSRs) and the significance of these SSR motifs in plant genes have received limited attention and remain inadequately comprehended [44]. It is noteworthy that numerous bacterial SSRs in the intergenic regions possess regulatory roles [45]. Further investigation is required to determine whether these SSR motifs in the intergenic regions of Cucurbita species contribute to specialization or gene regulation. The prevalence of repeat motifs was mostly characterized by a small quantity, with a significant contribution from AT-rich motifs, which constituted a substantial component of all di-nucleotide repeats in the Cucurbita species (Figure S1). The AT or AAT type is more prevalent in dicots [13], which aligns with our findings. A recent study on bitter gourd identified SSR markers, which predominantly consisted of tri-nucleotide repeat units. Among the several types of repeats, there was an overrepresentation of A/T, AT/AT, AAT/ATT, and AAAT/ATTT patterns [46]. This has also been detected in other genomes [11,47,48]. In contrast, the occurrence of the GC or CCG type was significantly less frequent in the genome [49,50], while the GC, TC, or GA types exhibit rather stable structures. The majority of AT types are found in non-genic areas, whereas the TC/GA types are mainly found in coding sequences [38].

4.2. Comparative analysis of chromosome synteny between *Cucurbita pepo* and other species in the Cucurbitaceae family

Chromosome synteny study has been performed in several species, including cucumber, watermelon, and melon. However, there have been few research on the chromosome synteny between different Cucurbita species or between Cucurbita species and other Cucurbitaceae crops. This study utilized the genomes of three Cucurbita species to construct SSR markers across the entire genome. These markers were then used to detect the syntenic relationships across the genomes by in silico PCR analysis, achieving a high-resolution level of analysis. While the length of the pumpkin genomes are comparable to those of other sequenced Cucurbitaceae species, the Cucurbita genus has a significantly greater number of cross-species SSR markers. In contrast to the numerous common markers found in prior research [14], we have discovered a significantly larger number of SSR markers in the Cucurbita genus that can be transferred across different species. These markers were then utilized for the investigation of chromosome synteny. The occurrence of Whole Genome Duplication (WGD) in Cucurbita, which has not been detected in other sequenced Cucurbitaceae species including cucumber [8], melon [10], and watermelon [9], could potentially explain the large number of SSR markers present. Based on the cross-species transferable SSR markers, we detected 52, 61, and 89 syntenic blocks dispersed across all chromosomes between *C. pepo* and cucumber, melon, and watermelon, respectively (Figure 3). The whole-genome comparison revealed the presence of similar homoeologous blocks [22]. This indicates that the cross-species transferable SSR markers are valuable and dependable for genome comparisons and chromosome synteny investigations. Multiple synteny blocks were commonly observed between *C. pepo* and other Cucurbitaceae species as a result of chromosome fission. The chromosome Cpe1 of *C. pepo* exhibited the most intricate syntenic pattern, being syntenic with seven watermelon chromosomes. This suggests that complex structural alterations had place following their separation from a shared ancestor. In Cucurbita, the proportion of collinear blocks to inversion blocks was approximately equal, with a ratio close to 1:1. This could be attributed to the random occurrence of genome duplication and inter-chromosomal exchanges throughout the evolution of chromosomes. Using cross-species transferable SSR markers, we found that Cucurbita species have more highly preserved syntenic blocks compared to melon, cucumber, and watermelon. Our research revealed that each block in three Cucurbita species from the same genus had a higher number of shared common SSR markers. Additionally, these homoeologous chromosomes were highly preserved, providing further evidence of their tight evolutionary ties within the Cucurbitaceae family. For instance, the *C. pepo* syntenic block had a greater number of markers compared to the melon syntenic block [12]. As a result of whole genome duplication (WGD) throughout the process of chromosome evolution and speciation, there was an increase in the number of chromosomes and cross-markers. Nevertheless, these blocks exhibited a significant degree of conservation during the process of chromosomal evolution across several species of the Cucurbitaceae family. Analysis of chromosomal pairs using cross-species SSR markers revealed the presence of eight significant inversions on various chromosomes between *C. pepo* and either *C. moschata* or *C. maxima*. This suggests that *C. pepo* underwent more intricate evolutionary processes, as depicted in Figure 4. Chr4 exhibited a mosaic region within the Cucurbita species, which is rather intriguing. Possible explanations include genome duplication, extensive inter-chromosomal transfers, or long-term evolutionary influences. Investigating the potential impact of the partial inversion of chromosome 4 in *C. pepo* on the mapping, cloning, and analysis of certain features is a worthwhile endeavor for future research.

4.3. The Genetic Diversity and Population Structure of *C. pepo* Germplasm

In the past, the limited availability of genomic sequences resulted in a lack of molecular markers for studying the genetic diversity and population structure of Cucurbita species. Although various methods such as sequence-related amplified polymorphism (SRAP), amplified fragment length polymorphism (AFLP), simple sequence repeats (SSR), random amplified polymorphic DNA (RAPD), and inter-simple sequence repeats (ISSRs) have been used to assess the genetic diversity of Cucurbita species, most of these markers exhibit high randomness, lack precise location information, and have limited genomic coverage and poor polymorphism. Consequently, their application is greatly restricted [18, 51, 52]. Using the draft genome of three cucurbit crops, we created 91,248 SSR markers with accurate positions on chromosomes. We then assessed the genetic diversity of 61 pumpkin accessions using 66 core SSR markers. The analysis of 61 accessions' population structure indicated that certain materials had a mixed history, consisting of genetic contributions from both group I and group II. This suggests that these accessions may have experienced gene exchange between two subspecies. The samples were gathered from several regions in China and were clearly categorized into two subspecies, namely subsp. ovifer (or subsp. texana) and subsp. pepo, in accordance with prior research [21,51]. Decker has defined the three subspecies of *C. pepo* as *C. pepo* subsp. fraterna (Bailey) Andres, *C. pepo* subsp. texana (Scheele) Filov, and *C. pepo* subsp. pepo [53]. The postulated progenitor of *C. pepo*, namely subsp. fraterna from northeastern Mexico, has been classified as a wild gourd [54]. The population structure and UPGMA results suggest that these accessions of *C. pepo* in China share a common ancestor. Therefore, there are significant opportunities for enhancing germplasm through improvement. The Cucurbita genus encompasses various economically significant crops, however its breeding progress has been slower compared to other Cucurbitaceous crops. The production needs cannot be met because to a scarcity of high-quality cultivars. Marker assisted selection can facilitate various breeding projects. The identification of whole-genome SSR markers in this study will enhance the progress and application of research on Cucurbita species.

AT A GLANCE

Simple sequence repeats (SSRs) have extensive use in mapping constructs as well as in comparative and genetic diversity assessments. A total of 103,056 SSR loci were identified in Cucurbita species using in silico PCR. Overall, the occurrence of these SSRs declined as the motif length increased, with di-nucleotide motifs being the most prevalent. The frequency of SSRs reduced significantly as the number of repeats increased, regardless of the repetition types. A significant proportion of the SSR loci were appropriate for the generation of markers, including 84.75% in *Cucurbita moschata*, 94.53% in *Cucurbita maxima*, and 95.09% in *Cucurbita pepo*. The researchers generated SSR markers that can be transferred between *C. pepo* and other Cucurbitaceae species. They next studied the complex mosaic relationships among these markers. The main syntenic links between *C. pepo* and *C. moschata* or *C. maxima* suggest that the chromosomes in the Cucurbita genomes have remained mostly unchanged throughout evolution. In addition, a total of 66 core SSR markers were chosen to assess the genetic variability in 61 *C. pepo* germplasms. These markers were then categorized into two groups using structural and unweighted pair group approach with arithmetic analysis. These findings will enhance the use of SSRs in both fundamental and practical studies of Cucurbita species.

REFERENCES

1. George, E.B.; Ronald, J.T. Toxic plants of North America. *Choice Curr. Rev. Acad. Libraries* 2013, *12*, 2202–2203.
2. Loy, J.B. Morpho-physiological aspects of productivity and quality in squash and pumpkins (*Culcurbita* spp.). *Crit. Rev. Plant Sci.* 2004, *23*, 337–363.
3. Savage, J.A.; Haines, D.F.; Holbrook, N.M. The making of giant pumpkins: How selective breeding changed the phloem of *Cucurbita maxima* from source to sink. *Plant Cell Environ.* 2015, *38*, 1543–1554.
4. Davis, A.R.; Perkins-Veazie, P.; Sakata, Y.; Lopez-Galarza, S.; Maroto, J.V.; Lee, S.G.; Huh, Y.C.; Sun, Z.Y.; Miguel, A.; King, S.R.; et al. Cucurbit grafting. *Crit. Rev. Plant Sci.* 2008, *27*, 50–74.
5. Lee, J.M.; Kubota, C.; Tsao, S.J.; Bie, Z.; Echevarria, P.H.; Morra, L.; Oda, M. Current status of vegetable grafting: Diffusion, grafting techniques, automation. *Sci. Hortic.* 2010, *127*, 93–105.
6. Lv, J.; Qi, J.J.; Shi, Q.X.; Shen, D.; Zhang, S.P.; Shao, G.J.; Li, H.; Sun, Z.Y.; Weng, Y.Q.; Shang, Y.; et al. Genetic Diversity and Population Structure of Cucumber (*Cucumis sativus* L.). *PLoS ONE* 2012, *7*, e46919.
7. Li, Y.; Wen, C.; Weng, Y. Fine mapping of the pleiotropic locus B for black spine and orange mature fruit color in cucumber identifies a 50 kb region containing a R2R3-MYB transcription factor. *Theor. Appl. Genet.* 2013, *126*, 2187–2196.
8. Huang, S.W.; Li, R.Q.; Zhang, Z.H.; Li, L.; Gu, X.F.; Fan, W.; Lucas, W.J.; Wang, X.W.; Xie, B.Y.; Ni, P.X.; et al. The genome of the cucumber, *Cucumis sativus* L. *Nat. Genet.* 2009, *41*, 1275–1281.
9. Guo, S.G.; Zhang, J.G.; Sun, H.H.; Salse, J.; Lucas, W.J.; Zhang, H.Y.; Zheng, Y.; Mao, L.Y.; Ren, Y.; Wang, Z.W.; et al. The draft genome of watermelon (*Citrullus lanatus*) and resequencing of 20 diverse accessions. *Nat. Genet.* 2013, *45*, 51–58.
10. Garcia-Mas, J.; Benjak, A.; Sanseverino, W.; Bourgeois, M.; Mir, G.; González, V.M.; Hénaff, E.; Câmara, F.; Cozzuto, L.; Lowy, E.; et al. The genome of melon (*Cucumis melo* L.). *Proc. Natl. Acad. Sci. USA* 2012, *109*, 11872–11877.
11. Cavagnaro, P.F.; Senalik, D.A.; Yang, L.; Simon, P.W.; Harkins, T.T.; Kodira, C.D.; Huang, S.; Weng, Y. Genome-wide characterization of simple sequence repeats in cucumber (*Cucumis sativus* L.). *BMC Genom.* 2010, *11*, 569.
12. Zhu, H.; Guo, L.; Song, P.; Luan, F.; Hu, J.; Sun, X.; Yang, L. Development of genome-wide SSR markers in melon with their cross-species transferability analysis and utilization in genetic diversity study. *Mol. Breed.* 2016, *36*, 153.

13. Zhu, H.; Song, P.; Koo, D.H.; Guo, L.; Li, Y.; Sun, S.; Weng, Y.; Yang, L. Genome wide characterization of simple sequence repeats in watermelon genome and their application in comparative mapping and genetic diversity analysis. *BMC Genom.* 2016, *17*, 557.

14. Li, D.W.; Cuevas, H.E.; Yang, L.M.; Li, Y.H.; Garcia-Mas, J.; Zalapa, J.; Staub, J.E.; Luan, F.S.; Reddy, U.; He, X.M.; et al. Syntenic relationships between cucumber (*Cucumis sativus* L.) and melon (*C. melo* L.) chromosomes as revealed by comparative genetic mapping. *BMC Genom.* 2011, *12*, 396.

15. Yang, L.M.; Koo, D.H.; Li, D.W.; Zhang, T.; Jiang, J.M.; Luan, F.S.; Renner, S.S.; Henaff, E.; Sanseverino, W.; Garcia-Mas, J.; et al. Next-generation sequencing, FISH mapping and synteny-based modeling reveal mechanisms of decreasing dysploidy in Cucumis. *Plant J.* 2014, *77*, 16–30.

16. Hi, L.Y.; Jeong, J.H.; Hong, K.H.; Dong, K.B. Use of Random Amplified Polymorphic DNAs for Linkage Group Analysis in Interspecific Hybrid F2 Generation of *Cucurbita*. *Hortic. Environ. Biotechnol.* 1995, *36*, 323–330.

17. Brown, R.N.; Myers, J.R. A genetic map of squash (*Cucurbita* sp.) with randomly amplified polymorphic DNA markers and morphological markers. *J. Am. Soc. Hortic.* 2002, *127*, 568–575.

18. Paris, H.S.; Yonash, N.; Portnoy, V.; Mozes-Daube, N.; Tzuri, G.; Katzir, N. Assessment of genetic relationships in *Cucurbita pepo* (Cucurbitaceae) using DNA markers. *Theor. Appl. Genet.* 2003, *106*, 971–978.

19. Zraidi, A.; Stift, G.; Pachner, M.; Shojaeiyan, A.; Gong, L.; Lelley, T. A consensus map for *Cucurbita pepo*. *Mol. Breed.* 2007, *20*, 375–388.

20. Gong, L.; Stift, G.; Kofler, R.; Pachner, M.; Lelley, T. Microsatellites for the genus *Cucurbita* and an SSR-based genetic linkage map of *Cucurbita pepo* L. *Theor. Appl. Genet.* 2008, *117*, 37–48.

21. Esteras, C.; Gomez, P.; Monforte, A.J.; Blanca, J.; Vicente-Dolera, N.; Roig, C.; Nuez, F.; Pico, B. High-throughput SNP genotyping in *Cucurbita pepo* for map construction and quantitative trait loci mapping. *BMC Genom.* 2012, *13*, 80.

22. Sun, H.H.; Wu, S.; Zhang, G.Y.; Jiao, C.; Guo, S.G.; Ren, Y.; Zhang, J.; Zhang, H.Y.; Gong, G.Y.; Jia, Z.C.; et al. Karyotype Stability and Unbiased Fractionation in the Paleo-Allotetraploid *Cucurbita* Genomes. *Mol. Plant* 2017, *10*, 1293–1306.

23. Montero-Pau, J.; Blanca, J.; Bombarely, A.; Ziarsolo, P.; Esteras, C.; Marti-Gomez, C.; Ferriol, M.; Gomez, P.; Jamilena, M.; Mueller, L.; et al. De novo assembly of the zucchini genome reveals a whole-genome duplication associated with the origin of the *Cucurbita* genus. *Plant Biotechnol. J.* 2018, *16*, 1161–1171.

24. Blanca, J.; Canizares, J.; Roig, C.; Ziarsolo, P.; Nuez, F.; Pico, B. Transcriptome characterization and high throughput SSRs and SNPs discovery in *Cucurbita pepo* (Cucurbitaceae). *BMC Genom.* 2011, *12*, 104.

25. Wyatt, L.E.; Strickler, S.R.; Mueller, L.A.; Mazourek, M. An acorn squash (*Cucurbita pepo* ssp. ovifera) fruit and seed transcriptome as a resource for the study of fruit traits in Cucurbita. *Hortic. Res.* 2015, *2*, 14070.

26. Xanthopoulou, A.; Psomopoulos, F.; Ganopoulos, I.; Manioudaki, M.; Tsaftaris, A.; Nianiou-Obeidat, I.; Madesis, P. De novo transcriptome assembly of two contrasting pumpkin cultivars. *Genom. Data* 2016, *7*, 200–201.

27. Aliki, X.; Ganopoulos, I.; Psomopoulos, F.; Manioudaki, M.; Moysiadis, T.; Kapazoglou, A.; Osathanunkul, M.; Michailidou, S.; Kalivas, A.; Tsaftaris, A.; et al. De novo comparative transcriptome analysis of genes involved in fruit morphology of pumpkin cultivars with extreme size difference and development of EST-SSR markers. *Gene* 2017, *622*, 50–66.

28. Vitiello, A.; Scarano, D.; D'Agostino, N.; Digilio, M.C.; Pennacchio, F.; Corrado, G.; Rao, R. Unraveling zucchini transcriptome response to aphids. *PeerJ. PrePrints* 2016, *4*.

29. Krzywinski, M.; Schein, J.; Birol, I.; Connors, J.; Gascoyne, R.; Horsman, D.; Jones, S.J.; Marra, M.A. Circos: An information aesthetic for comparative genomics. *Genome Res.* 2009, *19*, 1639–1645.

30. Murray, M.G.; Thompson, W.F. Rapid isolation of high molecular weight plant DNA. *Nucleic Acids Res.* 1980, *8*, 4321–4325.
31. Peakall, R.; Smouse, P.E. GenAlEx 6.5: Genetic analysis in Excel. Population genetic software for teaching and research—An update. *Bioinformatics* 2012, *28*, 2537–2539.
32. Tamura, K.; Stecher, G.; Peterson, D.; Filipski, A.; Kumar, S. MEGA6: Molecular Evolutionary Genetics Analysis Version 6.0. *Mol. Biol. Evol.* 2013, *30*, 2725–2729.
33. Pritchard, J.K.; Stephens, M.; Donnelly, P. Inference of population structure using multilocus genotype data. *Genetics* 2000, *155*, 945–959.
34. Evanno, G.; Regnaut, S.; Goudet, J. Detecting the number of clusters of individuals using the software STRUCTURE: A simulation study. *Mol. Ecol.* 2005, *14*, 2611–2620.
35. Lu, C.R.; Zou, C.S.; Zhang, Y.P.; Yu, D.Q.; Cheng, H.L.; Jiang, P.F.; Yang, W.C.; Wang, Q.L.; Feng, X.X.; Prosper, M.A.; et al. Development of chromosome-specific markers with high polymorphism for allotetraploid cotton based on genome-wide characterization of simple sequence repeats in diploid cottons (*Gossypium arboreum* L. and *Gossypium raimondii* Ulbrich). *BMC Genom.* 2015, *16*, 55.
36. Wang, Q.; Fang, L.; Chen, J.D.; Hu, Y.; Si, Z.F.; Wang, S.; Chang, L.J.; Guo, W.Z.; Zhang, T.Z. Genome-Wide Mining, Characterization, and Development of Microsatellite Markers in *Gossypium* Species. *Sci. Rep.* 2015, *5*, 10638.
37. Zhang, S.; Tang, C.J.; Zhao, Q.; Li, J.; Yang, L.F.; Qie, L.F.; Fan, X.K.; Li, L.; Zhang, N.; Zhao, M.C.; et al. Development of highly polymorphic simple sequence repeat markers using genome-wide microsatellite variant analysis in Foxtail millet [*Setaria italica* (L.) P. Beauv.]. *BMC Genom.* 2014, *15*, 78.
38. Wang, X.W.; Yang, S.; Chen, Y.D.; Zhang, S.M.; Zhao, Q.S.; Li, M.; Gao, Y.L.; Yang, L.; Bennetzen, J.L. Comparative genome-wide characterization leading to simple sequence repeat marker development for Nicotiana. *BMC Genom.* 2018, *19*, 500.
39. Weber, J.L. Informativeness of human (dC-dA)n.(dG-dT)n polymorphisms. *Genomics* 1990, *7*, 524–530.
40. Li, Y.C.; Korol, A.B.; Fahima, T.; Beiles, A.; Nevo, E. Microsatellites: Genomic distribution, putative functions and mutational mechanisms: A review. *Mol. Ecol.* 2002, *11*, 2453–2465.
41. Wang, Z.; Weber, J.L.; Zhong, G.; Tanksley, S.D. Survey of plant short tandem DNA repeats. *Theor. Appl. Genet.* 1994, *88*, 1–6.
42. Morgante, M.; Hanafey, M.; Powell, W. Microsatellites are preferentially associated with nonrepetitive DNA in plant genomes. *Nat. Genet.* 2002, *30*, 194–200.
43. Li, Y.C.; Korol, A.B.; Fahima, T.; Nevo, E. Microsatellites within genes: Structure, function, and evolution. *Mol. Biol. Evol.* 2004, *21*, 991–1007.
44. Varshney, R.K.; Graner, A.; Sorrells, M.E. Genic microsatellite markers in plants: Features and applications. *Trends Biotechnol.* 2005, *23*, 48–55.
45. Zhao, Z.X.; Guo, C.; Sutharzan, S.; Li, P.; Echt, C.S.; Zhang, J.; Liang, C. Genome-Wide Analysis of Tandem Repeats in Plants and Green Algae. *G3 Genes Genomes Genet.* 2014, *4*, 67–78.
46. Cui, J.; Cheng, J.; Nong, D.; Peng, J.; Hu, Y.; He, W.; Zhou, Q.; Dhillon, N.P.S.; Hu, K. Genome-Wide Analysis of Simple Sequence Repeats in Bitter Gourd (*Momordica charantia*). *Front. Plant Sci.* 2017, *8*, 1103.
47. Kim, T.S.; Booth, J.G.; Gauch, H.G., Jr.; Sun, Q.; Park, J.; Lee, Y.H.; Lee, K. Simple sequence repeats in Neurospora crassa: Distribution, polymorphism and evolutionary inference. *BMC Genom.* 2008, *9*, 31.
48. Cheng, J.; Zhao, Z.; Li, B.; Qin, C.; Wu, Z.; Trejo-Saavedra, D.L.; Luo, X.; Cui, J.; Rivera-Bustamante, R.F.; Li, S.; et al. A comprehensive characterization of simple sequence repeats in pepper genomes provides valuable resources for marker development in Capsicum. *Sci. Rep.* 2016, *6*, 18919.
49. Tangphatsornruang, S.; Somta, P.; Uthaipaisanwong, P.; Chanprasert, J.; Sangsrakru, D.; Seehalak, W.;

Sommanas, W.; Tragoonrung, S.; Srinives, P. Characterization of microsatellites and gene contents from genome shotgun sequences of mungbean (*Vigna radiata* (L.) Wilczek). *BMC Plant Biol.* 2009, *9*, 137.

50. Portis, E.; Lanteri, S.; Barchi, L.; Portis, F.; Valente, L.; Toppino, L.; Rotino, G.L.; Acquadro, A. Comprehensive Characterization of Simple Sequence Repeats in Eggplant (*Solanum melongena* L.) Genome and Construction of a Web Resource. *Front. Plant Sci.* 2018, *9*, 401.
51. Ferriol, M.; Pico, B.; Nuez, F. Genetic diversity of a germplasm collection of *Cucurbita pepo* using SRAP and AFLP markers. *Theor. Appl. Genet.* 2003, *107*, 271–282.
52. Ntuli, N.R.; Tongoona, P.B.; Zobolo, A.M. Genetic diversity in *Cucurbita pepo* landraces revealed by RAPD and SSR markers. *Sci. Hortic.* 2015, *189*, 192–200.
53. Decker, D.S. Origin (s), Evolution, and Systematics of *Cucurbita pepo* (Cucurbitaceae). *Econ. Bot.* 1988, *42*, 4–15.
54. Paris, H.S. Germplasm enhancement of *Cucurbita pepo* (pumpkin, squash, gourd: Cucurbitaceae): Progress and challenges. *Euphytica* 2015, *208*, 415–438.

12 Chapter

IDENTIFICATION AND CHARACTERIZATION OF HSF AND HSP GENE FAMILIES & GENE EXPRESSION ANALYSIS DURING HEAT STRESS IN EGGPLANT *(SOLANUM MELONGEMA* L.) AT A GENOME-WIDE SCALE

1. INTRODUCTION: AN OVERVIEW

Plants inhabit intricate settings where they may encounter many abiotic stresses, like salt, dehydration, and severe temperature, which can significantly impede their growth and development [1]. Due to their immobility, plants are unable to move to escape these pressures. Consequently, they have evolved mechanisms, such as increased expression of genes related to tolerance, to respond to heat stress [2,3]. Plants have developed self-defense systems over time to adapt and thrive under challenging environmental situations. Prior research has demonstrated that when exposed to heat stress (HS), plant cells promptly react to elevated temperatures by activating the expression of genes that encode heat shock proteins (Hsps). These proteins play a crucial role in mitigating heat-induced harm and enhancing the ability of plants to withstand high temperatures. This phenomenon has been observed in various plant species such as strawberry, walnut, barley, and grapevines [4,5,6,7]. Several Heat Shock Proteins (Hsps) serve as molecular chaperones to prevent protein misfolding and aggregation. This helps maintain protein balance in cells and triggers learned thermotolerance in plants [8]. Heat shock factors (Hsfs) are special types of transcription factors that control and regulate the expression of Hsps. These Hsfs often reside as inactive proteins [9].

At present, numerous plant Hsf and Hsp genes from different species have been identified and thoroughly examined. Hsps are categorized into five families, namely, small Hsp (sHsp), Hsp60s, Hsp70s, Hsp90s, and Hsp100s, based on their estimated molecular weights and similarities in their sequences [10]. There is a favorable correlation between the expression of sHsp and thermostability [11]. Hsp60 proteins serve as chaperones, aiding in the folding and aggregation of several proteins that are delivered to organelles, including chloroplasts and mitochondria [12]. Hsp70 chaperones, along with their co-chaperones, are a major group of cellular machinery that aids in various protein folding activities across practically all cellular compartments [13]. The downregulation of Hsp90 expression in Arabidopsis TU8 mutants results in increased heat sensitivity. Fungi that produce Hsp90 inhibitors stimulate the expression of the Hsp101 and Hsp70 genes in *Arabidopsis thaliana* seedlings, leading to an increase in plant heat tolerance [14]. Arabidopsis possesses a minimum of 21 Heat Shock Transcription Factors (Hsfs) [15]. HsfA1a, HsfA1b, and HsfA1d function as the primary promoters of the heat shock response [16], whereas HsfA2 can improve the ability of plants to withstand high temperatures [17]. Primarily, Hsfs and Hsps have essential functions in enhancing plant thermotolerance. The Hsf and Hsp gene families have undergone significant research in the model plant Arabidopsis thaliana, as well as in non-model plants including rice (*Oryza sativa*), poplar (*Populus trichocarpa*), maize (*Zea mays*), and Chinese cabbage (*Brassica rapa* L. ssp. pekinensis).

Eggplant, scientifically known as *Solanum melongena* L., is a significant solanaceous crop with economic importance. It holds the third position in terms of importance, following potato and tomato. Eggplant is predominantly cultivated in East Asia, South Asia, the Middle East, and northern Africa. The ideal temperature range for eggplant growth and development is between 22 °C and 30 °C. Due to global warming, the temperature in subtropical and tropical locations frequently exceeds 35 °C, leading to significant heat damage in eggplant. This includes stunted plant development, decreased yield, and compromised quality [2]. Thermotolerance is a crucial characteristic for eggplants in agriculture, although the specific molecular pathways underlying heat tolerant are still not well understood. Heat shock factors (Hsfs) and heat shock proteins (Hsps) have essential functions in the signal transduction pathways that mediate plant responses to heat stress. The study focused on investigating the role of Hsf and Hsp genes in eggplant's response to heat stress, as these genes play crucial regulatory activities in plant's ability to cope with high temperatures. The sequencing and assembly of the eggplant genome [22] allowed for the examination of the eggplant Hsf and Hsp families and their reactions to heat stressors at the molecular level. Consequently, a comprehensive analysis was performed to identify Hsf and Hsp genes throughout the entire genome of eggplant, with the aim of determining their expansion and evolutionary history. The expression differential between the thermotolerant line 05-4 and the thermosensitive line 05-1, caused by naturally elevated temperature, was investigated using RNA-seq data and quantitative real-time RT-PCR analysis. The findings offer a comprehensive overview of the Hsf and Hsp gene families in eggplant and clarify their association with thermotolerance. This serves as a basis for further investigations into the functional aspects of these genes in eggplant. Moreover, these discoveries have the potential to be valuable in comprehending the process of thermotolerance facilitated by Heat Shock Factors (Hsfs) and Heat Shock Proteins (Hsps) in eggplant.

2. MATERIALS AND METHODS

2.1. Identification and Classification of Hsf and Hsp Family Members in Eggplant

Published Arabidopsis *Hsf* and *Hsp* sequences [**23**] were retrieved and used as queries in BLAST searches against the eggplant genome database (**http://eggplant.kazusa.or.jp/**, accessed on 6 June 2021) to identify potential eggplant *Hsfs* and *Hsps*. All output genes identified according to Arabidopsis *Hsf* and *Hsp* sequences were collected and confirmed using Pfam (**http://pfam.xfam.org/search**, accessed on 6 June 2021) and SMART (**http://smart.embl-heidelberg.de/**, accessed on 6 June 2021). The isoelectric points and molecular weights were predicted using the Compute pI/Mw tool from ExPASy (**http://web.expasy.org/compute_pi**, accessed on 6 June 2021).

2.2. Phylogenetic Analysis

Alignments of the full eggplant Hsf and Hsp proteins were performed using clustal X2.1 [**24**]. Phylogenetic trees were constructed using the neighbor-joining (NJ) method in MEGA (version 5.0) [**25**] with bootstrap values from 1000 replicates indicated at each node. To identify signature domains, the Hsf and Hsp protein sequences were compared with Arabidopsis and tomato. *SmHsfs* and *SmHsps* (*sHsp, Hsp60s, Hsp70s, Hsp90s* and *Hsp100s*) were named based on the subfamily classification and their phylogenetic relationships with the corresponding *AtHsfs* and *AtHsps* and gene names of eggplant *sHsps* were revised according to their molecular weights in the eggplant genome database based on Hirakawa et al. [22].

2.3. Gene Structures, Conserved Motifs and Protein Functional Network Analysis

The exon and intron structures were illustrated using the Gene Structure Display Server (GSDS, **http://gsds.cbi.pku.edu.cn** accessed on 6 June **26**] (2021] by aligning the predicted cDNA sequences with their corresponding genomic DNA sequences. The conserved motifs in the encoded proteins were analyzed using the MEME online program (**http://meme.sdsc.edu**, v4.9.0, accessed on 6 June 2021) [**27**]. MEME was run locally with the following parameters: number of repetitions = any, maximum number of motifs = 20 and optimum motif width = 6–100 residues for *Hsf, sHsp, Hsp60, Hsp70* and *Hsp100*. The STRING protein interaction database (**http://string-**

db.org/, accessed on 6 June 2021) was used to analyze the interaction networks of Hsf and Hsp proteins in the highly specific protein and parameter selection model plant species *Arabidopsis thaliana*.

2.4. Plant Materials, Growth Conditions and Stress Treatments

In the present study, two inbred eggplant lines (selected by the Vegetable Research Institute, Guangdong Academy of Agricultural Sciences, Guangzhou, China), the thermotolerant line 05-4 and the thermosensitive line 05-1, were used. Eggplant seedlings were cultivated under 25/20 °C day/night conditions and a 16/8 h day/night photoperiod in a growth chamber until the four true leaves period for treatments. For the HS treatment, the seedlings of 05-1 and 05-4 with four leaves were directly placed in the 42 °C light incubator (RXZ-1000B3, Jiangnan Instrument Factory, Ningbo, China). For the heat treatment used for RNA-seq, the 3rd mature leaves of treated seedlings were collected at 0 and 6 h after HS treatment and 10 plants were used for each treatment. For the heat treatment used for the qRT-PCR, the 3rd mature leaves from two different lines were harvested at 0 and 6 h. The samples were harvested, immediately frozen in liquid nitrogen and stored at -80 °C for RNA extraction. Three biological replicates were performed and each replicate had 10 plants.

2.5. RNA Extraction and Quantitative Real-Time PCR Analysis

Total RNA was extracted using a TransZol Plant kit (TransGen Biotech/TransBionovo, Beijing, China) and the cDNA was synthesized according to the manufacturer's instructions (Takara, Dalian, China). Primers with amplicon lengths of 80–150 bp were designed using Primer5 software. All primer sequences are listed in **Table S14**. Real-time qRT-PCR was conducted on a Real-Time PCR Detection System (Bio-Rad, Hercules, CA, USA) using the SYBR Premix Ex Taq kit (TaKaRa, Dalian, China) according to the manufacturer's instructions. The 10 μL reaction system contained 5 μL of SYBR Green Supermix (2×), 4 μL of cDNA template (30 ng/μL) and 0.5 μL of each primer (10 μM). The qRT-PCR reaction was performed using the following parameters: pre-denaturation at 95 °C for 30 s, followed by 39 cycles of denaturation at 95 °C for 5 s, annealing at 60 °C for 15 s and extension at 72 °C for 15 s. The fluorescent signal was measured at the end of each cycle and the melting curve analysis was performed by heating the PCR product from 65 °C to 90 °C to verify the specificity of the primers. Three independent biological replicates were performed and the qPCR of each replicate was performed in triplicate. The relative expression levels of eggplant *Hsf* and *Hsp* genes were calculated using the -2ΔCt method [**28**]. The *SmEF1a* genes were used as internal controls.

3. RESULTS

3.1. Genome-Wide Identification and Analysis of Hsf and Hsp Gene Family Members in Eggplant

To search for *Hsf* and *Hsp* genes in eggplant, we used the conserved Hsf and Hsp domain consensus sequences of several proteins as BLASTP queries against the eggplant genome database (**http://eggplant.kazusa.or.jp/**, accessed on 6 June 2021). In addition, homology searches using identified protein sequences of *Arabidopsis thaliana* were performed. After automated database searching and a manual review, 24 and 117 genes were identified as members of the *Hsf* and *Hsp* families in eggplant, respectively, whose classification and naming were based on the rules of the *Hsp* gene families from Arabidopsis and tomato, including *sHsp, Hsp60, Hsp70, Hsp90* and *Hsp100.* The *Hsf* and *Hsp* gene families in eggplant were relatively large compared with those in Arabidopsis and those in tomato and rice, respectively. The numbers of identified genes in the *Hsf, sHsp, Hsp60, Hsp70, Hsp90* and *Hsp100* families of eggplant were ,30 ,21 ,39 ,24 17 and 10, respectively (**Table 1**).

Table 1. Numbers of *Hsf* and *Hsp* genes in Arabidopsis, eggplant, tomato and rice.

Family	Arabidopsis	Eggplant	Tomato	Rice
Hsf	22	24	23	25
Hsp20	27	39	23	39
Hsp60	18	21	16	20
Hsp70	19	30	22	24
Hsp90	7	17	8	9
Hsp100	8	10	13	10

The amino acid lengths for Hsfs ranged from 111 (*SmHsfA1c*) to 496 (*SmHsfA1b*), with deduced molecular weights from 12.2 kDa to 54.8 kDa and the predicted isoelectric points of Hsfs were divergent, ranging from 4.60 (*SmHsfA3*) to 9.64 (*SmHsfA1d*). The length of sHsp proteins ranged from 87 (*Sm10.2-sHsp*) to244 amino acids (*Sm27.2-sHsp*) and the predicted molecular weights were between 10.2 kDa (*Sm10.2-sHsp*) and 27.2 kDa (*Sm10.2-sHsp*). In addition, the predicted pI-values of sHsp proteins ranged from 4.56 (*Sm10.2-sHsp*) to 10.49 (*Sm12.7-sHsp*). The amino acids lengths were consistent with the molecular weights of Hsp60s. The amino acid number and molecular weight for Smcpn60-4 was the highest, while that for SmCpn60-7.3 was the lowest and the predicted pI-values ranged from 5.26 (*SmCpn60-a1*) to 10.29 (*SmCpn60-7.3*). The deduced length of the Hsp70 proteins ranged from 85 (*SmmtHsc70-3*) to 914 (*SmHsp70-18*) amino acids and the highest- and lowest-molecular-weight SmHsp70s were *SmHsp70-18* (103.1 kDa) and *SmHsp70-5* (11.5

kDa), respectively, while the pI values ranged from 4.52 (*SmmtHsc70-3*) to 9.35 (*SmHsp70-19*). The length of Hsp90 proteins ranged from 137 (*SmHsp90-4.4*) to 782 (*SmHsp90-6*) amino acids, the predicted molecular weights of Hsp90s were between 16.2 kDa (*SmHsp90-4.4*) and 89.6 kDa (*SmHsp90-7.1*) and the predicted isoelectric points ranged from 4.78 (*SmHsp90-5*) to 9.55 (*SmHsp90-2.1*). The longest amino acids lengths and highest molecular weights in Hsp100s were *SmHsp100-ClpB1*, with 979 amino acids and 110.2 kDa, respectively. In contrast, the smallest was *SmHsp100-ClpC3*; the predicted isoelectric points ranged from 5.38 (*SmHsp100-ClpB3*) to 9.07 (*SmHsp100-ClpC1*) and these proteins were distributed from the alkaline to acidic.

3.2. Phylogenetic and Sequence Structure Analysis of Hsf and Hsp Proteins in Eggplant

To evaluate the evolutionary relationship of the eggplant Hsf and Hsp proteins, a phylogenetic analysis of each family was performed based on the full-length amino acid sequences from Arabidopsis, eggplant and tomato and each family could be classified into different subfamilies. The *SmHsf* family contained three subfamilies: type A (18 genes), type B (5 genes) and type C (1 gene). Based on the phylogenetic tree, class HsfA had the maximum number of subclasses among the three classes and was closer to tomato Hsf proteins, which coincided with the botanical classification. A total of 39 *sHsp* genes could be grouped into 12 distinct subfamilies, containing 6 groups of cytosolic *sHsp* genes, C-I, C-II, C-III, C-IV, C-V and C-VI and 2 groups of mitochondrial *sHsp* genes, MT I and MT II. Notably, the C-I *sHsp* group in the eggplant genome was large, containing 24 genes, compared with 6 in Arabidopsis. The *Hsp60* family was divided into 4 subfamilies, including cytosol-localized Cpn12) 60 genes), mitochondrion-localized Hsp4) 60 genes) and chloroplast-localized Cpn-60a (2 genes) and Cpn-60b (3 genes). The *Hsp70* family contains genes encoding 19 cytosolic *Hsp70s*, 4 binding proteins (BIPs, Hsp70 homologs in the ER), 3 mitochondrial Hsp70s (mtHsc70s) and 2 chloroplastid Hsp70s (cpHsc70s) . Seventeen *Hsp90* family genes could be divided into cytoplasm (Cyt), mitochondrial (MT), endoplasmic reticulum (ER) and chloroplast, containing 8, 3, 2 and 1 proteins, respectively. The Hsp100 family can be classified into ClpB, C, D and X classes as follows: 3 ClpB proteins (designated as B1, B2 and B3), 4 ClpC proteins (C1, C2, C3 and C4), 1 ClpD protein (D1) and 2 ClpX proteins (X1 and X2) .

3.3. Structure of Hsf and Hsp Genes and Conserved Motifs of Hsf and Hsp Proteins in Eggplant

To obtained further insights into the structural diversity of *Hsf* and *Hsp* genes in eggplant, we used the Multiple Expectation maximization for Motif Elicitation (MEME) [27] to predict the conserved motifs shared among the related proteins within these families. In each family, 20 putative motifs were identified. . Most of the closely related members in the phylogenetic tree shared common motif compositions.

The exon/intron structures of eggplant *Hsf* and *Hsp* members were analyzed based on their coding sequences and the corresponding genome sequences. The eggplant *Hsfs* shared highly conserved exon/intron structures with 3–0 intron phases (**Figure 1**A). The intron phases were remarkably well conserved among family members. Most of the eggplant *sHsps* did not contain introns and only a few had 3–1 introns (**Figure 1**B). Interestingly, in the *Hsp60* family, two members, *SmCpn8-60* and *SmHsp3-60,* had no introns in their coding regions, while the other eggplant Hsp60s contained several introns (22–1) (**Figure 1**C). In the *Hsp70* family, cytosolic *Hsp70s* had 13–0 introns, ER-localized BIPs had 7–4 introns, mitochondrion-localized *mtHsc70s* had 4–0 introns and chloroplast-localized *cpHsc70s* had 6 introns, while truncated *Hsp70ts* had no introns (**Figure 2**A). With the exception of *SmHsp90-1.1*, each *Hsp90s* member contained 17–1 introns (**Figure 2**B). The number of exons and introns of *Hsp100* family members differed greatly. For example, *SmHsp-100ClpX1* contained up to 16 introns, but *SmHsp-100 ClpB3* and *SmHsp-100ClpC3* only had 4 introns (**Figure 2**C).

3.4. Expression Patterns of Eggplant Hsf and Hsp Genes

To examine the heat response for *Hsfs* and *Hsps* in eggplant, an RNA sequencing profile (data not shown) in leaves of thermosensitive line 1-05 and thermotolerant line 4-05, at 0 and 6 h after HS treatment, was used. *Hsf* and *Hsp* genes were selected according to annotations and their expression profiles were analyzed. We analyzed the transcription levels of 18 *Hsf*, 25 *sHsp*, 6 *Hsp60*, 18 *Hsp70*, 11 *Hsp90* and 6 *Hsp100* genes in the leaves. As shown in **Figure 3**, for the thermosensitive line 05-1, 16 genes (89%) of the *Hsf* family were upregulated and two members, *SmHsfA4e* and *SmHsfB3*, were downregulated under HS conditions, which were more than in the thermotolerant line 05-4, in which 17 *Hsf* genes (%94) were upregulated by HS and only *SmHsfA8* was downregulated. In contrast to line -05 1, *SmHsfA4e* and *SmHsfB3* were strongly induced in treated 4-05 leaves. In the leaves of the thermotolerant line 4-05, among the upregulated members, the expression levels of most A (A1a, A1b, A3, A4a, A4b, A4d, A4e, A5, A6a and A6b), B1 and B2a were higher than those of other members under HS.

Figure 1. Phylogenetic relationships, gene structures and motif compositions of *Hsf*, *sHsp* and *Hsp60* family members in eggplant. Multiple alignment of the Hsf (A), sHsp (B) and Hsp60 (C) proteins from eggplant (Sm) was performed with MEGA 5.0 using the neighbor-joining (NJ) method with 1000 bootstrap replicates (left panel). A schematic representation of conserved motifs (obtained using MEME) in the Hsf and sHsp proteins is displayed in the middle panel. Different motifs are represented by differently colored boxes. Details of the individual motifs are in Tables S8–S10. Exon/intron structures of the *Hsf* and *sHsp* genes are shown in the right panel. Green boxes represent exons and black lines represent introns.

Figure 2. Phylogenetic relationships, gene structures and motif compositions of the Hsp70, Hsp90 and Hsp100 family members in *S. melongena* (Sm). Multiple alignment of the Hsp70 (A), Hsp90 (B) and Hsp100 (C) proteins from *S. melongena* (Sm) was performed with MEGA 5.0 using the neighbor-joining (NJ) method with 1000 bootstrap replicates (left panel). A schematic representation of conserved motifs (obtained using MEME) in the Hsp70 (A), Hsp90 (B) and Hsp100 (C) proteins is displayed in the right panel. Different motifs are represented by differently colored boxes. Details of the individual motifs are in Tables S11–S13. The exon/intron structures of the Hsp70 (A), Hsp90 (B) and Hsp100 (C) genes are shown in the middle panel. Green boxes represent exons and black lines represent introns.

A strong response to HS in all of the 25 *sHsp* genes from both lines (1-05 and 4-05) was observed, in which a majority of these genes were upregulated and only *Sm-15.6sHsp* and *Sm-15.8sHsp* were downregulated. After high-temperature treatment, the expression of *SmHsp1-60*, *SmHsp60-2*, *SmCpn60-a2* and *SmCpn-60b3* was increased in the two inbred lines, while *SmHsp4-60* and *SmCpn-60b3* was increased in 4-05 and no significant difference could be observed in 1-05. Among the 18 *Hsp70* genes, the expression was remarkably changed in response to heat treatment in the thermosensitive 1-05 and thermotolerant 4-05 leaves and these genes were upregulated in both plants. Among these upregulated genes, the expression quantity of *SmHsp1-70*, *SmcpHsp70-2* and *SmHsp-70BIP4* was higher in 4-05, compared with 1-05.

However, *SmHsp14-70* and *SmHsp15-70* were increased in the thermotolerant line, but decreased in the thermosensitive line. Considering the Hsp90 genes, the expression levels of most genes (*SmHsp1.3-90*, *SmHsp90-1.4*, *SmHsp90-1.5*, *SmHsp90-2.3*, *SmHsp90-5*, *SmHsp90-6*, *SmHsp9-7.1* and *SmHsp7.2-90*) were increased and only *SmHsp90-2.2* and *SmHsp904.3* were downregulated in the two lines. Among the upregulated *Hsp90* genes, gene expression levels of six genes in 4-05 were obviously higher than in 1-05. After heat treatment, *SmHsp-100ClpB1*, *SmHsp100-ClpB2*, *SmHsp100-ClpB3*, *SmHsp100-ClpD1* and *SmHsp-100ClpX2* expressions in the two lines were significantly increased. Among these genes, *SmHsp-100ClpB1*, *SmHsp100-ClpB3* and *SmHsp-100ClpX2* showed higher expression in 4-05 than in 1-05, but *SmHsp-100ClpB2* was more abundant in the thermosensitive line.

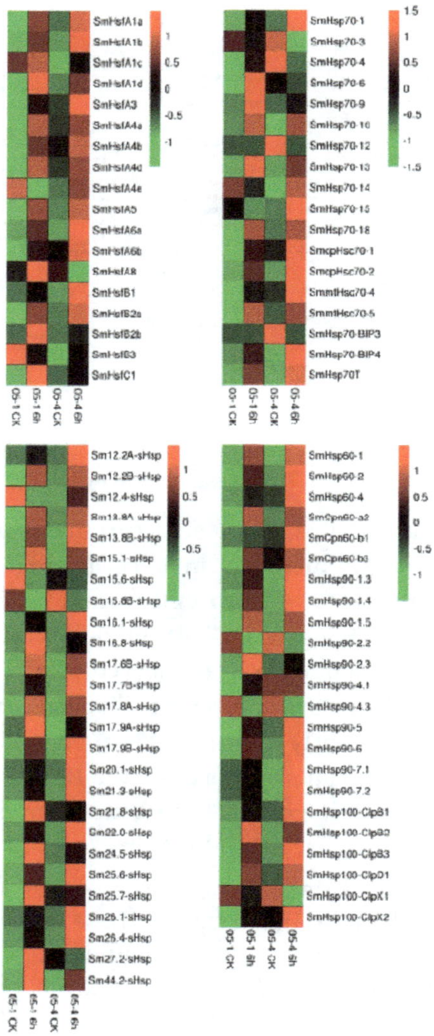

Figure 3. Expression analysis of eggplant *Hsf* and *Hsp* genes. Raw data were from RNA-seq data, in response to HS treatment in 05-1 and 05-4 leaves. HS treatment: 42 °C for 6 h; 05-1: eggplant thermosensitive line; 05-4: eggplant thermotolerant line.

3.5. Validation of Hsf and Hsp Gene Expression Levels by qRT-PCR

To verify the accuracy of the transcriptome sequencing, the expressions of 12 randomly selected genes were validated using quantitative real-time RT-PCR (qRT-PCR). The results showed that the expression pattern of each tested gene was similar to that of the transcriptome sequencing and the increase rate of all these *Hsf* and *Hsp* genes in the thermotolerant line 4-05 were significantly higher than those in thermosensitive line 1-05 (**Figure 4**).

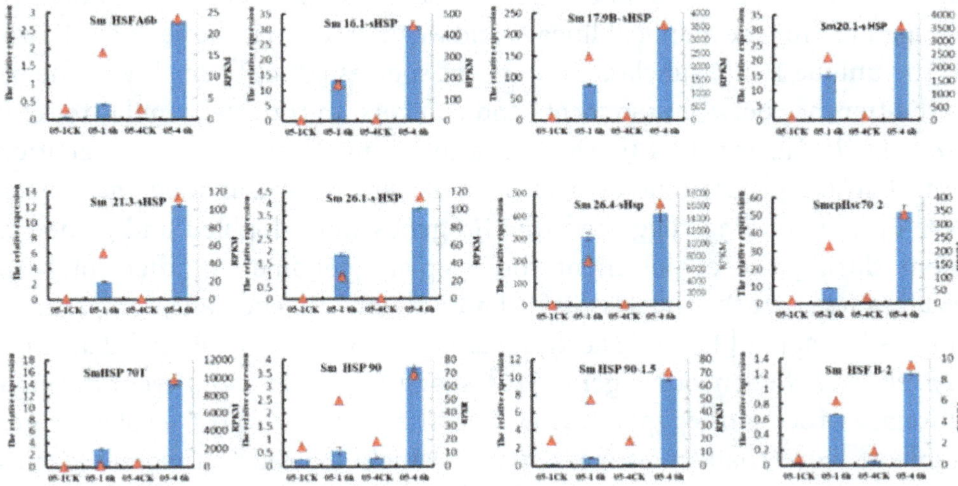

Figure 4. Comparison of transcripts expression results from RNA-seq and qRT-PCR analysis. Abscissa: Sample number; the ordinate (left): the relative expression of gene validated using qRT-PCR, represented by bar chart; coordinates (right): RPKM value obtained from the transcriptome sequencing, represented by triangle scatter diagram.

4. DISCUSSION

Many studies have suggested that *Hsfs* and *Hsps* play central roles in plant developmental and defense processes [**29,30**]. Benefiting from genome availability, the functions of the *Hsf* and *Hsp* family genes have been characterized in many plants. Although *Hsfs* and *Hsps* exist in all living organisms, their numbers vary in different plants. There are 22 *Hsfs* in Arabidopsis, 25 *Hsfs* in rice [**18**], 30 *Hsfs* in maize [**20**], 25 *Hsfs* in pepper [**31**] and 52 *Hsfs* in soybean [**32**]. Compared to the 27 *sHsp* genes in Arabidopsis [**33**], there are 35, 51 and 27 *sHsp* genes in pepper [**31**], soybean [**34**] and Chinese cabbage [**35**], respectively. Previous studies have identified 18 *Hsp70* genes in Arabidopsis and 32 genes in rice [**36**]. The grapevine genome contains at least seven genes encoding members of the *Hsp90* super family [**37**]. Zhang et al. (2015) reported 28 *Hsf*, 37 *sHsp*, 28 *Hsp60*, 20 *Hsp70* and 5 *Hsp100* genes in the poplar genome [**19**]. However, with the limited investigations into the molecular mechanism of heat tolerance, little is known about the *Hsf* family in eggplant.

In the present study, we identified 24 *Hsf* genes, 39 *sHsp* genes, 21 *Hsp60* genes, 30 *Hsp70* genes, 17 *Hsp90* genes and 10 *Hsp100* genes based on the eggplant genome (**Table 1**). Although the total number of *Hsf* and *Hsp* genes was similar to that of Arabidopsis [**18,38,39,40**], rice [**18,41**] and tomato [**42**], the members of some specific *Hsf* and *Hsp* subclasses in eggplant were different from the other three species. Two members were identified that belonged to subclass HsfC2 in rice, while no eggplant *Hsf* members were classified into subclass HsfC2 and the same events were also observed in *Arabidopsis thaliana* [**18**] and pepper [**31**]. Rice is the model plant use for the monocot lineage and we inferred that the gene duplications led to the unique HsfC2 subclass in monocot species [**17,42**], which was the most marked difference between monocots and eudicots. In contrast, similar to tomato and *Arabidopsis thaliana* [**18,43**], eggplant also has members that were partitioned into the HsfA6 subclass, but no rice *Hsf* members were classified into subclass HsfA**44**] 6]. This finding suggested that *Hsf* genes were doubled and gained new functions during the evolution of the eggplant genome. Another interesting observation was that the subclass *HsfA9* had 1 member in eggplant, compared with 4 members in pepper [**31**] and *Eucalyptos grandis* (Myrtaceae) contained at least 17 closely related *HsfA9*-encoding genes [**17**], suggesting a gene loss event during the evolutionary process of eggplant. However, there were two *HsfA4* subclass genes in eggplant, more than in pepper *CaHsfA4*, which showed that some *Hsfs* might have the similar functions, as in maize [**20**]. The reasons for the increase in the *HsfA9* genes need further investigation.

The phylogenetic analysis revealed that eggplant *Hsf* and *Hsp* members were more closely related to those from tomato than to those from Arabidopsis, which was consistent with the fact that both eggplant and tomato are members of the Solanaceae family [**45**]. Based on the previous analysis of the evolution of *Hsfs* and *Hsps* in Chinese cabbage [**21,35**], rice [**46**] and soybean [**47**], *Hsf* and *Hsp* genes essentially cover all the subfamilies and are relatively stable and conserved in the evolutionary process of eggplant and most of the *Hsf* and *Hsp* gene families were closely related to the evolutionary species.

Divergences in coding regions, particularly those that change the function of the gene, reflect amino acid altering substitutions and/or alterations in exon-intron structure [**19**]. The differences in intron and exon structure play important roles in the evolution of family genes. Structural analyses showed that the eggplant *Hsf* genes contained 7–0 introns and there were significant differences in the intron length; similar results were also obtained in cucumber [**48**], rice [**49**] and chickpea [**50**], but this result was different from that of pepper [**31**], for which all members have one intron. The number of introns of the *Hsp* gene family members in eggplant also showed differences, similar to the results of previous studies on poplar *sHsp*, *Hsp60*, *Hsp70* and *Hsp***19**] *100*]. Qiao (2015), researching the pear *Hsf* and Guo (2015), researching the pepper *Hsp20*, showed a lack of conserved motifs among all the family genes and none of these genes contained the whole sequence, consistent with the eggplant *Hsfs* and *Hsps* in the present study [**31,51**]. We speculated that the deletion of introns and domains leads to structural changes during evolution, leading to functional diversity in *Hsf* and *Hsp* genes in eggplant; however, this theory needs experimental confirmation.

Hsfs, as transcriptional activators of *Hsps*, cooperate with *Hsps* to form a network responding to various stresses. These factors play a broad role in the tolerance to multiple environmental stress treatments apart from heat stress [**52,53**]. The comprehensive analysis of the expression for individual *Hsf* and *Hsp* members under HS was necessary for further functional analyses in plant thermotolerance [**23,54**]. The present study showed that most members of the eggplant *sHsp*, *Hsp60*, *Hsp70*, *Hsp90* and *Hsp100* families were induced by HS treatment in lines 1-05 and 4-05 and only a few members were significantly downregulated. Several studies have indicated that the expression and accumulation of heat shock proteins and heat shock transcription factors can enhance the thermostability of tomato [**55**], wheat [**56**] and rice [**57**]. *Hsfs* are activated under HS conditions and subsequently bind the HSE elements of the promoters of the *Hsp* genes to regulate the expression of downstream *Hsp* genes [**17**]. The accumulation of the *Hsps* effectively reduces the damage from HS and enhances thermotolerance by binding denaturing proteins and preventing them from irreversible aggregation [**58**]. Thus far, only *sHsp* has been shown to play a major role in improving plant thermotolerance in the form of molecular chaperones and cell membrane stabilizing factors [**59**]. However, the specific mechanisms of other *Hsp* genes are less well established. Previous studies have shown that the response of plants to high temperature was a quantitative trait controlled by multiple genes; some normal genes were closed and some stress tolerance-related genes were induced under high-temperature stress, thus altering plant morphogenesis, physiological functions and biochemical and molecular structures, which in turn influenced the growth of plants [**60**]. In addition, heat shock proteins are different from other stress proteins and have their own unique characteristics. In the present study, *Hsps* (*sHsp*, *Hsp60*, *Hsp70*, *Hsp90* and *Hsp100*) showed species diversity, universal distribution and instantaneous response and structural conservation. For example, the synthesis of heat shock protein was fast, beginning between the first few minutes and tens of minutes and the expression lasted for up to several hours, occasionally continuing for 12 or more hours (**Figure 4**). Similar results were also observed in poplar [**19**] and grape [**61**].

Arabidopsis has four members belonging to the HsfA1 family, namely A1a, A1b, A1c, and A1d [62]. Research has demonstrated that HsfA1a has the ability to immediately detect heat stress and become active. Additionally, the same treatments were found to stimulate the binding to Hsp18.2 and Hsp70 promoters, as confirmed by chromatin immunoprecipitation [63]. Upregulating HsfA1a improves various stress resistance by stimulating the production of stress-induced genes Hsp18.2 and Hsp70 [64]. Furthermore, AtHsfA1 was found to be associated with both drought stress [65] and programmed cell death [66]. Therefore, in the case of eggplant, HsfA1 may likewise serve a comparable role to AtHsfA1 and concurrently interact with Hsps. Mounting evidence indicates that Hsp, regulated by Hsf, is a crucial heat stress protein that serves as the foundation for plant cells' reaction to damage caused by high temperatures [67,68,69]. Upon exposure to elevated temperatures, the majority of Hsf and Hsp genes in eggplant were promptly activated and the amount of gene expression in the thermotolerant line was significantly greater than that in the thermosensitive line. Thus, the Hsf–Hsp protein degradation pathway is the primary mechanism by which eggplants respond to high temperature stress and perhaps contributes significantly to the development of heat tolerance in eggplants. The results lay the groundwork for additional functional research on these genes in eggplant, which could possibly be valuable in understanding the mechanism of thermotolerance in eggplant, as well as in other solanaceous plants.

5. CONCISE SUMMARY

The current investigation discovered a total of 24 Heat Shock Factor (Hsf) genes and 117 Heat Shock Protein (Hsp) genes, which encompassed sHsp, Hsp60, Hsp70, and Hsp100, from the genome of the eggplant. An analysis was conducted on the phylogeny, gene structure, expression profiles, and heat stress responses of these genes. The eggplant has a comparable amount of Hsf and Hsp genes to Arabidopsis, tomato, and rice. These genes span all the subfamilies and exhibit a relatively stable and conserved gene structure and motif composition throughout the evolutionary process. SmHsf genes, acting as crucial transcriptional activators, governed distinct subfamilies of Hsps in eggplant. The majority of Hsf and Hsp genes in eggplant leaves are significantly upregulated in response to heat stress, suggesting their involvement in the heat stress response. The expression levels of these genes in the thermotolerant line were much greater than those in the thermosensitive line during heat stress (HS) in eggplant. This difference in gene expression may be the primary factor contributing to the strong thermotolerance seen in the thermotolerant eggplant. Based on the aforementioned findings, it is anticipated that the thermotolerance of various eggplant resources will be assessed by examining the alteration in expression of particular Hsf and Hsp genes under heat stress (HS). Additionally, the thermotolerance of additional solanaceae species resources will also be evaluated. The objective of this study was to provide a strong basis for functional research on the eggplant Hsf and Hsp gene families and enhance our comprehension of the mechanism by which Hsf and Hsp genes in solanaceous plants promote thermotolerance.

AT A GLANCE

When plant cells are exposed to high temperatures, a significant amount of proteins will undergo denaturation and become inactive. Simultaneously, the expression of Heat Shock Factors (Hsfs) and Heat Shock Proteins (Hsps) will be promptly stimulated to eliminate denatured proteins, thereby preventing programmed cell death and consequently improving the thermotolerance of plants. A thorough investigation was conducted to identify and analyze the Hsf and Hsp gene families in eggplant under conditions of heat stress. The eggplant genome was analyzed using the interolog from Arabidopsis, resulting in the identification of 24 Hsf-like genes and 117 Hsp-like genes. The genetic architecture and motif composition of Hsf and Hsp genes exhibited a high degree of conservation within each subfamily in eggplant. The RNA-seq data and qRT-PCR analysis demonstrated that the expressions of the majority of eggplant Hsf and Hsp genes were elevated in response to heat stress, particularly in the thermotolerant line. The thorough study revealed that distinct sets of SmHsps genes were engaged in the subsequent stages of specific SmHsfs genes. These results established a foundation for uncovering the functions of SmHsps and SmHsp in promoting thermotolerance in eggplant. This knowledge could potentially aid in comprehending the mechanism by which SmHsps and SmHsp contribute to thermotolerance in eggplant.

REFERENCES

1. Cramer, G.R.; Urano, K.; Delrot, S.; Pezzotti, M.; Shinozaki, K. Effects of abiotic stress on plants: A systems biology perspective. *BMC Plant Biol.* 2011, *11*, 163.
2. Li, Y.; Li, Z.; Luo, S.; Sun, B. Effects of heat stress on gene expression in eggplant (*Solanum melongema* L.) seedlings. *Afr. J. Biotechnol.* 2011, *10*, 18078–18084.
3. Zhang, A.; Zhu, Z.; Shang, J.; Zhang, S.; Shen, H.; Wu, X.; Zha, D. Transcriptome profiling and gene expression analyses of eggplant (*Solanum melongena* L.) under heat stress. *PLoS ONE* 2020, *15*.
4. Kesici, M.; Ipek, A.; Ersoy, F.; Ergin, S.; Gülen, H. Genotype-dependent gene expression in strawberry (*Fragaria x ananassa*) plants under high temperature stress. *Biochem. Genet.* 2020, *58*, 848–866.
5. Yang, G.; Gao, X.; Ma, K.; Li, D.; Jia, C.; Zhai, M.; Xu, Z. The walnut transcription factor *JrGRAS2* contributes to high temperature stress tolerance involving in Dof transcriptional regulation and HSP protein expression. *BMC Plant Biol.* 2018, *18*, 1.
6. Sadura, I.; Libik-Konieczny, M.; Jurczyk, B.; Gruszka, D.; Janeczko, A. HSP transcript and protein accumulation in brassinosteroid barley mutants acclimated to Low and high temperatures. *IJMS* 2020, *21*, 1889.
7. Zha, Q.; Xi, X.; Jiang, A.; Wang, S.; Tian, Y. Changes in the protective mechanism of photosystem II and molecular regulation in response to high temperature stress in grapevines. *Plant Physiol. Bioch.* 2016, *101*, 43–53.

8. Sung, D.-Y.; Kaplan, F.; Lee, K.-J.; Guy, C.L. Acquired tolerance to temperature extremes. *Trends Plant Sci.* 2003, *8*, 179–187.

9. Wang, F.; Dong, Q.; Jiang, H.; Zhu, S.; Chen, B.; Xiang, Y. Genome-wide analysis of the heat shock transcription factors in *Populus trichocarpa* and *Medicago truncatula*. *Mol. Biol. Rep.* 2012, *39*, 1877–1886.

10. Lindquist, S.; Craig, E.A. The heat-shock protein. *Annu. Rev. Genet.* 1988, *22*, 631–677.

11. Lopez-Matas, M.A.; Nuñez, P.; Soto, A.; Allona, I.; Casado, R.; Collada, C.; Guevara, M.A.; Aragoncillo, C.; Gomez, L. Protein cryoprotective activity of a cytosolic small heat shock protein that accumulates constitutively in chestnut stems and is up-regulated by low and high temperatures. *Plant Physiol.* 2004, *134*, 1708–1717.

12. Suzuki, K.; Nakanishi, H.; Bower, J.; Yoder, D.W.; Osteryoung, K.W.; Miyagishima, S.Y. Plastid chaperonin proteins Cpn60α and Cpn60β are required for plastid division in *Arabidopsis thaliana*. *BMC Plant Biol.* 2009, *9*, 38.

13. Usman, M.G.; Rafii, M.Y.; Ismail, M.R.; Malek, M.A.; Latif, M.A. Expression of target gene *Hsp70* and membrane stability determine heat tolerance in chili pepper. *J. Am. Soc. Hortic. Sci.* 2015, *140*, 144–150.

14. McLellan, C.A.; Turbyville, T.J.; Wijeratne, E.M.; Kerschen, A.; Vierling, E.; Queitsch, C.; Whitesell, L.; Gunatilaka, A.A.L. A rhizosphere fungus enhances Arabidopsis thermotolerance through production of an HSP90 inhibitor. *Plant Physiol.* 2007, *145*, 174–182.

15. Nover, L.; Bharti, K.; Döring, P.; Mishra, S.K.; Ganguli, A.; Scharf, K.D. Arabidopsis and the heat stress transcription factor world: How many heat stress transcription factors do we need? *Cell Stress Chaperon* 2001, *6*, 177–189.

16. Yoshida, T.; Ohama, N.; Nakajima, J.; Kidokoro, S.; Mizoi, J.; Nakashima, K.; Maruyama, K.; Kim, J.M.; Seki, M.; Todaka, D.; et al. Arabidopsis *HsfAl* transcription factors function as the main positive regulators in heat shock-responsive gene expression. *Mol. Genet. Genom.* 2011, *286*, 321–332.

17. Scharf, K.-D.; Berberich, T.; Ebersberger, I.; Nover, L. The plant heat stress transcription factor (Hsf) family: Structure function and evolution. *Biochim. Biophys. Acta.* 2012, *1819*, 104–119.

18. Guo, J.; Wu, J.; Ji, Q.; Wang, C.; Luo, L.; Yuan, Y.; Wang, Y.H.; Wang, J. Genome-wide analysis of heat shock transcription factor families in rice and Arabidopsis. *J. Genet. Genom.* 2008, *35*, 105–118.

19. Zhang, J.; Liu, B.; Li, J.; Zhang, L.; Wang, Y.; Zheng, H.; Lu, M.; Chen, J. Hsf and Hsp gene families in Populus: Genome-wide identification organization and correlated expression during development and in stress responses. *BMC Genom.* 2015, *16*, 181–200.

20. Lin, Y.; Jiang, H.; Chu, Z.; Tang, X.; Zhu, S.; Cheng, B. Genome-wide identification classification and analysis of heat shock transcription factor family in maize. *BMC Genom.* 2011, *12*, 76.

21. Song, X.; Liu, G.; Duan, W.; Liu, T.; Huang, Z.; Ren, J.; Li, Y.; Hou, X. Genome-wide identification classification and expression analysis of the heat shock transcription factor family in Chinese cabbage. *Mol. Genet. Genom.* 2014, *289*, 541–551.

22. Hirakawa, H.; Shirasawa, K.; Miyatake, K.; Nunome, T.; Negoro, S.; Ohyama, A.; Yamaguchi, H.; Sato, S.; Isobe, S.; Tabata, S.; et al. Draft Genome Sequence of eggplant (*Solanum melongena* L.): The representative solanum species indigenous to the old world. *DNA Res.* 2014, *21*, 649–660.

23. Swindell, W.R.; Huebner, M.; Weber, A.P. Transcriptional profiling of Arabidopsis heat shock proteins and transcription factors reveals extensive overlap between heat and non-heat stress response pathways. *BMC Genom.* 2007, *8*, 125.

24. Larkin, M.A.; Blackshields, G.; Brown, N.P.; Chenna, R.; McGettigan, P.A.; McWilliam, H.; Valentin, F.; Wallace, I.M.; Wilm, A.; Lopez, R.; et al. Clustal W and Clustal X version2.0. *Bioinformatics* 2007, *23*, 2947–2948.

25. Tamura, K.; Peterson, D.; Peterson, N.; Stecher, G.; Nei, M.; Kumar, S. MEGA5: Molecular evolutionary genetics analysis using maximum likelihood evolutionary distance and maximum parsimony methods. *Mol. Biol. Evol.* 2011, *28*, 2731–2739.

26. Guo, A.Y.; Zhu, Q.H.; Chen, X.; Luo, J.C. GSDS: A gene structure display server. *Yi Chuan* 2007, *29*, 1023–1026.

27. Bailey, T.L.; Williams, N.; Misleh, C.; Li, W.W. MEME: Discovering and analyzing DNA and protein sequence motifs. *Nucl. Acids Res.* 2006, *34*, W369–W373.

28. Schmittgen, T.D.; Livak, K.J. Analyzing real-time PCR data by the comparative C T method. *Nat Protoc.* 2008, 3, 1101–1108.

29. Giorno, F.; Wolters-Arts, M.; Grillo, S.; Scharf, K.D.; Vriezen, W.H.; Mariani, C. Developmental and heat stress-regulated expression of HsfA2 and small heat shock proteins in tomato anthers. *J. Exp. Bot.* 2010, *61*, 453–462.

30. Pérez-Salamó, I.; Papdi, C.; Rigó, G.; Zsigmond, L.; Vilela, B.; Lumbreras, V.; Nagy, I.; Horváth, B.; Domoki, M.; Darula, Z.; et al. The heat shock factor A4A confers salt tolerance and is regulated by oxidative stress and the mitogen-activated protein kinases MPK3 and MPK6. *Plant Physiol.* 2014, *165*, 319–334.

31. Guo, M.; Lu, J.; Zhai, Y.; Chai, W.; Gong, Z.; Lu, M. Genome-wide analysis expression profile of heat shock factor gene family (*CaHsfs*) and characterization of *CaHsfA2* in pepper (*Capsicum annuum* L.). *BMC Plant Biol.* 2015, *15*, 151.

32. Chung, E.; Kim, K.M.; Lee, J.H. Genome-wide analysis and molecular characterization of heat shock transcription factor family in *Glycine max*. *J. Genet. Genom.* 2013, 3, 127–135.

33. Siddique, M.; Gernhard, S.; von Koskull-Doring, P.; Vierling, E.; Scharf, K.D. The plant sHSP superfamily: Five new members in Arabidopsis thaliana with unexpected properties. *Cell Stress Chaperon* 2008, *13*, 183–197.

34. Lopescaitar, V.S.; Carvalho, M.C.D.; Darben, L.M.; Kuwahara, M.K.; Nepomuceno, A.L.; Dias, W.P.; Abdelnoor, R.V.; Marcelino-Guimarães, F.C. Genome-wide analysis of the *Hsp20* gene family in soybean: Comprehensive sequence genomic organization and expression profile analysis under abiotic and biotic stresses. *BMC Genom.* 2013, *14*, 577.

35. Tao, P.; Guo, W.; Li, B.; Wang, W.; Yue, Z.; Lei, J.; Zhong, X. Genome-wide identification classification and expression analysis of sHSP genes in Chinese cabbage (*Brassica rapa* ssp. pekinensis). *Genet. Mol. Res.* 2015, *14*, 11975–11993.

36. Sarkar, N.K.; Kundnani, P.; Grover, A. Functional analysis of Hsp70 superfamily proteins of rice (*Oryza sativa*). *Cell Stress Chaperon* 2013, *18*, 427–437.

37. Banilas, G.; Korkas, E.; Englezos, V.; Nisiotou, A.A.; Hatzopoulos, P. Genome-wide analysis of the heat shock protein 90 gene family in grapevine (*Vitis vinifera* L.). *Aust. J. Grape Wine Res.* 2012, *18*, 29–38.

38. Krishna, P.; Gloor, G. The *Hsp90* family of proteins in Arabidopsis thaliana. *Cell Stress Chaperon* 2001, *6*, 238–246.

39. Lee, U.; Rioflorido, I.; Hong, S.; Larkindale, J.; Waters, E.R.; Vierling, E. The Arabidopsis ClpB/Hsp100 family of proteins: Chaperones for stress and chloroplast development. *Plant J.* 2007, *49*, 115–127.

40. Scharf, K.D.; Siddique, M.; Vierling, E. The expanding family of Arabidopsis thaliana small heat stress proteins and a new family of proteins containing alpha-crystallin domains (Acd proteins). *Cell Stress Chaperon* 2001, *6*, 225–237.

41. Singh, A.; Singh, U.; Mittal, D.; Grover, A. Genome-wide analysis of rice *ClpB/HSP100 ClpC* and *ClpD* genes. *BMC Genom.* 2010, *11*, 95.

42. Wang, G.D.; Kong, F.Y. Research advancement of heat shock factors in tomato. *Plant Physiol. J.* 2013, *49*, 217–224.

43. Chauhan, H.; Khurana, N.; Agarwal, P.; Khurana, P. Heat shock factors in rice (*Oryza sativa* L.): Genome-wide expression analysis during reproductive development and abiotic stress. *Mol. Genet. Genom.* 2011, *286*, 171–187.

44. Mueller, L.A.; Tanksley, S.D. The SOL Genomics Network: A comparative resource for Solanaceae biology and beyond. *Plant Physiol.* 2005, *138*, 1310–1317.

45. Xu, G.; Guo, C.; Shan, H.; Kong, H. Divergence of duplicate genes in exon-intron structure. *Proc. Natl. Acad. Sci. USA* 2012, *109*, 1187–1192.

46. Wang, Y.; Lin, S.; Song, Q.; Tao, H.; Huang, J.; Chen, X.; Que, S.; He, H. Genome-wide identification of heat shock proteins (Hsps) and Hsp interactors in rice: Hsp70s as a case study. *BMC Genom.* 2014, *15*, 344.

47. Zhang, L.; Zhao, H.K.; Dong, Q.L.; Zhang, Y.Y.; Wang, Y.M.; Li, H.Y.; Xing, G.J.; Li, Q.Y.; Dong, Y.S. Genome-wide analysis and expression profiling under heat and drought treatments of *HSP70* gene family in soybean (*Glycine max* L.). *Front Plant Sci.* 2015, *6*, 773.

48. Zhou, S.; Zhang, P.; Jing, Z.; Shi, J. Genome-wide identification and analysis of heat shock transcription factor family in cucumber (*Cucumis sativus* L.). *Plant Omics.* 2013, *6*, 449–455.

49. Wan, B.L.; Zha, Z.P.; Du, X.S. Expression profile analysis of rice heat shock transcription factor (*HSF*) genes in response to plant growth regulators and abiotic stresses. *China Biotechnol.* 2010, *30*, 22–32.

50. Zafar, S.A.; Hussain, M.; Raza, M.; Ahmed, M.D.; Rana, I.A.; Sadia, B. Genome wide analysis of heat shock transcription factor (*HSF*) family in chickpea and its comparison with Arabidopsis. *Plant Omics* 2016, *9*, 136–141.

51. Qiao, X.; Li, M.; Li, L.; Yin, H.; Wu, J.; Zhang, S. Genome-wide identification and comparative analysis of the heat shock transcription factor family in Chinese white pear (*Pyrus bretschneideri* L.) and five other Rosaceae species. *BMC Plant Biol.* 2015, *15*, 12. **[Google Scholar] [CrossRef] [Green Version]**

52. Wahid, A.; Gelani, S.; Ashraf, M.; Foolad, M.R. Heat tolerance in plants: An overview. *Environ. Exp. Bot.* 2007, *61*, 199–223.

53. Hahn, A.; Bublak, D.; Schleiff, E.; Scharf, K.D. Crosstalk between *Hsp90* and *Hsp70* chaperones and heat stress transcription factors in tomato. *Plant Cell* 2011, *23*, 741–755.

54. Hu, W.; Hu, G.; Han, B. Genome-wide survey and expression profiling of heat shock proteins and heat shock factors revealed overlapped and stress specific response under abiotic stresses in rice. *Plant Sci.* 2009, *176*, 583–590.

55. Mishra, S.K.; Tripp, J.; Winkelhaus, S.; Tschiersch, B.; Theres, K.; Nover, L.; Scharf, K.D. In the complex family of heat stress transcription factors *HsfA1* has a unique role as master regulator of thermotolerance in tomato. *Genes Dev.* 2002, *16*, 1555–1567.

56. Chauhan, H.; Khurana, N.; Nijhavan, A.; Khurana, J.P.; Khurana, P. The wheat chloroplastic small heat shock protein (sHsp26) is involved in seed maturation and germination and imparts tolerance to heat stress. *Plant Cell Environ.* 2012, *35*, 1912–1931.

57. Kim, S.R. and An, G. Rice chloroplast-localized heat shock protein 70 OsHsp70CP1 is essential for chloroplast development under high-temperature conditions. *J. Plant Physiol.* 2013, *170*, 854–863.

58. Park, H.S.; Jeong, W.J.; Kim, E.C.; Jung, Y.; Lim, J.M.; Hwang, M.S.; Park, E.J.; Ha, D.S.; Choi, D.W. Heat shock protein gene family of the porphyra seriata and enhancement of heat stress tolerance by *PsHsp70* in Chlamydomonas. *Mar. Biotechnol.* 2012, *14*, 332–342.

59. Li, D.; Yang, F.; Lu, B.; Chen, D.; Yang, W. Thermotolerance and molecular chaperone function of the small heat shock protein HSP20 from hyperthermophilic archaeon *Sulfolobus solfataricus* P2. *Cell Stress Chaperon* 2012, *17*, 103–108.

60. Bondino, H.G.; Valle, E.M.; Ten, H.A. Evolution and functional diversification of the small heat shock protein/alpha-crystallin family in higher plants. *Planta* 2012, *235*, 1299–1313.

61. Liu, G.; Wang, J.; Cramer, G.; Dai, Z.; Duan, W.; Xu, H.; Wu, B.; Fan, P.; Wang, L.; Li, S. Transcriptomic analysis of grape (*Vitis vinifera* L.) leaves during and after recovery from heat stress. *BMC Plant Biol.* 2012, *12*, 174.

62. Liu, H.; Liao, H.; Charng, Y. The role of class A1 heat shock factors (*HSFA1s*) in response to heat and other stresses in Arabidopsis. *Plant Cell Environ.* 2011, *34*, 738–751.

63. Liu, Y.; Zhang, C.; Chen, J.; Guo, L.; Li, X.; Li, W.; Yu, Z.; Deng, J.; Zhang, P.; Zhang, K.; et al. Arabidopsis heat shock factor *HsfA1a* directly senses heat stress pH changes and hydrogen peroxide via the engagement of redox state. *Plant Physiol. Bioch.* 2013, *64*, 92–98.

64. Qian, J.; Chen, J.; Liu, Y.; Yang, L.; Li, W.; Zhang, L. Overexpression of Arabidopsis *HsfA1*a enhances diverse stress tolerance by promoting stress-induced *Hsp* expression. *Genet Mol. Res.* 2014, *13*, 1233–1243.

65. Bechtold, U.; Albihlal, W.S.; Lawson, T.; Fryer, M.J.; Sparrow, P.A.C.; Richard, F. Arabidopsis heat shock transcription factor *A1b* overexpression enhances water productivity resistance to drought and infection. *J. Exp. Bot.* 2013, *64*, 3467–3481.

66. Li, N.; Zeng, J.; Xu, Y.; Guo, L. Effects of Heat Shock Factor *AtHsfA1a* on programmed cell death in Arabidopsis thaliana under cold stress. *Agric. Biotechnol.* 2016, *5*, 57–59.

67. Wang, W.; Vinocur, B.; Shoseyov, O.; Altman, A. Role of plant heat-shock proteins and molecular chaperones in the abiotic stress response. *Trends Plant Sci.* 2004, *9*, 244–252.

68. Ostling, P.; Björk, J.K.; Roos-Mattjus, P.; Mezger, V.; Sistonen, L. Heat shock factor 2 (HSF2) contributes to inducible expression of *hsp* genes through interplay with HSF1. *J. Biol. Chem.* 2007, *282*, 7077–7086.

69. Singh, A.; Mittal, D.; Lavania, D.; Agarwal, M.; Mishra, R.C.; Grover, A. *OsHsfA2c* and *OsHsfB4b* are involved in the transcriptional regulation of cytoplasmic *OsClpB* (*Hsp100*) gene in rice (*Oryza sativa* L.). *Cell Stress Chaperones* 2012, *17*, 243–254.

13 Chapter

ADVANCEMENTS IN THE STUDY OF ALLELOPATHY USING VOLATILE ORGANIC COMPOUNDS (VOCS) EMITTED BY PLANTS

1. INTRODUCTION: AN OVERVIEW

The term "allelopathy" was initially introduced by Austrian scientist Hans Molisch in 1937 and primarily pertained to the chemical interactions between plants. Allelopathy is a significant ecological phenomena that plays a crucial role in the ecological adaptation of plants [1,2]. The allelopathic effects might have both beneficial and detrimental consequences. Multiple studies have documented the benefits of allelopathic effects in agricultural systems, including the control of weeds [3,4,5,6], the suppression of pests [7,8,9,10], the prevention of diseases [11,12], the enhancement of soil nutrition [13,14], and the promotion of microbial interactions [15,16]. In conclusion, the allelopathy of the majority of plants has a significant impact on plant growth [10,17,18]. Plants have the ability to produce a wide range of secondary metabolites as they grow and develop. Plant volatile organic compounds (VOCs) differ depending on the species of plants and are influenced by the presence and composition of nearby plant species [19,20]. Secondary metabolites can have either positive or negative effects on other organisms when they are stored or released into the environment. For example, plants store secondary metabolites that can deter animals from feeding on them and protect against microbial infestation. On the other hand, when these metabolites are released into the air, they can attract insect pollinators [21]. Plants establish communication with species in their environment by emitting volatile organic compounds (VOCs). This allows them to carry out various ecological roles, including influencing their own growth, development, defense mechanisms, reproduction, and life cycle [22]. In 1984, Rice defined allelopathy as the harmful effect that one plant (including microorganisms) has on another through the release of chemical compounds into the environment. These compounds, known as allelochemicals, can be found in various parts of the plant (leaves, stems, flowers, seeds, fruits, and/or roots) and can be released through different means: volatilization, foliar leaching, root exudations, and decomposition of plant residue (Figure 1).

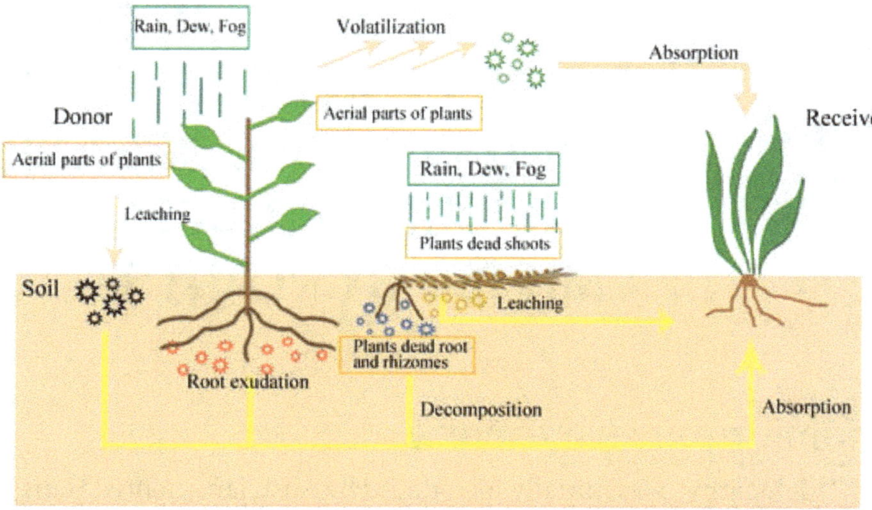

Figure 1. **The allelopathy pathways of plants.**

Volatile organic compounds (VOCs) are plant-derived secondary metabolites that are released into the air and are commonly found as allelochemicals in plants [24]. The four primary synthesis pathways of volatile organic compounds (VOCs) in plants are the shikimate/phenylalanine, mevalonic acid (MVA), methylerythritol phosphate (MEP), and lipoxygenase (LOX) pathways. Through these pathways, plants are capable of synthesizing and emitting a wide range of VOCs, including terpenoids, phenylpropanoids/benzenoids, and fatty acid derivatives [25]. The volatile organic compounds (VOCs) emitted by these plants serve various ecological purposes, including chemical signaling, kin identification, insect attraction or repulsion, and numerous other impacts [21,26,27,28,29]. While the majority of research on plant VOCs focuses on aboveground chemical signals, an increasing number of studies indicate that VOCs also play a crucial role in belowground plant-plant interactions [30]. Indeed, the observation that plants emit allelochemicals through the volatile pathway has been long-standing. Molisch conducted one of the initial empirical investigations on allelopathy, specifically focusing on the effects of volatile organic compounds (VOCs). The study revealed that VOCs emitted by apples and pears have the ability to hinder the germination of potatoes [31]. Volatile organic compounds (VOCs) have been extensively shown to principally protect against herbivorous insects [32,33], bacteria, and diseases [34,35,36], consequently decreasing severe environmental stress [37,38] and enhancing nutritional uptake [11]. Muller et al. [39] conducted a study on the volatile compounds emitted by annual grassland species in Salvia leucophylla Greene and Artemisia californica communities. Their research revealed that these volatile allelochemicals had interspecific allelopathic effects on the woody herbaceous plants, resulting in negative impacts on the recipient plant species [40,41] and alterations in soil microorganisms [42,43]. In addition to the volatile organic compounds (VOCs) emitted by plant shoots, root volatiles can also have allelopathic effects on nearby plants. For instance, the VOCs generated by the roots of huge sagebrush (Artemisia tidentata Nutt.) were found to hinder the germination of wild tobacco (Nicotiana attenuata) seeds [44]. The majority of allelochemicals produced by plant roots are classified as "root exudates" [45]. However, a small number of allelochemicals are released by the process of volatilization from the roots, known as VOCs. These VOCs have a significant ecological impact on the soil ecosystem and have not yet been extensively researched [46,47].

Allelopathy has garnered significant interest and emerged as a prominent scientific issue in the field of ecology [48]. Allelopathy is being developed as a distinct scientific discipline, and we are conducting thorough and extensive study from both theoretical and practical perspectives. Plant-derived volatile organic compounds (VOCs) are a primary mechanism for achieving allelopathic effects. Utilizing the allelopathic effect of plant natural products to enhance agricultural productivity is a cost-effective, eco-friendly, and efficient approach [49]. While there have been several comprehensive reviews on the potential applications of VOCs [20,22,49], there is a lack of systematic review and reporting on the allelopathy of plant VOCs. Volatile Organic Compounds (VOCs) are naturally occurring chemical molecules that are released into the air by plants. These compounds serve as natural herbicides and fungicides, helping to protect nearby plants from stress and enhance agricultural productivity [49]. We believe that volatile organic compounds (VOCs) have a significantly wider spectrum of possible applications. Therefore, in this specific situation, doing research on the allelopathy of volatile organic compounds (VOCs) is of utmost significance for the advancement of sustainable agriculture. The review primarily examines recent studies on the allelopathic effects of plant volatile organic compounds (VOCs). These studies investigate the ability of VOCs to resist diseases, prevent pests, inhibit weed growth, break dormancy, regulate plant growth, affect reactive oxygen species (ROS) levels and enzyme activity, modulate plant respiration and photosynthesis, and act as signaling molecules. We provide the data from the references to elucidate these roles and enhance the comprehension of allelopathy in plants through the study of volatile organic compounds (VOCs).

2. ALLELOPATHY OF VOCS OF PLANTS

Experts are paying more attention to the allelopathy of volatile compounds, which has led to the gradual discovery of the possible role of volatile organic compounds (VOCs) in agriculture. The extant literature summarizes the allelopathic effects of volatile organic compounds (VOCs) on plants, specifically focusing on the primary allelopathic interactions involving plant VOCs. (See Figure 2).

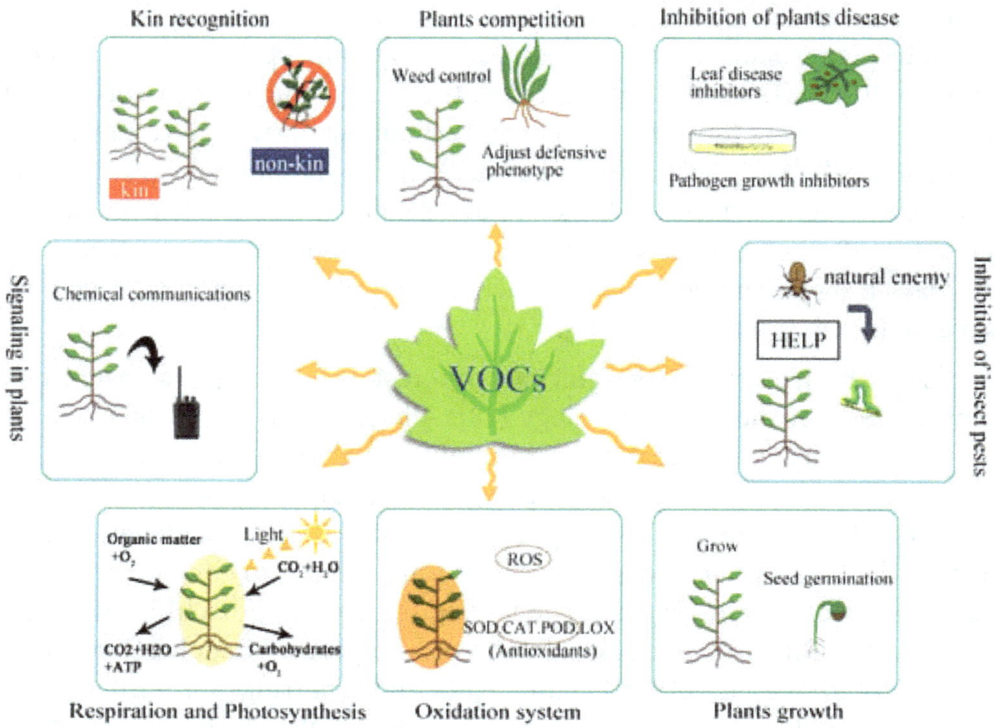

Figure 2. The main allelopathy of plants VOCs. (Note: This figure is widely inspired and adapted from Brilli et al. (2019) [49]).

2.1. VOCs and Plants Growth

Numerous results showed that VOCs have an effect on plant cell growth and differentiation [**50,51,52,53**], such as diallyl disulfide (DADS) of garlic VOCs, which can affect mitotic activity and cell length of tomato roots by impacting on cell division, endogenous plant hormone levels, expansin gene expression, and sulfate assimilation and glutathione (GSH) metabolism [**53,54**].

Similarly, many studies confirm that plant volatiles can inhibit seed germination and the growth of root and seedling [**50,55,56,57,58,59,60**]. β-terpineol, linalool, eugenol, and tetradecanoic acid are the VOCs released from tomato (*Solanum lycopersicon* Mill.) foliage, and they could inhibit seed germination of the tropical plant *Amaranthus mangostanus* L. [**55**]. The brassica species exude allelochemicals, which are glucosinolates [**61**] that could break down into several biological action compounds, such as isothiocyanates, which have biologically active and inhibit germination and growth of exposed plant species [**57,62**]. VOCs released from pine needles and the roots of *Pinus halepensis* L. mainly inhibited the seed germination and root growth of two herbaceous target species *Lactuca sativa* L. and *Linum strictum* L. [**63**]. VOCs affect plant growth, mainly to change plant morphology and reduce plant biomass. Therefore, seed germination, seedling root length, and seedling height are often used as intuitive indicators to evaluate allelopathy. Low concentrations of DADS promoted the growth of cucumber roots and induce elongation of the main roots by up-regulating the expression of CsCDKA and CsCDKB genes and regulating the hormone balance of the roots [**64**]. The VOCs released by *Atriplex cana* Ledeb. (Amaranthaceae) significantly inhibited seedling growth of *Amaranthus retroflexus* L. and *Poa annua* L., and 5 μg/mL essential oil completely inhibited the seed germination of *A. retroflexus*, *Medicago sativa* L., *P. annua*, and *Echinochloa crusgalli* L. [**65**]. Besides, Effah et al. found that plant VOCs mediate multiple ecological networks, and they may mediate the allelopathic effects of the germination or growth of competitors seeds [**66**]. Monoterpenoids are considered as effective inhibitors of seed germination and seedling growth [**67,68**].

VOCs affect plant growth, mainly to inhibit plant growth, but some studies have found that VOCs have dual effects on germination and plant growth, both promoting and inhibiting; for example, Arroyo et al. found that volatile chemicals from *Artemisia herba-alba* Asso. inhibited the germination of *Pinus halepensis* Mill. seeds, promoted the growth of *P. halepensis* seedlings, and reduced the root biomass of *Salsola vermiculata* L. seedlings [**69**]. VOCs also have an effect on plants growth direction. Runyon et al. found that *Cuscuta pentagona* seedlings were favored by not only the growth of host tomato plants, but also the direction of tomato VOCs [**70**]. The extracted VOCs of tomato and wheat were placed on both sides of *C. pentagona* seedlings, and the *C. pentagona* seedlings continued to grow in the direction of tomato VOCs. It was also found that β-phellandrene and β-myrcene of tomato VOCs could significantly attract the growth of *C. pentagona* seedlings.

2.2. VOCs and Weed Control

Allelopathy gain extensive attention in biological weed control [**3,4,5,6,71**]. Boydston et al. [**3**] found that mustard seed meal has the potential for weed control in organic production systems. The release of volatile organic compounds from the leaves can cause allelopathic effects and damage the growth of other competitive plants [**72**]. The allelopathy of the volatile allelochemicals may perform a vital part in inhibiting the competitive ability of weed species, be one of the alternatives to control weed infestation, and it has excited the greatest interest [**67,73,74,75**]. Therefore, the best way to control weeds is to use the crops' own competition and allelopathy. Wei et al. reported that the volatile oil released by *Atriplex cana* Ledeb. significantly inhibited the growth of seedlings of four weed species, including *Amaranthus retroflexus* L. and *Poa annua* L., and it had a high value for further use as a biological herbicide [**65**].

Brassicas produce the allelochemicals glucosinolates throughout their plant parts and released them into the environment by volatilization [**76**]. In the natural environment, glucosinolates are broken down into several compounds, and the most important compound of them is isothiocyanate [**77**], which can inhibit the growth and development of plant or weed [**78**]. *Digitaria sanguinalis* is a common non-irrigated weed that severely affects crop yields. Pardo-Muras et al. showed that the oxygenated monoterpenes in VOCs produced and released by both *Ulex europaeus* and *Cytisus scoparius* inhibited the germination and early growth of two weeds, *A. retroflexus* and *D. sanguinalis* [**79**]. Many studies have reported that in addition to independent effects of VOCs, VOCs also have synergistic or antagonistic effects. These synergistic and antagonistic effects lead to the final allelopathy [**80,81**].

Besides, VOCs from invasive species can also reduce interspecific performance and improve the performance of invasive species [**82**]. VOCs might also be perceived by neighboring plants to adjust their defensive phenotype according to the present risk of attack. Allelochemicals as natural herbicide have an attractive prospect, and some plants may be expected to develop into a new generation of herbicides or fungicides.

2.3. VOCs and Plants Dormancy

Dormancy is a physiological state in which plants respond to stress [**83**]. DADS is considered as main allelochemical of VOCs in garlic [**53**]. Hosoki et al. reported that the sulfur-containing compounds from garlic VOCs could break the bud dormancy in some corms, tubers, and ornamental trees [**84,85**]. Kubota et al. found that exposure to volatile diallyl disulfides and trisulfides was the most effective treatment to promote the bud break of single-bud cuttings of "Kyoho" (*Vitis vinifera × labruscana Bailey*) [**86**]. So, the allelochemicals in garlic that break the dormancy of vine buds are sulfur compounds, particularly DADS. In addition to DADS, more and more studies can prove that VOCs can affect plant dormancy. In a recent study, Shukla et al. studied the breaking dormancy of potato tubers effects on 20 essential oils from medicinal and aromatic plants [**87**]. The essential oils could induce or inhibit the sprouting process of potato tubers by altering the accumulation of reducing sugars, ethylene production, and expression of genes, thereby affecting the dormancy of plants [**87**]. Besides, eugenol from clove essential oil and carvone from caraway and dill essential oils have been reported to inhibit potato tubers sprouting [**88,89,90**]. The results showed that the essential oils of lemon grass and clove were the most effective VOCs for breaking dormancy and inducing germination of potato tubers. The oils of palmarosa and ajwain inhibited the sprout of potato tubers [**87**]. Owolabi et al. found the essential oils of *Lippia multiflora*, *Cymbopogon citratus*, and *Zingiber officinale* could control potato tubers dormancy [**91**]. They are suitable for application as sprout suppressants. At present, there are few studies on the effects of VOCs on plant dormancy, and the specific mechanism of action needs to be studied in the future.

2.4. VOCs and the Inhibition of Plants Diseases and Insect Pests

2.4.1. Inhibition of Plants Disease

VOCs not only have an inhibitory effect on plants, but also on pathogenic bacteria. A number of experimental trials showed that leaves' VOCs inhibited germination and the growth of plant pathogens and had stronger activity than commercial fungicides [**92,93,94**]. The essential oil of oregano triggers the expression of hundreds of genes involved in the grapevine immune system, so it can prevent *Plasmopara viticola* infection in grapevine (*Vitis Vinifera*) and primes plant immunity mechanisms [**95**]. The inhibition rate of volatile allelochemicals from leaves of *Ocimum adscendens* to 29 different kinds of mycelium was over %98, and the inhibition rate of mycelium growth of Aspergillus reached %100; especially, the activity of *A. lavus* was 100~10 times higher than another five commercial fungicides [**96**]. Chaturvedi et al. found that volatiles released by *Adenocalymma allicea* could effectively control leaf spot in rice and kill the pathogenic fungus *Drechslera oryzae*. When volatile oil obtained from *A.alicea* plants compared with the activity of the synthetic fungicides blitox50- and m45-, the activity of volatile oil to *D. oryzae* was up to four times higher than those synthetic fungicides [**97**]. In particular, this volatile oil had no harmful effect on the growth and development of rice seedlings. The volatile oil released from the leaves of the same genus *Ocimum basilicum* also inhibited the growth of other fungal mycelia by more than %85, and the dosage was only 4/1 of the commercial fungicide [**98**]. Therefore, using VOCs to replace commercial fungicide is no longer a dream. Phenols are a kind of allelochemicals that have been concerned and studied. Eugenol is an important class of phenolic allelochemicals. Cloves have a long history of use, and the main volatile component eugenol has a strong inhibitory effect on fungi and bacteria [**99,100**]. In a recent study, Quintana-Rodriguez et al. performed a screening on the efficacy of 22 VOCs, which were known to be volatilized from infected plants leaves, against the fungal pathogens *Colletotrichum lindemuthianum*, *Fusarium oxysporum*, and *Botrytis cinerea*. The work results showed that nonanal, (+)-carvone, citral, trans-2-decenal, L-linalool, nerolidol, and eugenol significantly inhibited the growth of the three fungal species, and eugenol had the most active among them. Therefore, the VOCs of plants have the disease resistance function [**101**].

In addition to eugenol, there are many VOCs that can also inhibit pathogens. DADS is a volatile organosulfur compound derived from garlic (*Allium sativum* L.) bulbs, and it is known as an allelochemical because of the potential allelopathy of garlic. A large number of study results show that it has a strong inhibitory effect on a variety of pathogenic bacteria [**102**]. In a recent study, Yang et al. demonstrated that the VOCs (DADS) from green garlic (*Allium sativum* L.) increased the accumulation of H2O2 and the disease resistance of cucumber [**103**]. Sekine et al. reported that other VOCs such as cuminaldehyde and *p*-cymene also have been demonstrated to possession antifungal activity against *B. cinerea*, *F. oxysporum*, *Verticillium dahliae*, and *Alternaria mali* [**104**]. According to the work of Mandal and Mandal, linalool, a substance with antifungal and antioxidant potential, was found in the volatile oil of coriander (*Coriandrum sativum* L.) [**105**].

2.4.2. Inhibition of Plants Insect Pests

Numerous studies showed that plants not only produce toxins and hormones directly in response to insect feeding, but also release VOCs to attract predators [**106,107**]. A class of VOCs produced as a response to herbivore attacking are mainly terpenoids, predominantly monoterpenoids, and sesquiterpenoids [**107**], and they benefited the host plant by interacting with herbivores. For example, when the larva of *Spodoptera exigua* Hübner. were feeding on corn, the corn released volatile terpenoids to attract the parasitic wasp, which was the natural enemy of *S. exigua*. If it was not mechanically damaged, the corn did not release VOCs to attract the natural enemy. Further studies found that the oral saliva of the beet moth contained volicitin, when the beet moth bit corn, its messenger jasmonic acid derivatives were activated by volicitin to release terpenoids to lure *Cotesia marginiventris* Cresson [**108**]. Kessler and Baldwin found that wild tobacco could release VOCs to attract mealoptera, the natural predators of caterpillars, when caterpillars ate them. Additionally, the VOCs released by wild tobacco also could prevent caterpillars from laying eggs on the leaves [**21**].

Under natural conditions, plants can also achieve effective control of related pests by releasing volatile substances to attract natural enemies. It is confirmed that the indirect chemical defense of plants by releasing VOCs to attract natural enemies is a chemical relationship in nature. *Ageratum conyzoides* L. released volatile terpenoid into the air by stalks, leaves, and flowers to attract predatory mites (*Amblyseius* SPP.) and maintain their population density [**26**]. Predatory mites are the most effective natural enemies of red mites (*Panonychus citri*). So, the population density of red mites was reduced [**26**]. Degenhardt et al. found that VOCs from corn root (E)-b-caryophyllene attracted insect-killing nematodes to control a major pest [**109**]. Therefore, the sustainable control of pests and diseases can be achieved through the natural chemical mechanisms that exist in the field to regulate plant–organism interactions. As a natural fungicide, VOCs have no harmful effect on plant growth and development, and using VOCs is a more environmentally friendly and economical way to kill bacteria.

2.5. VOCs and Plants Respiration and Photosynthesis

Previous studies showed that allelochemical can affect plant respiration by interfering in various stages of respiration, including the generation of carbon dioxide (CO_2) by electron transport, oxidative phosphorylation, and the activity of ATPase, and it has the potential to inhibit plant growth and development [**41,110,111**]. VOCs released from the leaves of *A. tridentata* Nutt. var. *vaseyana* and *Sasa cernua* Makino. inhibited the respiration of germinating seeds [**110,112**]. In fact, studies of allelopathy processes in shrubs in the 1960s found that volatile terpenes could reduce respiration in the young leaves of some plants and increase respiration in mature leaves [**113**]. Similarly, terpenes in eucalyptus volatile oil could affect target plants by inhibiting cellular respiration [**114**]. The influence mechanism of these allelochemicals on plant respiration deserves more in-depth study.

In addition, photosynthesis plays an important role in realizing the energy conversion in nature and maintains the carbon–oxygen balance in the atmosphere. The high concentration of allelochemicals involved in multiple metabolic steps may lead to the inhibition of plant photosynthesis, or even block the mechanism of photosynthesis by inhibiting electron receptors, energy coupling, or destroying photosynthetic pigments and enzymes [**115**]. Isoprene volatilized from the foliage of many woody species was thought to increase the rate of photosynthesis by stabilizing thylakoid membranes, so that adjacent plants could tolerate high temperatures [**116**]. Kaur et al. found the volatile oils α-Pinene and 1,8-cineole from *Eucalyptus tereticornis* Sm. could significantly reduce the respiration and photosynthetic pigment content of *Amaranthus viridis* Linn. Seedlings; thereby, the negative effects of the oil on photosynthetic machinery was explained [**117**]. These studies can confirm that olefin compounds can affect the photosynthesis of plants. In addition, Tsubo et al. found that exposure to low concentrations of the volatile chemicals released by *A. adamsii* Besser stimulated the photosynthetic rates of *Stipa krilovii* Roshev [**118**]. The volatile oil of *Artemisia ordosica* Krasch. inhibited the growth and photosynthesis of *Palmellococcus miniatus* through oxidative damage [**119**]. Zhao et al. studied the effects of eucalyptol and limonene, the main terpenoids in cyanobacteria VOCs on the photosynthetic capacity of *Chlorella vulgaris* [**120**]. The results showed that the compounds could induce the degradation of photosynthetic pigments and reduce the photosynthetic abilities of other algae. These studies can confirm that VOCs have the ability to affect plant photosynthesis.

2.6. VOCs and Plants ROS Content and Enzymatic Activity

Reactive oxygen species (ROS) play a vital role in the plant defense against stresses. The balance between ROS generation and scavenging is considered as paramount in cellular homeostasis. In recent years, more and more attention has been paid to the effectiveness and feasibility of monomer organic sulfide extracted from garlic as an anti-tumor drug, and the research and development of monomer organic sulfide has become an important research topic [**121**]. Similarly, DADS had an effect on the ROS content of plants. Yang et al. used cucumber and garlic as test materials to study the allelopathy of VOCs from green garlic on the scavenging of cucumber ROS [**103**]. The results showed that DADS, a volatile substance in garlic, reduced superoxide anions and increased hydrogen peroxide accumulation in cucumber seedlings. The effects of VOCs on antioxidant enzymatic activities were species dependent. They can regulate the activity of antioxidant enzymes (SOD, CAT, and POD) of cucumber seedlings in response to oxidative stress. VOCs released from *Acacia dealbata* Link. leaves increased the activity of superoxide dismutase (SOD) and peroxidase (POD) in *L. multiflorum* flowers, but decreased SOD activity in *T. subterraneum* [**60**]. The volatile allelochemical myrcene rapidly induced ROS production and significantly increased the activity of lipoxygenase (LOX) in rice roots [**122**]. In other cases, Mutlu et al. reported that the aerial parts of *Nepeta meyeri* Benth. contained two volatile oils, Germacrene-d and Caryophyllene oxide, and they could reduce the SOD activity of six weed species [**123**]. Jin et al. reported that the essential oil such as carvacrol, cinnamaldehyde, perillaldehyde, and linalool enhanced the SOD and POD activities of Chinese bayberries, and carvacrol had the best effect [**124**].

2.7. VOCs and Plants Signal Transduction

2.7.1. Chemical Communications

Plants, similar to animals, do not exist in isolation. Plant individuals and populations maintain population relationships and resist external stress through chemical communication. VOCs volatilized from some plants are involved in plant–plant communication. Through more than 20 years of research, it has been found that when plants were stressed by insect feeding, microbial infection, and mechanical damage, they could use volatile organic substances to carry out inter-chemical and intra-chemical chemical communication [**125,126**]. Plants sent VOCs signals under attack or stress, then neighboring plants received these VOCs signals directly or indirectly to turn on the chemical defense mechanisms and produce phenolic alkaloids chemical defense compounds including directly terpenoids and other defense substances. Indirect chemical defense the use of VOCs to attract the natural enemies of pests, through the methods of predation and parasitism to eliminate pests. For example, the VOCs signals released by injured *A. tidentata* could induce direct chemical defense by protease inhibitors in *Nicotiana attenuate*, and the VOC signals released by maize leaves when insects fed could induce plants to rapidly release monoterpenoids that attracted insect predators for indirect defense

[**127,128**]. Wild lima bean quickly synthesized and secreted excess nectar to attract natural enemies after receiving VOC signals from plants that have been harmed by foraging [**129**]. In recent years, more and more research has gone deep into the mechanism of action of these chemical communication signaling molecules. Baldwin et al. [**125**] found when in response to the attack, infested leaves released (E)-β-ocimene (typical signaling chemicals of volatile terpenoids) to increase the resistance of un-infested leaves and induce the expression of defense-related genes in neighboring un-infected leaves [**130**]. In future research, based on the clarification of the biochemical and metabolic processes of the phytochemical signal substances that have been discovered, the response mechanism of the recipient plants to these signal substances should be further explored. Therefore, although the research of phytochemical communication faces huge challenges, the clarification of phytochemical identification and communication relationships will broaden the horizons of the interspecific and intraspecific relationships of plants.

2.7.2. Plant Kin Recognition

Kin recognition is simply the ability of an individual to distinguish the relationship between genetically close related kin and non-kin. The kin recognition of plants has very important ecological and evolutionary significance. Increasing evidence shows that plants' recognition of neighboring allogenous and heterogeneous plants is mostly mediated by chemicals [**131,132**]. When plants are attacked by herbivores, they will emit volatile signals to surrounding plants. Some plants, such as *A. tidentata*, suffered less damage than other plants that receive volatile signals from non-self-wounds [**133,134**]. This shows that VOCs play a role in plant self- and non-self-recognition. Because VOCs are the simplest and fastest chemical signal that can send to neighboring plants, plant recognition of volatiles helps plants to establish corresponding response strategies in the early stages of competition. This can avoid wasting resources in the competition between self and relatives as much as possible [**130**]. After a series of studies recently, although the kin recognition of plants is ubiquitous, most studies show that some plants do not have kin recognition behaviors in order to avoid meaningless competition [**27,135,136,137**]. Plants' kin recognition has always been a hot topic in the study of behavioral ecology and evolution. From the perspective of genetic recognition, it is of great theoretical and practical significance to re-examine the ecological interaction of individual crops and groups and the environment.

3. METHOD OF VOCS COLLECTION AND IDENTIFICATION

The allelopathy of VOCs has attracted widespread attention in recent years and a lot of research work has been done. With the advancement of organic separation and identification technology and the participation of more and more experts, the collection and identification technology of plants VOCs is no longer a difficult issue in research.

3.1. Collection of Plants VOCs

The volatiles mostly are organic compounds with a molecular weight between 100 and 200, such as hydrocarbons, alcohols, ketones, organic acids, nitrogen compounds, and organic sulfur [138]. Most of them have high chemical activity. Different collection methods may directly affect the type and proportion of VOCs, so it is particularly important to choose the appropriate method.

Traditional distillation collection techniques include steam distillation (SD), simultaneous distillation and solvent extraction (SDE), microwave-assisted hydrodistillation extraction (MWHD), ultrasound-assisted extraction (USE), and solid-phase trapping solvent extraction (SPTE). They have certain disadvantages in the isolation and purification of chemical constituents from plants tissues, such as long extraction time, high volumes solvent, and low efficiency [139]. In addition, many natural products are thermally unstable and may degrade during thermal extraction or distillation.

The most mainstream approach is headspace solid-phase microextraction (HS-SPME). It has some advantages over SD, SDE, and SPTE, such as rapid solvent-free extraction, no apparent thermal degradation, less laborious manipulation and sample requirement, and so on [140]. Moreover, due to the relatively low temperature and short headspace solid-phase extraction time, the risk of thermal artifacts is extremely low compared to other techniques [141]. Additionally, it is easy to standardize and fully integrate into the analysis system [142]. Thus, HS-SPME is an ideal technology of plants' VOC collection.

3.2. Identification of Plants VOCs

The process of identifying allelochemicals requires both quantitative and qualitative assessment. Qualitative identification refers to the process of determining the specific kind and structure of allelochemicals. Qualitative identification encompasses techniques such as gas chromatography (GC), mass spectrometry (MS), nuclear magnetic resonance (NMR), Fourier transform infrared (FT-IR) spectroscopy, and various other methods [143,144,145]. These are the methodologies used for analyzing volatile organic compounds (VOCs), however, the challenges in doing the analysis and the necessary instruments vary significantly. Quantitative identification refers to the process of determining the concentration of allelochemicals by specifically identifying their kind. Chromatography is employed to determine the concentration of known compounds. Distinct techniques must be chosen for qualitative and quantitative identification, with the selection criteria based on the attributes of VOCs. The current methods of identification encompass gas chromatography mass spectrometry (GC-MS), high-performance liquid chromatography (HPLC), proton-transfer-reaction mass spectrometry (PTR-MS), and other similar techniques. PTR-MS offers the capability to directly sample volatile organic compounds (VOCs) in real-time and perform rapid quantitative analysis without the need for any sample preparation [146,147]. GC-MS [148] is the most commonly employed method for identification among these techniques. While PTR-MS is capable of achieving more accurate quantitative identification, the majority of the volatiles that are identified are still considered preliminary [149]. Gas chromatography-mass spectrometry (GC-MS) has superior selectivity and sensitivity when it comes to identifying volatile organic compounds (VOCs) and effectively separating and identifying the substances being analyzed.

4. CONCISE SUMMARY

This article provides a comprehensive overview of the allelopathic effects of volatile organic compounds (VOCs) produced by plants. It covers several aspects such as the impact of VOCs on plant growth, competition, dormancy, resistance to pests and diseases, respiration, photosynthesis, reactive oxygen species (ROS) content, enzyme activity, and signaling. Additionally, it provides a concise overview of the primary techniques used for gathering and distinguishing volatile organic compounds (VOCs). The research of allelopathy is a complex endeavor that encompasses multiple disciplines, including chemistry, ecology, biology, microbiology, and others. Collaboration among scientists in these disciplines is necessary to carry out research. The investigation of allelopathy on plants' volatile organic compounds (VOCs) is a nascent area of inquiry. The majority of research efforts currently concentrate on investigating the expression of the allelopathic phenomenon. However, these studies lack sufficient depth and breadth. For instance, there is a dearth of research on the mechanisms underlying allelopathy in plants, as well as the interplay between chemical recognition, communication mechanisms, and allelopathy. Recently, there has been an increasing number of publications on volatile organic compounds (VOCs). The study of VOC transmission, emission, and accumulation is currently a prominent area of research that warrants more investigation. There remain numerous unresolved matters that require additional investigation. Further investigation is necessary for the study of volatile organic compounds (VOCs) emitted by plants.

5. VOCS MAY HAVE A WIDE RANGE OF POTENTIAL ALLELOPATHIC EFFECTS IN AGRICULTURE

There are numerous possible allelopathic consequences of volatile organic compounds (VOCs) in agriculture. In addition to controlling weeds and pests in their natural state, volatile organic compounds (VOCs) can control plant growth, competition, dormancy, respiration and photosynthesis, ROS concentration, enzyme activity, and resistant resistance to disease. The fact that they are derived from plants and satisfy our criteria for creating green agriculture is what matters most. Utilizing the allelopathic effects of natural plant products to advance the agricultural business is a more cost-effective, environmentally responsible, and efficient strategy. However, there are certain drawbacks to employing plant allelopathy of volatile organic compounds (VOCs); for instance, VOCs might be difficult to manage in field experiments. Plant species, cultivars, genotypes, organs, and habitats all have an impact on how quickly volatile organic compounds (VOCs) volatilize. Actually, a complex combination of VOCs rather than a single VOC is released by plants [150]. Temperature, humidity, wind direction, and other climatic and meteorological factors all have an impact on the emission of volatile organic compounds (VOCs) from plants during agricultural production. Controlling the release of VOCs is challenging due to these considerations. Furthermore, compared to laboratory studies, open field investigations frequently have lower VOC concentrations. Therefore, a greenhouse with easy management is a better place for plant VOCs that exhibit allelopathy in agricultural production. The identification of allelochemicals is no longer a challenging issue in the study of allelopathy thanks to advancements in allelochemicals identification technologies and the growing involvement of chemists. Nevertheless, little is known about the kinetics of the active release pathway of volatile organic compounds (VOCs) in plants, and the molecular mechanism of VOCs has not been investigated. It is also unclear how plants perceive VOCs. Consequently, future studies will primarily investigate the dynamic process of volatile organic compound (VOC) emission as well as the type and regularity of plant VOC allelopathy in order to understand how VOCs are perceived within plant tissues. In the last 10 years, numerous studies have demonstrated the involvement of volatile organic compounds (VOCs) in plant signal transduction, and numerous allelochemicals that are involved in chemical communication have been found. The interaction between the mechanisms of allelopathy and chemical recognition and communication is currently the issue we are confronting. It is vital to find solutions for the general and specific chemical identification as well as the research on the identification and transfer method of the communicative material. Determining the natural VOC content is another significant issue we encounter since plant-based volatile organic compounds (VOCs) are difficult to regulate.

Currently holding a prominent position in agricultural growth, volatile organic compounds (VOCs) will only become more competitive in the future, particularly in the greenhouse, due to the growing demands for environmental protection and sustainable development. Allelopathy of plants VOCs is a theoretical field with significant practical applications. It is important for maintaining and enhancing natural resources, building efficient planting systems, rationally arranging rotation and intercropping, and improving the efficiency with which natural resources are utilized. It also plays a significant role in controlling pests and weeds and promoting sustainable agricultural production. Furthermore, the usage of plant volatile allelopathy can influence the activity of plant life by controlling the development, dormancy, respiration and photosynthesis, ROS concentration, enzyme activity, and chemical communication between plants. In conclusion, plant allelopathy of volatile organic compounds (VOCs) has enormous promise for agricultural advancement. In light of this, developing and putting into practice the allelopathic potential of plant volatile organic compounds is crucial.

AT A GLANCE

Allelopathy is an ecological phenomenon characterized by the interference of organisms with one another. Allelopathy is a management method in agricultural systems that primarily aims to control weeds, fight pests and diseases, and enhance the interaction between soil nutrition and microorganisms. Volatile organic compounds (VOCs) are chemical compounds released into the air by plants and are known to have many ecological functions. This study presents the latest developments in the allelopathic effects of volatile organic compounds (VOCs) on plants, including their impact on growth, competition, dormancy, disease and insect resistance, reactive oxygen species (ROS) levels, enzyme activity, respiration, and photosynthesis. Volatile organic compounds (VOCs) also play a role in inter-plant communication by acting as signaling molecules. This article provides a concise overview of the primary techniques used for the gathering and characterization of volatile organic compounds (VOCs). Additionally, it highlights the drawbacks of volatile organic compounds (VOCs) and proposes possible avenues to advance research and address unresolved questions in this developing field. Studying the allelopathic mechanisms of plant VOCs is essential in order to establish a theoretical foundation for the use of VOCs. To summarize, the use of allelopathy, namely the production of volatile organic compounds (VOCs) by plants, is a cost-effective, ecologically acceptable, and efficient approach to promote the growth of the agricultural business. This method harnesses the natural allelopathic effects of plant-derived substances.

REFERENCES

1. Farooq, M.; Jabran, K.; Cheema, Z.A.; Wahid, A.; Siddique, K.H.M. The role of allelopathy in agricultural pest management. *Pest Manag. Sci.* 2011, *67*, 493–506.

2. Pan, L.; Li, X.-Z.; Yan, Z.-Q.; Guo, H.-R.; Qin, B. Phytotoxicity of umbelliferone and its analogs: Structure-activity relationships and action mechanisms. *Plant Physiol. Biochem.* 2015, *97*, 272–277.

3. Boydston, R.A.; Morra, M.J.; Borek, V.; Clayton, L.; Vaughn, S.F. Onion and weed response to mustard (*Sinapis alba*) seed meal. *Weed Sci.* 2011, *59*, 546–552.

4. Awan, F.K.; Rasheed, M.; Ashraf, M.; Khurshid, M.Y. Efficacy of brassica sorghum and sunflower aqueous extracts to control wheat weeds under rainfed conditions of pothwar. *Pakistan J. Anim. Plant Sci.* 2012, *22*, 715–721.

5. Bajwa, A.A.; Mahajan, G.; Chauhan, B.S. Nonconventional weed management strategies for modern agriculture. *Weed Sci.* 2015, *63*, 723–747.

6. Tavella, L.B.; Lima e Silva, P.S.; Monteiro, A.L.; de Oliveira, V.R.; de Oliveira Fernandes, P.L. *Gliricidia sepium* intercropping for weed management in immature corn ear production. *Rev. Cienc. Agron.* 2017, *48*, 650–656.

7. Avato, P.; D'Addabbo, T.; Leonetti, P.; Argentieri, M.P. Nematicidal potential of brassicaceae. *Phytochem. Rev.* 2013, *12*, 791–802.

8. Liu, T.; Cheng, Z.; Meng, H.; Ahmad, I.; Zhao, H. Growth, yield and quality of spring tomato and physicochemical properties of medium in a tomato/garlic intercropping system under plastic tunnel organic medium cultivation. *Sci. Hortic.* 2014, *170*, 159–168.

9. Glinwood, R.; Ninkovic, V.; Pettersson, J. Chemical interaction between undamaged plants—Effects on herbivores and natural enemies. *Phytochemistry* 2011, *72*, 1683–1689.

10. Singh, A.; Weisser, W.W.; Hanna, R.; Houmgny, R.; Zytynska, S.E. Reduce pests, enhance production: Benefits of intercropping at high densities for okra farmers in Cameroon. *Pest Manag. Sci.* 2017, *73*, 2017–2027.

11. Bertin, C.; Yang, X.; Weston, L.A. The role of root exudates and allelochemicals in the rhizosphere. *Plant Soil* 2003, *256*, 67–83.

12. Abdel-Monaim, M.F.; Abo-Elyousr, K.A.M. Effect of preceding and intercropping crops on suppression of lentil damping-off and root rot disease in New Valley—Egypt. *Crop Prot.* 2012, *32*, 41–46.

13. Ma, Y.-h.; Fu, S.-l.; Zhang, X.-p.; Zhao, K.; Chen, H.Y.H. Intercropping improves soil nutrient availability, soil enzyme activity and tea quantity and quality. *Appl. Soil Ecol.* 2017, *119*, 171–178.

14. Ndungu-Magiroi, K.W.; Wortmann, C.S.; Kibunja, C.; Senkoro, C.; Mwangi, T.J.K.; Wamae, D.; Kifuko-Koech, M.; Msakyi, J. Maize-bean intercrop response to nutrient application relative to maize sole crop response. *Nutr. Cycl. Agroecosyst.* 2017, *109*, 17–27.

15. Bressan, M.; Roncato, M.-A.; Bellvert, F.; Comte, G.; Haichar, F.e.Z.; Achouak, W.; Berge, O. Exogenous glucosinolate produced by *Arabidopsis thaliana* has an impact on microbes in the rhizosphere and plant roots. *Isme J.* 2009, 3, 1243–1257.

16. Zhao, M.; Jones, C.M.; Meijer, J.; Lundquist, P.-O.; Fransson, P.; Carlsson, G.; Hallin, S. Intercropping affects genetic potential for inorganic nitrogen cycling by root-associated microorganisms in *Medicago sativa* and *Dactylis glomerata*. *Appl. Soil Ecol.* 2017, *119*, 260–266.

17. Farooq, M.; Bajwa, A.A.; Cheema, S.A.; Cheema, Z.A. Application of allelopathy in crop production. *Int. J. Agric. Biol.* 2013, *15*, 1367–1378.

18. Alemayehu, A.; Tamado, T.; Nigussie, D.; Yigzaw, D.; Kinde, T.; Wortmann, C.S. Maize-common bean intercropping to optimize maize-based crop production. *J. Agric. Sci.* 2017, *155*, 1124–1136.

19. Vivaldo, G.; Masi, E.; Taiti, C.; Caldarelli, G.; Mancuso, S. The network of plants volatile organic compounds. *Sci. Rep.* 2017, *7*, 11050.

20. Kigathi, R.N.; Weisser, W.W.; Reichelt, M.; Gershenzon, J.; Unsicker, S.B. Plant volatile emission depends on the species composition of the neighboring plant community. *BMC Plant Biol.* 2019, *19*, 58.

21. Kessler, A.; Baldwin, I.T. Defensive function of herbivore-induced plant volatile emissions in nature. *Science* 2001, *291*, 2141–2144.

22. Bouwmeester, H.; Schuurink, R.C.; Bleeker, P.M.; Schiestl, F. The role of volatiles in plant communication. *Plant J.* 2019, *100*, 892–907.

23. Rice, E.L. *Allelopathy*, 2nd ed.; Academic Press: Orlando, FL, USA, 1984.

24. Adebesin, F.; Widhalm, J.R.; Boachon, B.; Lefevre, F.; Pierman, B.; Lynch, J.H.; Alam, I.; Junqueira, B.; Benke, R.; Ray, S.; et al. Emission of volatile organic compounds from petunia flowers is facilitated by an ABC transporter. *Science* 2017, *356*, 1386–1388.

25. Dudareva, N.; Klempien, A.; Muhlemann, J.K.; Kaplan, I. Biosynthesis, function and metabolic engineering of plant volatile organic compounds. *New Phytol.* 2013, *198*, 16–32.

26. Kong, C.; Hu, F.; Xu, X.; Zhang, M.; Liang, W. Volatile allelochemicals in the *Ageratum conyzoides* intercropped citrus orchard and their effects on mites *Amblyseius newsami* and Panonychus citri. *J. Chem. Ecol.* 2005, *31*, 2193–2203.

27. McNickle, G.G.; St Clair, C.C.; Cahill, J.F., Jr. Focusing the metaphor: Plant root foraging behaviour. *Trends Ecol. Evol.* 2009, *24*, 419–426.

28. Dicke, M.; Baldwin, I.T. The evolutionary context for herbivore-induced plant volatiles: Beyond the "cry for help". *Trends Plant Sci.* 2010, *15*, 167–175.

29. Erb, M.; Veyrat, N.; Robert, C.A.M.; Xu, H.; Frey, M.; Ton, J.; Turlings, T.C.J. Indole is an essential herbivore-induced volatile priming signal in maize. *Nat. Commun.* 2015, *6*.

30. Gfeller, V.; Huber, M.; Foerster, C.; Huang, W.; Koellner, T.G.; Erb, M. Root volatiles in plant-plant interactions I: High root sesquiterpene release is associated with increased germination and growth of plant neighbours. *Plant Cell Environ.* 2019, *42*, 1950–1963.

31. Molish, H. *Der Einfluss Einer Pflanze auf die Andere-Allelopathie*; Gustav Fischer Verlag: Jena, Germany, 1937.

32. Simms, E.L.; Rausher, M.D. Costs and benefits of plant resistance to herbivory. *Am. Nat.* 1987, *130*, 570–581.

33. Kim, J.; Felton, G.W. Priming of antiherbivore defensive responses in plants. *Insect Sci.* 2013, *20*, 273–285.

34. Langenheim, J.H. Higher plant terpenoids: A phytocentric overview of their ecological roles. *J. Chem. Ecol.* 1994, *20*, 1223–1280.

35. Ameye, M.; Audenaert, K.; De Zutter, N.; Steppe, K.; Van Meulebroek, L.; Vanhaecke, L.; De Vleesschauwer, D.; Haesaert, G.; Smagghe, G. Priming of wheat with the green leaf volatile Z-3-hexenyl acetate enhances defense against fusarium graminearum but boosts deoxynivalenol production. *Plant Physiol.* 2015, *167*, 1671–1684.

36. Siri-Udom, S.; Suwannarach, N.; Lumyong, S. Applications of volatile compounds acquired from *Muscodor heveae* against white root rot disease in rubber trees (*Hevea brasiliensis* Mull. Arg.) and relevant allelopathy effects. *Fungal Biol.* 2017, *121*, 573–581.

37. Lerdau, M.; Gray, D. Ecology and evolution of light-dependent and light-independent phytogenic volatile organic carbon. *New Phytol.* 2003, *157*, 199–211.

38. Cofer, T.M.; Engelberth, M.; Engelberth, J. Green leaf volatiles protect maize (*Zea mays*) seedlings against damage from cold stress. *Plant Cell Environ.* 2018, *41*, 1673–1682.

39. Muller, C.H.; Muller, W.H.; Haines, B.L. Volatile growth Inhibitors produced by aromatic shrubs. *Science* 1964, *143*, 471–473.

40. Muller, C.H. Inhibitory terpenes volatilized from salvia shrubs. *Bull. Torrey Bot. Club* 1965, *92*, 38–45.
41. Abrahim, D.; Braguini, W.L.; Kelmer-Bracht, A.M.; Ishii-Iwamoto, E.L. Effects of four monoterpenes on germination, primary root growth, and mitochondrial respiration of maize. *J. Chem. Ecol.* 2000, *26*, 611–624.
42. Norton, J.M.; Harman, G.E. Responses of soil microorganisms to volatile exudates from germinating pea seeds. *Can. J. Bot.* 1985, *63*, 1040–1045.
43. Won Yun, K.; Kil, B.S.; Han, D.M. Phytotoxic and antimicrobial activity of volatile constituents of *Artemisia princeps* var. orientalis. *J. Chem. Ecol.* 1993, *19*, 2757–2766.
44. Jassbi, A.R.; Zamanizadehnajari, S.; Baldwin, I.T. Phytotoxic volatiles in the roots and shoots of *Artemisia tridentata* as detected by headspace solid-phase microextraction and gas chromatographic-mass spectrometry analysis. *J. Chem. Ecol.* 2010, *36*, 1398–1407.
45. Hütsch, B.W.; Augustin, J.; Merbach, W. Plant rhizodeposition—An important source for carbon turnover in soils. *J. Plant Nutr. Soil Sci.* 2002, *165*, 397–407.
46. Lin, C.; Owen, S.M.; Peñuelas, J. Volatile organic compounds in the roots and rhizosphere of *Pinus* spp. *Soil Biol. Biochem.* 2007, *39*, 951–960.
47. Delory, B.M.; Delaplace, P.; Fauconnier, M.-L.; du Jardin, P. Root-emitted volatile organic compounds: Can they mediate belowground plant-plant interactions? *Plant Soil* 2016, *402*, 1–26.
48. Fitter, A. Making allelopathy respectable. *Science* 2003, *301*, 1337–1338.
49. Brilli, F.; Loreto, F.; Baccelli, I. Exploiting plant volatile organic compounds (VOCs) in agriculture to Improve sustainable defense strategies and productivity of crops. *Front. Plant Sci.* 2019, *10*.
50. Oleszek, W. Allelopathic effects of volatiles from some Cruciferae species on lettuce, barnyard grass and wheat growth. *Plant Soil* 1987, *102*, 271–273.
51. Romagni, J.G.; Allen, S.N.; Dayan, F.E. Allelopathic effects of volatile cineoles on two weedy plant species. *J. Chem. Ecol.* 2000, *26*, 303–313.
52. Schmidt-Silva, V.; Pawlowski, Â.; Kaltchuk-Santos, E.; Zini, C.; Soares, G. Cytotoxicity of essential oils from two species of Heterothalamus (Asteraceae). *Aust. J. Bot.* 2011, *59*, 682–691.
53. Cheng, F.; Cheng, Z.; Meng, H.; Tang, X. The garlic allelochemical diallyl disulfide affects tomato root growth by influencing cell division, phytohormone balance and expansin gene expression. *Front. Plant Sci.* 2016, *7*, 1199.
54. Cheng, F.; Cheng, Z.-H.; Meng, H.-W. Transcriptomic insights into the allelopathic effects of the garlic allelochemical diallyl disulfide on tomato roots. *Sci. Rep.* 2016, *6*, 38902.
55. Kim, Y.S.; Kil, B.-S. Allelopathic effects of some volatile substances from the tomato plant. *J. Crop Prod.* 2001, *4*, 313–321.
56. Alves, M.d.C.S.; Medeiros Filho, S.; Innecco, R.; Torres, S.B. Alelopatia de extratos voláteis na germinação de sementes e no comprimento da raiz de alface. *Pesqui. Agropecuária Bras.* 2004, *39*, 1083–1086.
57. Norsworthy, J.K.; Meehan, J.T. Use of isothiocyanates for suppression of Palmer amaranth (*Amaranthus palmeri*), pitted morningglory (*Ipomoea lacunosa*), and yellow nutsedge (*Cyperus esculentus*). *Weed Sci.* 2017, *53*, 884–890.
58. Horiuchi, J.-i.; Badri, D.V.; Kimball, B.A.; Negre, F.; Dudareva, N.; Paschke, M.W.; Vivanco, J.M. The floral volatile, methyl benzoate, from snapdragon (*Antirrhinum majus*) triggers phytotoxic effects in *Arabidopsis thaliana*. *Planta* 2007, *226*, 1–10.
59. Silva, E.R.; Overbeck, G.E.; Soares, G.L.G. Phytotoxicity of volatiles from fresh and dry leaves of two Asteraceae shrubs: Evaluation of seasonal effects. *S. Afr. J. Bot.* 2014, *93*, 14–18.
60. Souza-Alonso, P.; Novoa, A.; González, L. Soil biochemical alterations and microbial community responses under *Acacia dealbata* Link invasion. *Soil Biol. Biochem.* 2014, *79*, 100–108.

61. Haramoto, E.R.; Gallandt, E.R. Brassica cover cropping for weed management: A review. *Renew. Agric. Food Syst.* 2007, *19*, 187–198.

62. Morra, M.J.; Kirkegaard, J.A. Isothiocyanate release from soil-incorporated *Brassica tissues*. *Soil Biol. Biochem.* 2002, *34*, 1683–1690.

63. Santonja, M.; Bousquet-Mélou, A.; Greff, S.; Ormeño, E.; Fernandez, C. Allelopathic effects of volatile organic compounds released from *Pinus halepensis* needles and roots. *Ecol. Evol.* 2019, *9*, 8201–8213.

64. Ren, K.; Hayat, S.; Qi, X.; Liu, T.; Cheng, Z. The garlic allelochemical DADS influences cucumber root growth involved in regulating hormone levels and modulating cell cycling. *J. Plant Physiol.* 2018, *230*, 51–60.

65. Wei, C.; Zhou, S.; Li, W.; Jiang, C.; Yang, W.; Han, C.; Zhang, C.; Shao, H. Chemical composition and allelopathic, phytotoxic and pesticidal activities of *Atriplex cana* Ledeb. (Amaranthaceae) essential oil. *Chem. Biodivers.* 2019, *16*, e1800595.

66. Effah, E.; Holopainen, J.K.; McCormick, A.C. Potential roles of volatile organic compounds in plant competition. *Perspect. Plant Ecol. Evol. Syst.* 2019, *38*, 58–63.

67. Macias, F.A.; Molinillo, J.M.; Varela, R.M.; Galindo, J.C. Allelopathy—A natural alternative for weed control. *Pest Manag Sci* 2007, *63*, 327–348.

68. Luiza Ishii-Iwamoto, E.; Marusa Pergo Coelho, E.; Reis, B.; Sebastiao Moscheta, I.; Moacir Bonato, C. Effects of monoterpenes on physiological processes during seed germination and seedling growth. *Curr. Bioact. Compd.* 2012, *8*, 50–64.

69. Arroyo, A.I.; Pueyo, Y.; Pellissier, F.; Ramos, J.; Espinosa-Ruiz, A.; Millery, A.; Alados, C.L. Phytotoxic effects of volatile and water soluble chemicals of *Artemisia herba-alba*. *J. Arid Environ.* 2018, *151*, 1–8.

70. Runyon, J.B.; Mescher, M.C.; De Moraes, C.M. Volatile chemical cues guide host location and host selection by parasitic plants. *Science* 2006, *313*, 1964–1967.

71. Hunt, N.D.; Hill, J.D.; Liebman, M. Reducing freshwater toxicity while maintaining weed control, profits, and productivity: Effects of increased crop rotation diversity and reduced herbicide usage. *Environ. Sci. Technol.* 2017, *51*, 1707–1717.

72. Arimura, G.-i.; Shiojiri, K.; Karban, R. Acquired immunity to herbivory and allelopathy caused by airborne plant emissions. *Phytochemistry* 2010, *71*, 1642–1649.

73. Verdeguer, M.; Blázquez, M.; Boira, H. Phytotoxic effects of *Lantana camara*, *Eucalyptus camaldulensis* and *Eriocephalus africanus* essential oils in weeds of Mediterranean summer crops. *Biochem. Syst. Ecol.* 2009, *37*, 362–369.

74. Benvenuti, S.; Cioni, P.L.; Flamini, G.; Pardossi, A. Weeds for weed control: Asteraceae essential oils as natural herbicides. *Weed Res.* 2017, *57*, 342–353.

75. Mushtaq, W.; Ain, Q.; Siddiqui, M.B.; Alharby, H.; Hakeem, K.R. Allelochemicals change macromolecular content of some selected weeds. *S. Afr. J. Bot.* 2020, *130*, 177–184.

76. Fahey, J.W.; Zalcmann, A.T.; Talalay, P. The chemical diversity and distribution of glucosinolates and isothiocyanates among plants. *Phytochemistry* 2001, *56*, 5–51.

77. Halkier, B.A.; Gershenzon, J. Biology and biochemistry of glucosinolates. *Annu. Rev. Plant Biol.* 2006, *57*, 303–333.

78. Petersen, J.; Belz, R.; Walker, F.; Hurle, K. Weed suppression by release of Isothiocyanates from turnip-rape mulch. *Agron. J.* 2001, *93*, 37–43.

79. Pardo-Muras, M.; Puig, C.G.; López-Nogueira, A.; Cavaleiro, C.; Pedrol, N. On the bioherbicide potential of *Ulex europaeus* and *Cytisus scoparius*: Profiles of volatile organic compounds and their phytotoxic effects. *PLoS ONE* 2018, *13*, e0205997.

80. Pardo-Muras, M.; Puig, C.G.; Pedrol, N. *Cytisus scoparius* and *Ulex europaeus* produce volatile organic compounds with powerful synergistic herbicidal effects. *Molecules* 2019, *24*, 4539.

81. Vokou, D.; Douvli, P.; Blionis, G.J.; Halley, J.M. Effects of monoterpenoids, acting alone or in pairs, on seed germination and subsequent seedling growth. *J. Chem. Ecol.* 2003, *29*, 2281–2301.

82. Barney, J.N.; Sparks, J.P.; Greenberg, J.; Whitlow, T.H.; Guenther, A. Biogenic volatile organic compounds from an invasive species: Impacts on plant-plant interactions. *Plant Ecol.* 2009, *203*, 195–205.

83. Campbell, M.; Segear, E.; Beers, L.; Knauber, D.; Suttle, J. Dormancy in potato tuber meristems: Chemically induced cessation in dormancy matches the natural process based on transcript profiles. *Funct. Integr. Genom.* 2008, *8*, 317–328.

84. Hosoki, T.; Hiura, H.; Hamada, M. Breaking bud dormancy in corms, tubers, and trees with sulfur-containing compounds. *HortScience* 1985, *20*, 290–291.

85. Hosoki, T. Breaking bud dormancy in corms and trees with sulfide compounds in garlic and horseradish. *HortScience* 1986, *21*, 114–116.

86. Kubota, N.; Yamane, Y.; Toriu, K.; Kawazu, K.; Higuchi, T.; Nishimura, S. Identification of active substances in garlic responsible for breaking bud dormancy in grapevines. *J. Jpn. Soc. Hortic. Sci.* 1999, *68*(6), 1111–1117.

87. Shukla, S.; Pandey, S.S.; Chandra, M.; Pandey, A.; Bharti, N.; Barnawal, D.; Chanotiya, C.S.; Tandon, S.; Darokar, M.P.; Kalra, A. Application of essential oils as a natural and alternate method for inhibiting and inducing the sprouting of potato tubers. *Food Chem.* 2019, *284*, 171–179.

88. Hartmans, K.J.; Diepenhorst, P.; Bakker, W.; Gorris, L.G.M. The use of carvone in agriculture: Sprout suppression of potatoes and antifungal activity against potato tuber and other plant diseases. *Ind. Crop. Prod.* 1995, *4*, 3–13.

89. Song, X.; Bandara, M.S.; Tanino, K.K. Potato dormancy regulation: Use of essential oils for sprout suppression in potato storage. *Fruit Veg. Cereal Sci. Biotechnol* 2009, *2*, 110–117.

90. Finger, F.L.; Santos, M.M.d.S.; Araujo, F.F.; Lima, P.C.C.; Costa, L.C.d.; França, C.d.F.M.; Queiroz, M.d.C. Action of essential oils on sprouting of non-dormant potato tubers. *Braz. Arch. Biol. Technol.* 2018, *61*.

91. Owolabi, M.S.; Olowu, R.A.; Lajide, L.; Oladimeji, M.O.; Padilla-Camberos, E.; Flores-Fernández, J.M. Inhibition of potato tuber sprouting during storage by the controlled release of essential oil using a wick application method. *Ind. Crop. Prod.* 2013, *45*, 83–87.

92. Komai, K.; Tang, C.-S. A chemotype of Cyperus rotundus in Hawaii. *Phytochemistry* 1989, *28*, 1883–1886.

93. Neri, F.; Mari, M.; Brigati, S.; Bertolini, P. Fungicidal activity of plant volatile compounds for controlling *Monilinia laxa* in stone fruit. *Plant Dis.* 2007, *91*, 30–35.

94. Da Silva, A.C.; de Souza, P.E.; Amaral, D.C.; Zeviani, W.M.; Brasil Pereira Pinto, J.E. Essential oils from *Hyptis marrubioides*, *Aloysia gratissima* and *Cordia verbenacea* reduce the progress of Asian soybean rust. *Acta Sci. Agron.* 2014, *36*, 159–166.

95. Rienth, M.; Crovadore, J.; Ghaffari, S.; Lefort, F. Oregano essential oil vapour prevents *Plasmopara viticola* infection in grapevine (*Vitis Vinifera*) and primes plant immunity mechanisms. *PLoS ONE* 2019, *14*, e0222854.

96. Asthana, A.; Tripathi, N.N.; Dixit, S.N. Fungitoxic and phytotoxic studies with essential oil of *Ocimum adscendens*. *J. Phytopathol.* 1986, *117*, 152–159.

97. Chaturvedi, R.; Dikshit, A.; Dixit, S.N. *Adenocalymma allicea*, a new source of a natural fungitoxicant. *Trop. Agric.* 1987, *64*, 318–322.

98. Dube, S.; Upadhyay, P.D.; Tripathi, S.C. Antifungal, physicochemical, and insect-repelling activity of the essential oil of Ocimum basilicum. *Can. J. Bot.* 1989, *67*, 2085–2087.

99. Wang, C.; Zhang, J.; Chen, H.; Fan, Y.; Shi, Z. Antifungal activity of eugenol against *Botrytis cinerea*. *Trop. Plant Pathol.* 2010, *35*, 137–143.

100. Rao, P.V.; Gan, S.H. Cinnamon: A multifaceted medicinal plant. *Evid Based Complement Altern. Med.* 2014, *2014*, 642942.

101. Quintana-Rodriguez, E.; Rivera-Macias, L.E.; Adame-Alvarez, R.M.; Torres, J.M.; Heil, M. Shared weapons in fungus-fungus and fungus-plant interactions? Volatile organic compounds of plant or fungal origin exert direct antifungal activity in vitro. *Fungal Ecol.* 2018, 33, 115–121.

102. Mohammad, S.F.; Woodward, S.C. Characterization of a potent inhibitor of platelet aggregation and release reaction isolated from *Allium sativum* (Garlic). *Thromb. Res.* 1986, *44*, 793–806.

103. Yang, F.; Liu, X.; Wang, H.; Deng, R.; Yu, H.; Cheng, Z. Identification and allelopathy of green garlic (*Allium sativum* L.) volatiles on scavenging of cucumber (*Cucumis sativus* L.) reactive oxygen species. *Molecules* 2019, *24*, 3263.

104. Sekine, T.; Sugano, M.; Majid, A.; Fujii, Y. Antifungal effects of volatile compounds from black zira (*Bunium persicum*) and other spices and herbs. *J. Chem. Ecol.* 2007, 33, 2123–2132.

105. Mandal, S.; Mandal, M. Coriander (*Coriandrum sativum* L.) essential oil: Chemistry and biological activity. *Asian Pac. J. Trop. Biomed.* 2015, 5, 421–428.

106. Friberg, M.; Schwind, C.; Roark, L.C.; Raguso, R.A.; Thompson, J.N. Floral scent contributes to interaction specificity in coevolving plants and their insect pollinators. *J. Chem. Ecol.* 2014, *40*, 955–965.

107. Yazaki, K.; Arimura, G.-i.; Ohnishi, T. 'Hidden' terpenoids in plants: Their biosynthesis, localization and ecological roles. *Plant Cell Physiol.* 2017, *58*, 1615–1621.

108. Alborn, H.; Turlings, T.; Jones, T.H.; Stenhagen, G.; Loughrin, J.; Tumlinson, J. An elicitor of plant volatiles from beet armyworm oral secretion. *Science* 1997, *276*, 945–949.

109. Degenhardt, J.; Hiltpold, I.; Köllner, T.G.; Frey, M.; Gierl, A.; Gershenzon, J.; Hibbard, B.E.; Ellersieck, M.R.; Turlings, T.C.J. Restoring a maize root signal that attracts insect-killing nematodes to control a major pest. *Proc. Natl. Acad. Sci. USA* 2009, *106*, 13213–13218.

110. Li, H.-H.; Nishimura, H.; Hasegawa, K.; Mizutani, J. Allelopathy of *Sasa cernua*. *J. Chem. Ecol.* 1992, *18*, 1785–1796.

111. Cheng, F.; Cheng, Z. Research progress on the use of plant allelopathy in agriculture and the physiological and ecological mechanisms of allelopathy. *Front. Plant Sci.* 2015, *6*.

112. Weaver, T.; Klarich, D. Allelopathic effects of volatile substances from *Artemisia tridentata* Nutt. *Am. Midl. Nat.* 1977, *92*, 508–512.

113. Einhellig, F.A. *Mechanism of Action of Allelochemicals in Allelopathy*; ACS Publications: Washington, DC, USA, 1995.

114. Kohli, R.K.; Batish, D.R.; Singh, H.P. Eucalypt oils for the control of Parthenium (*Parthenium hysterophorus* L.). *Crop Prot.* 1998, *17*, 119–122.

115. Einhellig, F.A.; Rasmussen, J.A. Effects of three phenolic acids on chlorophyll content and growth of soybean and grain sorghum seedlings. *J. Chem. Ecol.* 1979, 5, 815–824.

116. Sharkey, T.D.; Yeh, S. Isoprene emission from plants. *Annu. Rev. Plant Biol.* 2001, *52*, 407–436.

117. Kaur, S.; Singh, H.P.; Batish, D.R.; Kohli, R.K. Chemical characterization and allelopathic potential of volatile oil of *Eucalyptus tereticornis* against *Amaranthus viridis*. *J. Plant Interact.* 2011, 6, 297–302.

118. Tsubo, M.; Nishihara, E.; Nakamatsu, K.; Cheng, Y.; Shinoda, M. Plant volatiles inhibit restoration of plant species communities in dry grassland. *Basic Appl. Ecol.* 2012, *13*, 76–84.

119. Yang, X.; Deng, S.; De Philippis, R.; Chen, L.; Hu, C.; Zhang, W. Chemical composition of volatile oil from *Artemisia ordosica* and its allelopathic effects on desert soil microalgae, *Palmellococcus miniatus*. *Plant Physiol. Biochem.* 2012, *51*, 153–158.

120. Zhao, J.; Yang, L.; Zhou, L.; Bai, Y.; Wang, B.; Hou, P.; Xu, Q.; Yang, W.; Zuo, Z. Inhibitory effects of eucalyptol and limonene on the photosynthetic abilities in *Chlorella vulgaris* (Chlorophyceae). *Phycologia* 2016, *55*, 696–702.

121. Tsai, C.W.; Yang, J.J.; Chen, H.W.; Sheen, L.Y.; Lii, C.K. Garlic organosulfur compounds upregulate the expression of the pi class of glutathione S-transferase in rat primary hepatocytes. *J. Nutr.* 2005, *135*, 2560–2565.

122. Hsiung, Y.-C.; Chen, Y.-A.; Chen, S.-Y.; Chi, W.-C.; Lee, R.-H.; Chiang, T.-Y.; Huang, H.-J. Volatilized myrcene inhibits growth and activates defense responses in rice roots. *Acta Physiol. Plant.* 2013, *35*, 2475–2482.

123. Mutlu, S.; Atici, Ö.; Esim, N.; Mete, E. Essential oils of catmint (*Nepeta meyeri* Benth.) induce oxidative stress in early seedlings of various weed species. *Acta Physiol. Plant.* 2011, *33*, 943–951.

124. Jin, P.; Wu, X.; Xu, F.; Wang, X.; Wang, J.; Zheng, Y. Enhancing antioxidant capacity and reducing decay of Chinese bayberries by essential oils. *J. Agric. Food Chem.* 2012, *60*, 3769–3775.

125. Baldwin, I.T.; Halitschke, R.; Paschold, A.; von Dahl, C.C.; Preston, C.A. Volatile signaling in plant-plant interactions: "Talking trees" in the genomics era. *Science* 2006, *311*, 812–815.

126. Heil, M.; Silva Bueno, J.C. Within-plant signaling by volatiles leads to induction and priming of an indirect plant defense in nature. *Proc. Natl. Acad. Sci. USA* 2007, *104*, 5467–5472.

127. Kessler, A.; Halitschke, R.; Diezel, C.; Baldwin, I.T. Priming of plant defense responses in nature by airborne signaling between *Artemisia tridentata* and *Nicotiana attenuata*. *Oecologia* 2006, *148*, 280–292.

128. Ton, J.; D'Alessandro, M.; Jourdie, V.; Jakab, G.; Karlen, D.; Held, M.; Mauch-Mani, B.; Turlings, T.C.J. Priming by airborne signals boosts direct and indirect resistance in maize. *Plant J.* 2007, *49*, 16–26.

129. Heil, M.; Ton, J. Long-distance signalling in plant defence. *Trends Plant Sci.* 2008, *13*, 264–272.

130. Biedrzycki, M.L.; Bais, H.P. Kin recognition in plants: A mysterious behaviour unsolved. *J. Exp. Bot.* 2010, *61*, 4123–4128.

131. Broz, A.K.; Broeckling, C.D.; De-la-Peña, C.; Lewis, M.R.; Greene, E.; Callaway, R.M.; Sumner, L.W.; Vivanco, J.M. Plant neighbor identity influences plant biochemistry and physiology related to defense. *BMC Plant Biol.* 2010, *10*, 115.

132. Chen, B.J.; During, H.J.; Anten, N.P. Detect thy neighbor: Identity recognition at the root level in plants. *Plant Sci. Int. J. Exp. Plant Biol.* 2012, *195*, 157–167.

133. Karban, R.; Shiojiri, K. Self-recognition affects plant communication and defense. *Ecol. Lett.* 2009, *12*, 502–506.

134. Masclaux, F.; Hammond, R.L.; Meunier, J.; Gouhier-Darimont, C.; Keller, L.; Reymond, P. Competitive ability not kinship affects growth of Arabidopsis thaliana accessions. *New Phytol.* 2010, *185*, 322–331.

135. Milla, R.; Forero, D.M.; Escudero, A.; Iriondo, J.M. Growing with siblings: A common ground for cooperation or for fiercer competition among plants? *Proc. R. Soc. B Biol. Sci.* 2009, *276*, 2531–2540.

136. Yang, X.-F.; Li, L.-L.; Xu, Y.; Kong, C.-H. Kin recognition in rice (*Oryza sativa*) lines. *New Phytol.* 2018, *220*, 567–578.

137. Torices, R.; Gómez, J.M.; Pannell, J.R. Kin discrimination allows plants to modify investment towards pollinator attraction. *Nat. Commun.* 2018, *9*, 2018.

138. Brooks, G.T. Comprehensive insect physiology, biochemistry and pharmacology: Edited by G. A. Kerkut and L. I. Gilbert. Pergamon Press, Oxford. 1985. 13 Volumes. 8200 pp approx. £1700.00/$2750.00. ISBN 0 08 026850 1. *Insect Biochem.* 1985, *15*, i–xiv.

139. Kimparis, A.; Siatis, N.; Daferera, D.; Tarantilis, P.; Pappas, C.; Polissiou, M. Comparison of distillation and ultrasound-assisted extraction methods for the isolation of sensitive aroma compounds from garlic (*Allium sativum*). *Ultrason. Sonochem.* 2006, *13*, 54–60.

140. Lee, S.N.; Kim, N.S.; Lee, D.S. Comparative study of extraction techniques for determination of garlic flavor components by gas chromatography-mass spectrometry. *Anal. Bioanal. Chem.* 2003, *377*, 749–756.

141. Mutarutwa, D.; Navarini, L.; Lonzarich, V.; Compagnone, D.; Pittia, P. GC-MS aroma characterization of vegetable matrices: Focus on 3-alkyl-2-methoxypyrazines. *J. Mass Spectrom.* 2018, *53*, 871–881.

142. Sgorbini, B.; Cagliero, C.; Liberto, E.; Rubiolo, P.; Bicchi, C.; Cordero, C. Strategies for accurate quantitation of volatiles from foods and plant-origin materials: A challenging task. *J. Agric. Food Chem.* 2019, *67*, 1619–1630.

143. Verpoorte, R.; Choi, Y.H.; Kim, H.K. NMR-based metabolomics at work in phytochemistry. *Phytochem. Rev.* 2007, *6*, 3–14.

144. Marshall, T.L.; Chaffin, C.T.; Makepeace, V.D.; Hoffman, R.M.; Hammaker, R.M.; Fateley, W.G.; Saarinen, P.; Kauppinen, J. Investigation of the effects of resolution on the performance of classical least-squares (CLS) spectral interpretation programs when applied to volatile organic compounds (VOCs) of interest in remote sensing using open-air long-path Fourier transform infrared (FT-IR) spectrometry. *J. Mol. Struct.* 1994, *324*, 19–28.

145. Stierlin, É.; Nicolè, F.; Fernandez, X.; Michel, T. Development of a headspace solid-phase microextraction gas chromatography-mass spectrometry method to study volatile organic compounds (VOCs) emitted by lavender roots. *Chem. Biodivers.* 2019, *16*, e1900280.

146. Danner, H.; Samudrala, D.; Cristescu, S.M.; Van Dam, N.M. Tracing hidden herbivores: Time-resolved non-invasive analysis of belowground volatiles by proton-transfer-reaction mass spectrometry (PTR-MS). *J. Chem. Ecol.* 2012, *38*, 785–794.

147. Capozzi, V.; Lonzarich, V.; Khomenko, I.; Cappellin, L.; Navarini, L.; Biasioli, F. Unveiling the molecular basis of mascarpone cheese aroma: VOCs analysis by SPME-GC/MS and PTR-ToF-MS. *Molecules* 2020, *25*, 1242.

148. Lebanov, L.; Ghiasvand, A.; Paull, B. Data handling and data analysis in metabolomic studies of essential oils using GC-MS. *J. Chromatogr. A* 2021, *1640*, 461896.

149. Majchrzak, T.; Wojnowski, W.; Lubinska-Szczygeł, M.; Różańska, A.; Namieśnik, J.; Dymerski, T. PTR-MS and GC-MS as complementary techniques for analysis of volatiles: A tutorial review. *Anal. Chim. Acta* 2018, *1035*, 1–13.

150. Hare, J.D. Ontogeny and season constrain the production of herbivore-inducible plant volatiles in the field. *J. Chem. Ecol.* 2010, *36*, 1363–1374.

14 Chapter

AN ANALYSIS OF GENE EXPRESSION IN PEPPER (*CAPSICUM ANNUUM* L.) SEEDLINGS WAS CONDUCTED TO INVESTIGATE THE EFFECTS OF HEAT STRESS

1. INTRODUCTION: AN OVERVIEW

Pepper, scientifically known as *Capsicum annuum* L., is a significant plant belonging to the Solanaceae family. It holds great importance as both a spice and a vegetable crop in numerous nations [1]. The substance has a high concentration of capsaicin, capsanthin, and vitamins, which have the potential to enhance hunger and promote overall well-being [2]. Pepper thrives in warm regions but is vulnerable to extreme heat, with the optimal temperature range for growth and development being 20–30 °C [3]. When the temperature goes beyond 35 °C, the plant will experience heat stress (HS) and display signs of high temperature damage throughout its growth stage. This will have negative effects on the plant's physical appearance, as well as its physiological and biochemical processes, among other aspects [3,4]. The greenhouse effect has caused a significant increase in global temperatures, which has negatively affected the growth and development of crops. This has been a major challenge for agricultural regions worldwide, resulting in a significant decrease in economic yields and quality [5]. Hence, it is crucial to study the molecular mechanisms that govern the response of pepper to heat stress (HS) in order to enhance the development of varieties that can thrive in harsher environments.

HS impacts the cellular architecture of plants, causing protein denaturation and disruption of lipid transport. This leads to the breakdown of the plasma membrane structure and ultimately the demise of particular cells or tissues. HS induces excessive water loss through plant transpiration, leading to a reduction in the photosynthetic rate and disrupting normal metabolic processes. These effects have a negative impact on the growth and development of plants [4]. Photosynthesis is a very temperature-sensitive physiological process that can be readily hindered by HS, impacting several aspects of photosynthesis such as photosystem II, photosystem I, electron transport chain, ATP synthesis, and carbon fixation [6,7]. Furthermore, alongside the decline in net photosynthetic efficiency and photosystem activity, there is an accumulation of reactive oxygen species (ROS). This accumulation leads to the degradation of D1 protein and antenna pigment, which can severely impair the plants' capacity to absorb and utilize light energy and store carbon. This has been observed in severe cases [8,9]. Furthermore, the presence of reactive oxygen species (ROS) induced by heat stress (HS) in plants leads to cellular oxidative damage. Elevated temperatures induce metabolic disruptions and trigger the generation of reactive oxygen species (ROS) in plants. This exacerbates the oxidation of lipids and the denaturation of proteins in the cell membrane, thereby impacting the integrity and functionality of biofilms. In severe instances, this can result in cellular harm and the demise of plants [10]. Elevated temperatures have a significant impact on plant metabolism. Specifically, high temperatures trigger the activation and suppression of numerous genes involved in the anthocyanin production pathway of eggplant. Consequently, this leads to a reduction in the accumulation of anthocyanin [11]. Several non-living factors, such as heat stress (HS), have a direct or indirect impact on the production, concentration, processing, movement, and retention of sugars. Soluble sugars have the potential to operate as signal molecules that interact with light, nitrogen, and abiotic stressors in order to regulate plant growth and development [12,13,14].

Perceiving and transmitting thermal signals is crucial for plants to resist stress. This process involves several pathways, including calcium-dependent protein kinases (CDPKs) and mitogen-activated protein kinases (MAPK/MPKs), as well as signal molecules like reactive oxygen species (ROS) and plant hormones. These components are essential for relaying external stimuli to generate internal cellular responses in various signaling networks. The sensing and transduction of thermal signals actively control the expression of genes and the function of proteins in response to different stresses, leading to adaptation to environmental challenges [15,16,17,18,19]. For instance, tomato mutants with SlMAPK3 gene edited using CRISPR/Cas9 technology exhibited greater tolerance to heat compared to normal plants. These mutants showed reduced wilting and damage to cell membranes, lower levels of reactive oxygen species (ROS), increased activity of antioxidant enzymes, and higher levels of gene expression [20]. The 47 kD MBP-phosphorylated protein SlMPK1, which is activated by heat, has a deleterious impact on the heat tolerance of tomato plants. It achieves this by facilitating antioxidant defense and regulating redox metabolism. SlSPRH1, a protein similar to SlMPK1 and abundant in serine and proline, is the specific target of SlMPK1 and can be phosphorylated by it. Increased expression of SlSPRH1 diminishes the heat resistance and antioxidant capability of plants and is associated with the phosphorylation of SlSPRH1. The SlMPK1-SlSPRH1 module has a negative regulatory effect on the high-temperature signal during the high-temperature response process and works in conjunction with the oxidative stress system [21]. Research indicates that HS is accompanied by a specific level of oxidative stress, and there is an interaction between the signals of heat and oxidative stressors. According to a study, H_2O_2 is released rather quickly in the presence of HS due to the action of NADPH oxidase [22]. This outbreak was associated with the activation of heat shock response genes [23]. Pretreatment with H2O2 or menadione can enhance heat tolerance in plants [24]. BZR1, a crucial controller of the brassinoid (BR) response, governs the heat shock (HS) response in tomato by means of RBOH1-dependent reactive oxygen species (ROS) signaling, partially by regulating FER2 and FER3 [25].

Plant heat shock transcription factors (Hsfs) play a crucial role in signal transduction by regulating the transcription of heat shock factors (HSPs) and other genes triggered by heat shock [26]. HsfA1a controls the first reaction, although both HsfA1a and HsfB1 are consistently expressed at a low level of mRNA. throughout heat stress (HS), there is a significant increase in the levels of HsfA2 mRNA and protein. As a result, HsfA2 becomes the predominant heat shock factor (Hsf) and plays a crucial role in regulating heat tolerance throughout the recovery phase or after repeated episodes of HS [27,28,29]. Under conditions without stress, the excessive production of HsfB1 triggers the simultaneous activation of HsfB1, leading to the buildup of proteins related to heat shock (HS) and an improvement in heat tolerance [30,31]. Hsps, such as Hsp100/ClpB, Hsp90/HtpG, Hsp70/DnaK, Hsp60/GroEL, and small Hsp (sHsp), are regulated by Hsfs. These Hsfs are crucial molecular chaperones that help maintain and restore protein homeostasis. This function is essential for the survival of plants under high temperature stress (HS) conditions [32,33]. Aside from Hsfs, other prominent families of transcription factors in plants, including WRKY, bZIP, MYB, and NAC, are also implicated in heat shock responses. CaWRKY27 acts as a downstream negative regulator of the H2O2-mediated heat shock response, preventing inappropriate responses during both heat shock and recovery [3]. CabZIP63, a member of the bZIP family in pepper, controls the expression of CaWRKY40 both at the transcriptional and post-transcriptional levels. It also establishes a positive feedback loop with CaWRKY40 during the pepper's response to ralstonia solanacearum inoculation or high temperature–high humidity [34]. Increased expression of SlAN2 resulted in the activation of multiple structural genes involved in the anthocyanin biosynthesis pathway and led to the accumulation of anthocyanin in tomato plants. This accumulation improved the plant's ability to tolerate high temperatures.

When plants are exposed to heat stress (HS) in their natural environment, their capacity to recuperate becomes crucial. This is because the more robust their ability to recover, the quicker the plant can regain its metabolic equilibrium and sustain its regular growth. Nevertheless, the specific molecular processes and networks that regulate pepper have not been discovered. Hence, in this investigation, we conducted transcriptome analysis of the heat-tolerant variety 17-03 and the heat-sensitive variety H1023 during heat stress recovery, with the aim of identifying candidate genes that exhibited changes in transcription levels in the leaves of pepper plants. Overall, our research results establish a theoretical foundation for the development of superior, heat-tolerant cultivars.

2. MATERIALS AND METHODS

2.1. Plant Materials and Heat Treatments

Two pepper varieties, heat-tolerant 17-03 and heat-sensitive H1023, were obtained from the Hubei Key Laboratory of Vegetable Germplasm Enhancement and Genetic Improvement, Hubei Academy of Agricultural Sciences, for transcriptome analysis. Seeds were sown in 50-hole trays and grown in a growth chamber under cool white fluorescent lights (approximately 200 $\mu mol/m^2/s$) at 25 ± 2 °C with a photoperiod of 16 h light/8 h dark and 70–80% relative humidity. At two weeks, seedlings with uniform growth were transplanted to plastic pots (10 × 10 × 10 cm), containing peat, vermiculite, and soil (v/v/v = 1:1:1), until they reached the stage of 6–8 true leaves. For heat treatment, seedlings were well watered and cultivated at 42 °C for 3 d before recovering at 25 °C for 1 d. The control seedlings were placed at 25 °C for 4 d. The lighting conditions and humidity were not changed. The first fully expanded leaf from the top of each plant was sampled from eight seedlings (three replicates) in the control and treatment groups. All samples were immediately frozen in liquid nitrogen and stored at -80 °C for RNA sequencing. The samples were named CK1 (HT1_1-HT1_3) (control group of 17-03), T1 (HT2_1-HT2_3) (heat treatment group of 17-03), M1 (HT3_1-HT3_3) (recovered group of 17-03), CK2 (HS1_1-HS1_3) (control group of H1023), T2 (HS2_1-HS2_3) (heat treatment group of H1023), and M2 (HS3_1-HS3_3) (recovered group of H1023).

2.2. Measurement of Relative Electrolyte Leakage and Proline Content

Relative electrolyte leakage, which measures cellular membrane integrity, is frequently used to evaluate plant stress tolerance [36]. The upper third of the fully expanded leaves from treated and non-treated plants of 17-03 and H1023 were excised and used to generate leaf discs (9 mm in diameter). Three replicates were used for each line, with 20 leaf discs per replicate. The leaf discs were placed into 50 mL centrifuge tubes containing 25 mL of distilled deionized water and shaken at 60 rpm for 12 h in the dark at 25 °C. The electrolyte leakage (R1) of the solution was measured using a portable magnetic conductivity meter (DDB-303A, Shanghai, China) at 25 °C. The solutions were boiled for 30 min and then cooled to room temperature. The electrolyte leakage in the boiled solution (R2) was then determined using the same method. The relative electrolyte leakage (%) was calculated as (R1/R2) × 100.

The proline content was measured as described by Ben et al. [37]. Briefly, 200 mg of ground leaf sample was extracted with 2 mL of 3% sulfosalicylic acid in boiling water for 10 min. The samples were cooled and centrifuged at 4000 rpm for 15 min. Then, 1 mL of the supernatant was transferred to a new 15 mL test tube. Then, 1 mL of glacial acetic acid and 1 mL of acid-ninhydrin reagent were added. The mixture was boiled for 30 min and cooled in an ice bath to terminate the reaction. Next, the reaction mixture was partitioned by adding toluene (3 mL). After static delamination, the upper liquid was centrifuged at 4000 rpm for 15 min. The absorbance of the organic phase was measured in a spectrophotometer (Schimadzu, Japan) at a 520 nm wavelength. The proline content of the leaf samples was calculated using a standard curve constructed with known amounts of proline.

2.3. RNA Extraction, Library Construction, and Transcriptome Sequencing

According to the manufacturer's instructions, total RNA was extracted from 18 pepper leaf samples using TRIzol reagent (Life Technologies, California, CA, USA). The mRNA with poly (A) in the total RNA was enriched using Oligo (dT) magnetic beads and divided into fragments approximately 300 bp in length using ion interruption. First-strand cDNA was synthesized using a M-MuLV reverse transcriptase system, using these RNA fragments as templates and random hexamer primers, while the second strand cDNA was synthesized using the first-strand cDNA as a template. Subsequently, the cDNA libraries were constructed after polymerase chain reaction (PCR) amplification and selected according to the fragment length of 450 bp. Then, the quality of the cDNA libraries was checked using an Agilent 2100 Bioanalyzer system (Agilent Technologies, Inc., Santa Clara, CA, USA). Based on the Illumina sequencing platform, the qualified libraries were sequenced using a double terminal (paired-end, PE) sequencer (Illumina, Foster, CA, USA).

2.4. Read Alignment and Mapping Reads to the Reference Genome

To analyze the sequencing results effectively and accurately, low-quality raw data or connectors in the sequencing data were filtered. Cutadapt was used to remove the 3′ sequencing adapter (https://cutadapt.readthedocs.io/en/stable/) (accessed on 10 March 2021), and reads with an average mass fraction lower than Q20 were removed [38]. The high-quality clean reads from each library were mapped to the pepper reference genome CM334 (https://ftp.solgenomics.net/genomes/Capsicum_annuum/C.annuum_cvCM334/) (accessed on 10 March 2021) using HISAT2 software (http://ccb.jhu.edu/software/hisat2/index.shtml) (accessed on 10 March 2021). The read count value of each gene was mapped using HTSeq as the original expression of the gene [39]. Fragments per kilobase of transcript per million mapped reads (FPKM) was used to standardize the gene expression levels based on Cufflinks software [40].

2.5. Functional Enrichment Analysis of Differentially Expressed Genes

Genes with an absolute value of |log2fold change| > 1 and a false discovery rate (FDR) < 0.05 were identified as representing significantly differentially expressed genes (DEGs), using DESeq2 in the four comparisons of CK1_vs_T1, CK1_vs_M1, CK2_vs_T2, and CK2_vs_M2 [**41**]. To study the putative functions and pathways of the DEGs in the above four comparisons, gene ontology (GO) functional enrichment analysis was conducted using Blast2GO (version 3.0; **https://www.blast2go.com/**) (accessed on 10 March 2021) [**42**]. Kyoto Encyclopedia of Genes and Genomes (KEGG) pathway annotation of DEGs was performed using Cytoscape software (version 3.2.0) (**https://cytoscape.org/**) (accessed on 10 March 2021) with the ClueGO plugin using a hypergeometric test and Benjamini-Hochberg FDR correction (FDR ≤ 0.05) [**43**]. Transcription factors were predicted using PlantTFDB [**44**].

2.6. Reverse Transcription and Quantitative Reverse-Transcription Polymerase Chain Reaction Analysis

RNA was first treated with DNase I and then reverse-transcribed to cDNA using a HiScript II 1st Strand cDNA Synthesis Kit (+gDNA wiper), according to the manufacturer's instructions (Vazyme, Nanjing, China). Then, the concentration of cDNA was diluted to 100 ng/μL, and the internal reference gene *CaUBI-3* forward and reverse primers were used for PCR to detect whether the reverse transcription was successful [**45**]. For the quantitative reverse-transcription polymerase chain reaction (qRT-PCR) assay, all primers were designed using Primer3 (**http://primer3.ut.ee/**) (accessed on 10 March 2021), and the specificity of the designed fragment was tested using SGN (**https://solgenomics.net/**) (accessed on 10 March 2021) (**Table S8**). qRT-PCR was performed using SYBR® Premix Ex Taq™ (Vazyme), with three technical replicates in 10 μL volumes containing 5 μL Fast SYBR™ Green Master Mix (2×), 2 μL cDNA template (100 ng/μL), 0.2 μL of each primer (10 μM), and 2.6 μL ddH2O. The PCR cycling conditions were as follows: 95 °C for 30 s, followed by 40 cycles of 95 °C for 5 s, and finally 60 °C for 20 s. The relative expression level of each gene was calculated using the 2–ΔΔCT method [**46**].

3. RESULTS

3.1. Phenotypic and Physiological Responses of 17-03 and H1023 under HS

To accurately evaluate the heat resistance of pepper in the seedling stage under HS, 17-03 (heat-tolerant variety) and H1023 (heat-sensitive variety) were treated at 42 °C for 3 d and then recovered for 1 d at 25 °C. Compared with the leaves of seedlings of 17-03 and H1023 that did not undergo HS (Figure 1a,d), the leaves of seedlings 17-03 after heat treatment were slightly bent (Figure 1b), while the leaves of H1023 were severely sagged, and the lower leaves were severely wilted. Additionally, the growth points of the plants were necrotic after treatment at 42 °C for 3 d (Figure 1e). After recovering at 25 °C for 1 d, the down-bent leaves of 17-03 were completely extended (Figure 1c), while those of H1023 only partially recovered and did not stretch out completely. Additionally, the edges of the leaves had different degrees of withering (Figure 1f).

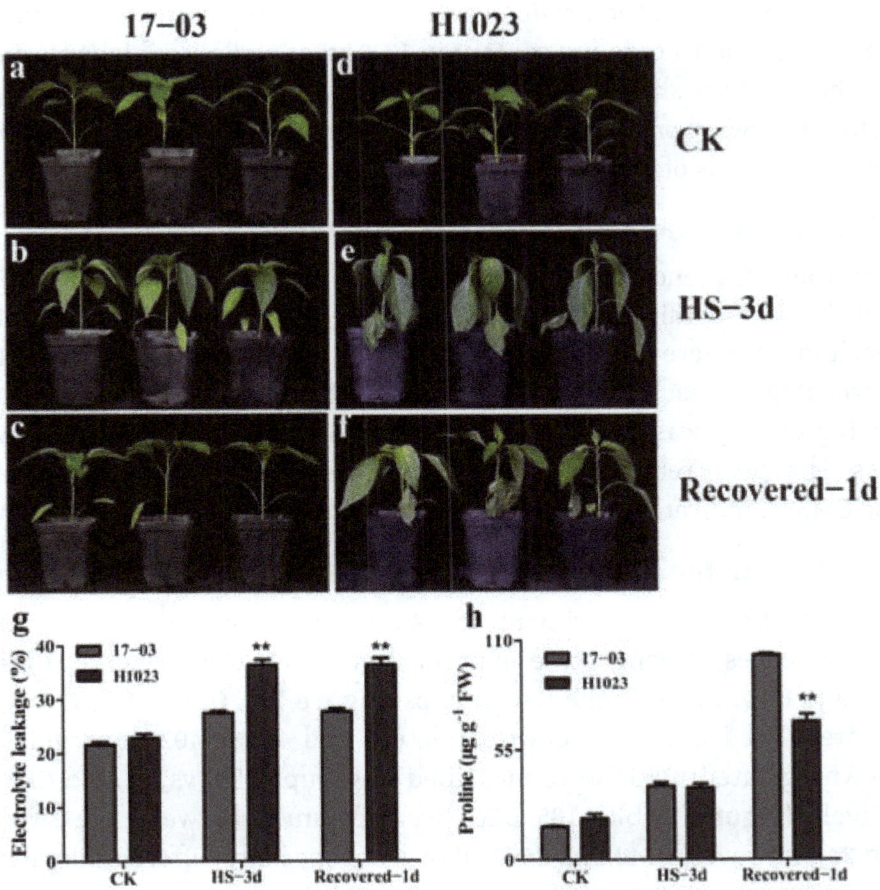

Figure 1. Phenotypic and physiological responses of 03-17 and H1023 under heat stress (HS). Seedlings of 03-17 and H1023 in control (a,d), treated at 42 °C for 3 d (b,e), and recovered for 1 d after 3 d of heat treatment (c,f). Relative electrolyte leakage (g) and proline content (h) in leaves treated at 42 °C for 0 d and 3 d and recovered for 1 d. Three independent biological replicates were used in each treatment, with 9 plants (6–8 true leaves) per replicate. Data are presented as mean ±SEM of three independent biological replicates. Asterisks indicate statistically significant differences between tolerant and sensitive genotypes. **, $p < 0.01$, Student's *t* test.

In this study, relative electrolyte leakage and proline content were measured in the treated plants to evaluate heat tolerance. Under normal growth conditions, there was no significant difference in the relative electrolyte leakage and proline content between the two varieties (Figure 1g,h). However, after 3 d of heat treatment at 42 °C and 1 d recovery at 25 °C, there was a remarkable increase in relative electrolyte leakage in the two varieties, with levels being significantly lower in 17-03 than in H1023 (Figure 1g). The proline content in plants increased after HS and recovery in both varieties (Figure 1h). After 3 d of heat treatment, the proline content increased, but there was no significant difference between the two varieties. However, the proline content further increased significantly in the recovery stage, and the proline content in 17-03 was obviously higher than that in H1023. These results show that 17-03 was more heat-tolerant than H1023, as the cell membranes were protected from damage, and osmotic stress was alleviated by increasing the levels of proline, an important osmotic protectant.

3.2. Overview of Transcriptomic Data for 17-03 and H1023

Transcriptome sequencing yielded 815.2 M raw reads. After filtering, 724.31 M valid reads were obtained in 18 libraries. The average effective data obtained from each sample were 6.03 G, accounting for 88.85% of the original data. The Q30 base percentage of each library was above 92.81%). The results showed that the sequencing quality was reliable, and the data were suitable for the subsequent analyses. The clean reads (82.23–85.35%) were mapped to the pepper reference genome CM334, indicating that the data could be used for subsequent analysis.

3.3. Identification of Differentially Expressed Genes (DEGs)

Based on the RNA-seq experiment, 24,448 expressed genes were identified in the pepper leaves. Among these expressed genes, 11,633 DEGs were identified in 17-03 and H1023 among the four groups (**Figure** 2a). Of the 11,633 DEGs, 7327 (3435 upregulated and 3892 downregulated) and 7778 (4025 upregulated and 3753 downregulated) DEGs were identified in groups CK1_vs_T1 and CK2_vs_T2, respectively (**Figure** 2a,b); 5185 DEGs were common between the two groups (**Figure** 2a); however, 2142 and 2593 DEGs were specially differentially expressed in CK1_vs_T1 and CK2_vs_T2, respectively (**Figure** 2a). Approximately, 3338 (2021 upregulated and 1317 downregulated) and 4822 (2256 upregulated and 2566 downregulated) DEGs were identified in groups CK1_vs_M1 and CK2_vs_M2, respectively (**Figure** 2b); there were 1934 common DEGs and 1404 and 2888 DEGs specially differentially expressed DEGs in CK1_vs_M1 and CK2_vs_M2, respectively (**Figure** 2a). Interestingly, 1229 common DEGs were identified among all four groups (**Figure** 2a).

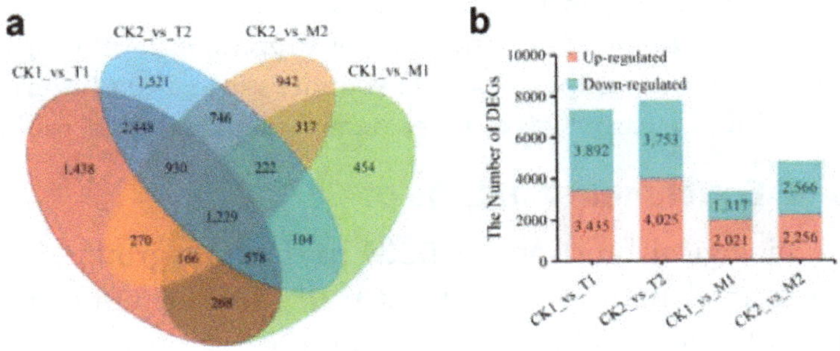

Figure 2. Expression analysis of differentially expressed genes (DEGs) in 17-03 and H1023 leaves after 3 d heat treatment at 42 °C and 1 d recovery at 25 °C. Numbers of DEGs in 17-03 and H1023 at different times (a). Numbers of up- and down regulated DEGs in 17-03 and H1023 at different times (b).

3.4. Validation of RNA-Seq Data Using qRT-PCR

To confirm the accuracy of the RNA-seq data, transcriptional levels of 14 randomly selected DEGs, representing a wide range of expression levels and patterns, were detected using qRT-PCR. All 14 DEGs participated in the process of HS response, including small heat shock protein (CA03g21390), HS transcription factor (CA03g16300), bZIP (CA08g12820), MYB (CA04g16680, CA06g27890), WRKY genes (CA03g32070, CA08g08240, and CA09g11940), NAC genes (CA05g04410, CA07g18020, CA09g12970, and CA11g04440), EG45-like domain-containing protein (CA07g00930), and universal stress protein A-like protein (CA11g00890) (**Figure 3**). The fold changes varied in the RNA-Seq and qRT-PCR analyses. Generally, the expression patterns determined using qRT-PCR were consistent with those obtained using RNA-Seq (**Figure 3**), which confirmed the accuracy of the RNA-Seq results reported in this study.

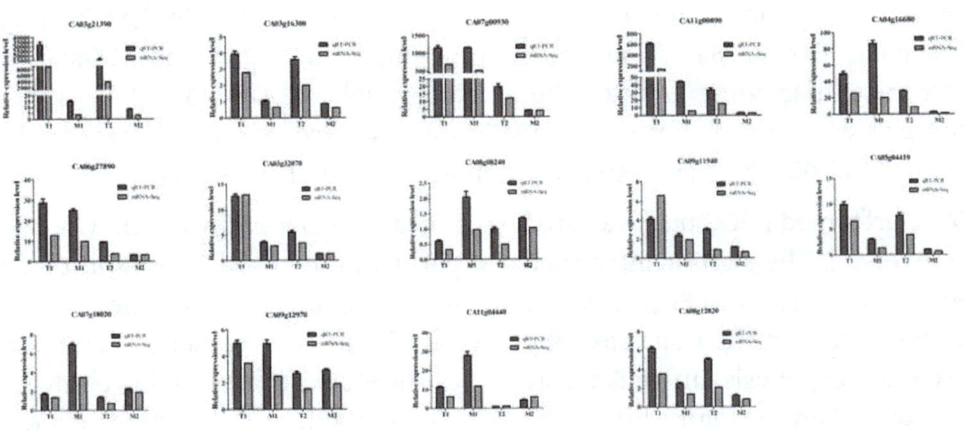

Figure 3. Quantitative reverse-transcription polymerase chain reaction (qRT-PCR)-based validation of differentially expressed genes (DEGs) in response to heat stress (HS)

at different time intervals. Ordinate represents fold changes of RNA-Seq data and the relative expression level of qRT-PCR. The relative expression level of each gene under stress at each time point was compared with that under normal conditions. qRT-PCR data are presented as mean ± SEM of three independent technical replicates.

3.5. Functional Enrichment Analysis of DEGs

To explore the biological functions of DEGs in the four groups, GO enrichment analysis was performed. In total, 5133 of 11,633 DEGs in the four comparisons were annotated with GO terms and assigned to three categories: molecular function (MF), biological process (BP), and cellular component (CC) (**Figure 4**). Under HS, the GO terms "photosynthesis, light harvesting" (GO:0009765), "DNA conformation change" (GO:0071103), "nucleosome assembly" (GO:0006334), "DNA packaging" (GO:0006323), "chromatin assembly" (GO:0031497), "nucleosome organization" (GO:0034728), and "chromatin assembly or disassembly" (GO:0006333) were the most commonly enriched components in the BP category in 17-03 and H1023 (**Figure 4**). The most commonly enriched components were "cell wall" (GO:0005618) and "external encapsulating structure" (GO:0030312) in the CC category and "nucleosome binding" (GO:0031491) in the MF category in 17-03 and H1023 (**Figure 4**). However, "DNA replication" (GO:0006260), "DNA replication initiation" (GO:0006270), "protein folding" (GO:0006457), and "protein-DNA complex assembly" (GO:0065004) in the BP category, "protein-DNA complex" (GO:0032993) in the CC category, and "nucleosomal DNA binding" (GO:0031492) in the MF category were only enriched in 17-03. For the recovery stage after HS, the most significantly enriched GO terms differed between 17-03 and H1023. In 17-03, DEGs were most enriched in "translation" (GO:0006412), "peptide biosynthetic process" (GO:0043043), "peptide metabolic process" (GO:0006518), "amide biosynthetic process" (GO:0043604), "cellular amide metabolic process" (GO:0043603), and "organonitrogen compound biosynthetic process" (GO:1901566) in the BP category. "Ribosome" (GO:0005840), "non-membrane-bounded organelle" (GO:0043228), and "intracellular non-membrane-bounded organelle" (GO:0043232) were the top three CC categories. The terms of the MF category were "structural constituent of ribosome" (GO:0003735) and "structural molecule activity" (GO:0005198) .

We performed KEGG pathway analysis to examine the pathways in which DEGs were involved. The significantly enriched pathways involved in HS and recovery responses are shown in **Figure 5**. The common significantly enriched pathways under HS were identified, such as "photosynthesis–antenna proteins" (cann00196), "ribosome biogenesis in eukaryotes" (cann03008), "fatty acid elongation" (cann03008), "anthocyanin biosynthesis" (cann00942), and "glycine, serine and threonine metabolism" (cann00260) (**Figure 5**). However, the significantly enriched KEGG pathways during recovery were different. The term "ribosome" (cann03010) was the most significantly enriched in 17-03, whereas "photosynthesis–antenna proteins" (cann00196), "carbon fixation in photosynthetic organisms" (cann00710), "porphyrin and chlorophyll metabolism" (cann00860), and "carotenoid biosynthesis" (cann00906) were the most significantly enriched in H1023 (**Figure 5**).

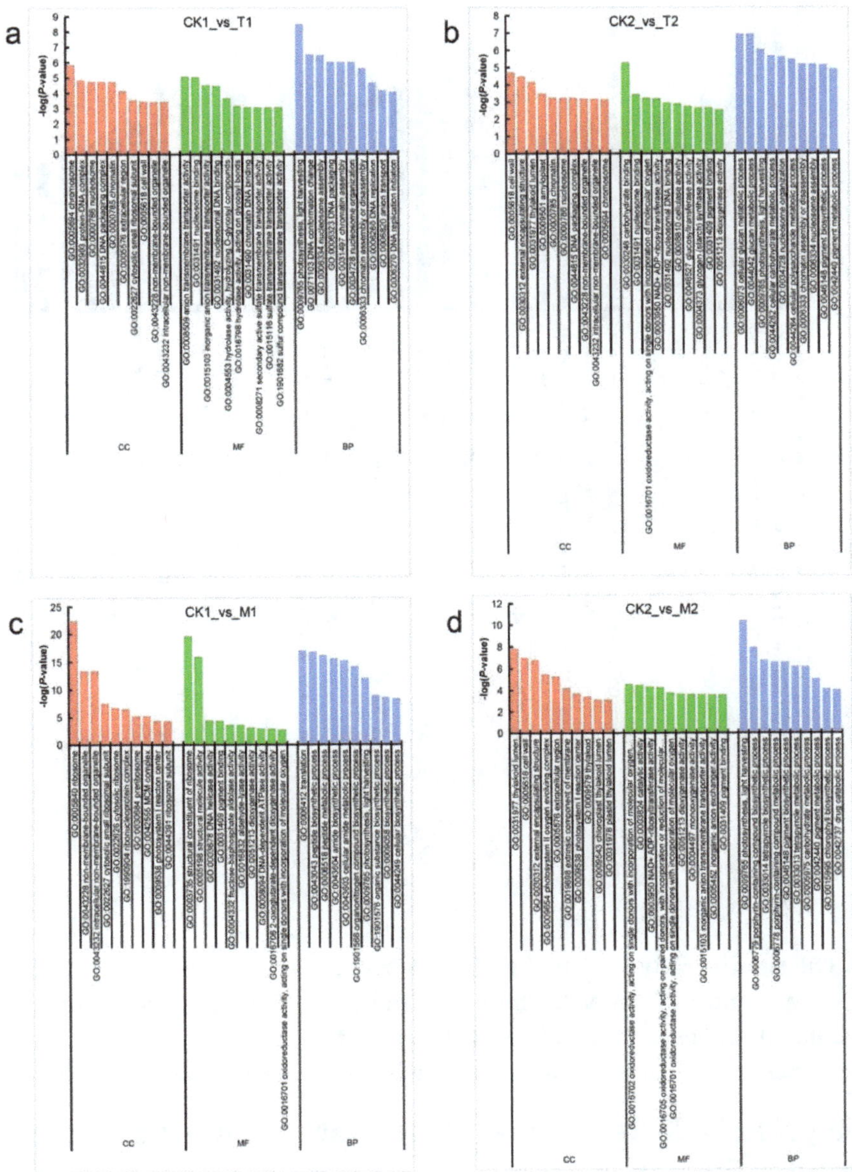

Figure 4. Gene ontology (GO) classifications of differentially expressed genes (DEGs) in the comparison groups CK1_vs_T1 (a), CK2_vs_T2 (b), CK1_vs_M1 (c), and CK2_vs_M2 (d). The DEGs were assigned to three categories: biological process (BP), cellular component (CC), and molecular function (MF). The X-axis indicates the top ten most significantly enriched BP, CC, and MF categories. The Y-axis indicates -log10 (*p*-value).

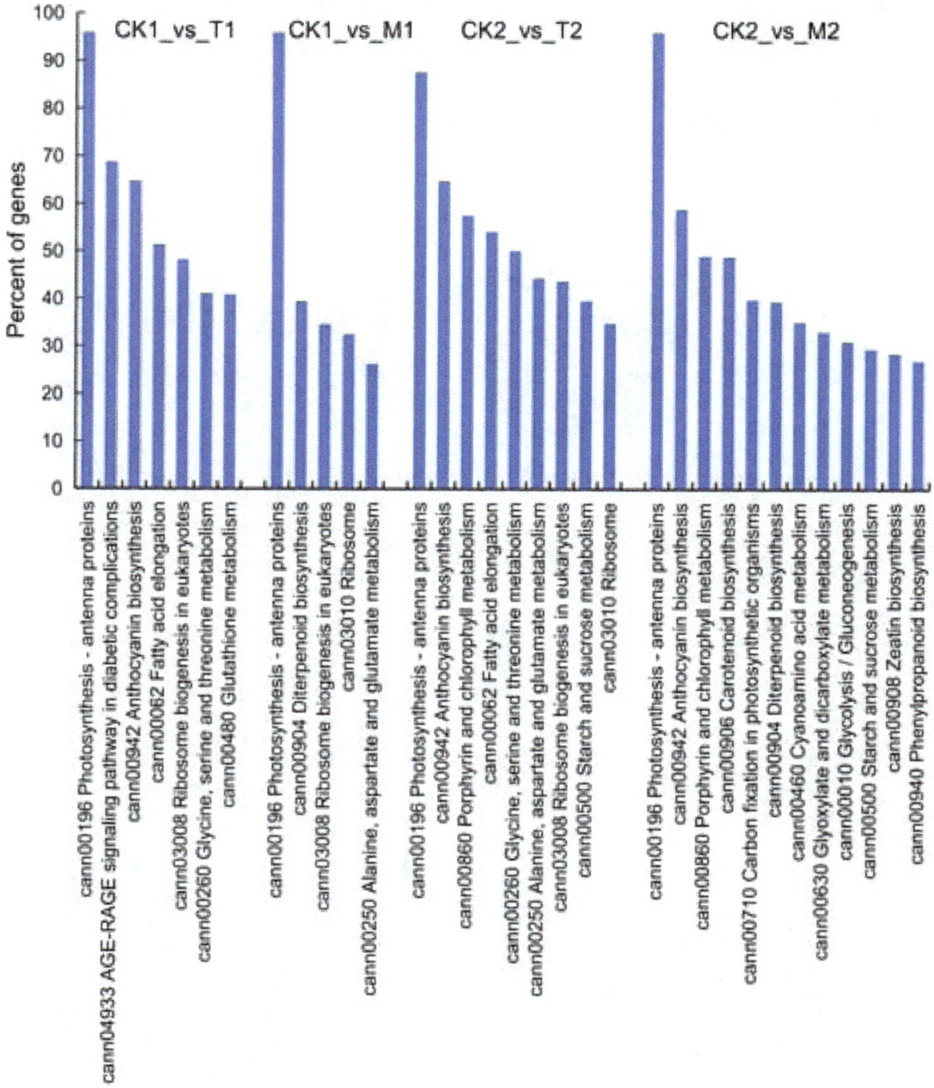

Figure 5. The significantly enriched pathways in comparison groups CK1_vs_T1, CK2_vs_T2, CK1_vs_M1, and CK2_vs_M2. The *X*-axis indicates the significantly enriched Kyoto Encyclopedia of Genes and Genomes (KEGG) pathways, and the *Y*-axis indicates the percentage of a specific category of genes in the main category.

3.6. Analysis of HS-Responsive Heat Shock Proteins and Heat Transcription Factors

Hsps and Hsfs are sensitive to HS, indicating that they play important roles in the HS response. Based on the results of transcriptome sequencing, 47 Hsp members were identified as differentially expressed in at least one of the four comparison groups (Figure 6). Most of these were dramatically upregulated in pepper leaves after 3 d of heat treatment and were more highly expressed in H1023 than in 17-03 (Figure 6). Moreover, seven Hsps were expressed at higher levels after recovery from HS. CA01g13220, CA02g11030, CA09g06120, and CA10g10840 were highly expressed in H1023 cells (Figure 6), and CA01g31330 was highly expressed in 17-03 and H1023 cells (Figure 6). Furthermore, the expression levels of five Hsps (CA04g02800, CA09g03220, CA09g06120, CA11g13160, and CA11g13170) markedly decreased after 3 d of heat treatment (Figure 6). Similarly, 17 significantly differentially expressed Hsfs were identified, most of which were upregulated during and after recovery from HS (Figure 6). Among them, six (CA02g11030, CA03g06850, CA05g00840, CA06g08710, CA07g15920, and CA10g20440) were significantly highly expressed in H1023 after recovery from HS (Figure 6). These results indicate that these significantly differentially expressed Hsps and Hsfs might play an important role in plant protection in the long-term HS response of pepper.

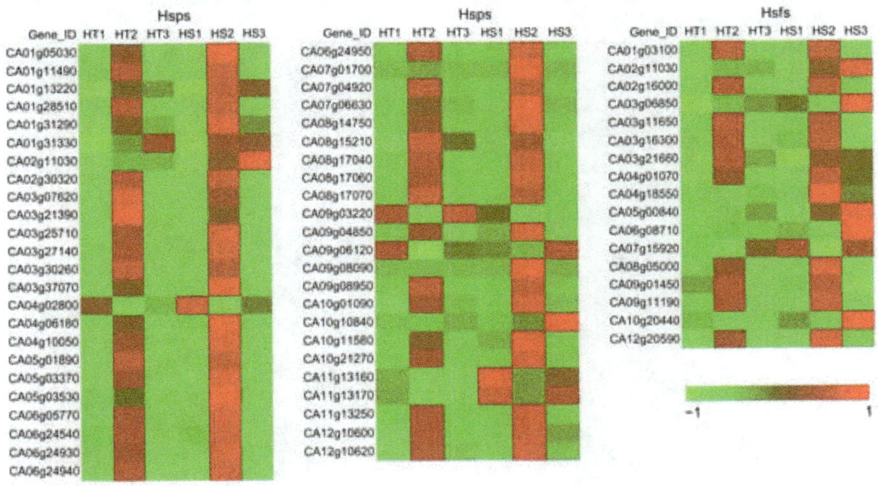

Figure 6. Heatmap of differentially expressed genes (DEGs) encoding Hsps and Hsfs in the comparison groups CK1_vs_T1, CK2_vs_T2, CK1_vs_M1, and CK2_vs_M2. The color gradient represents the normalized fragments per kilobase of transcript per million mapped reads (FPKM) value (Z-score) of DEGs. The redder the bars, the higher the gene expression level.

3.7. Analysis of HS-Responsive Transcription Factors

Transcription factors (TFs) play an important role in plant growth and development, as well as in biotic and abiotic stress response networks. Transcriptome analysis showed that many TFs in pepper were regulated by high temperatures and participated in plant recovery. A total of 49 TF families of 635 TFs were differentially expressed during heat treatment and recovery in 17-03 and H1023, including HSF, NAC, WRKY, ERF, bHLH, MYB, C2H2, B3, GRAS, bZIP, and HD-ZIP (Table S7). In this study, 38 DEGs encoding NAC proteins were identified. Among them, most were upregulated during and after recovery from HS. Moreover, the expression levels of most NAC TFs were higher in H1023 than in 17-03 during and after recovery from HS (Figure 7). WRKY proteins have also been reported to play important roles in heat response. Here, 35 DEGs encoding WRKY proteins were identified, with almost half of them upregulated in both H1023 and 17-03 during and after recovery from HS (Figure 7). Moreover, five WRKY genes (CA01g01280, CA01g34460, CA09g05110, CA09g11940, and CA12g09290) were upregulated in 17-03 but downregulated in H1023 after recovery from HS (Figure 7). Furthermore, some WRKY genes, such as CA11g05370, CA02g18540, CA09g08120, CA11g03750, and CA01g22410, were significantly highly expressed in H1023 after recovery from HS, but not at other times in H1023 or in 17-03 (Figure 7).

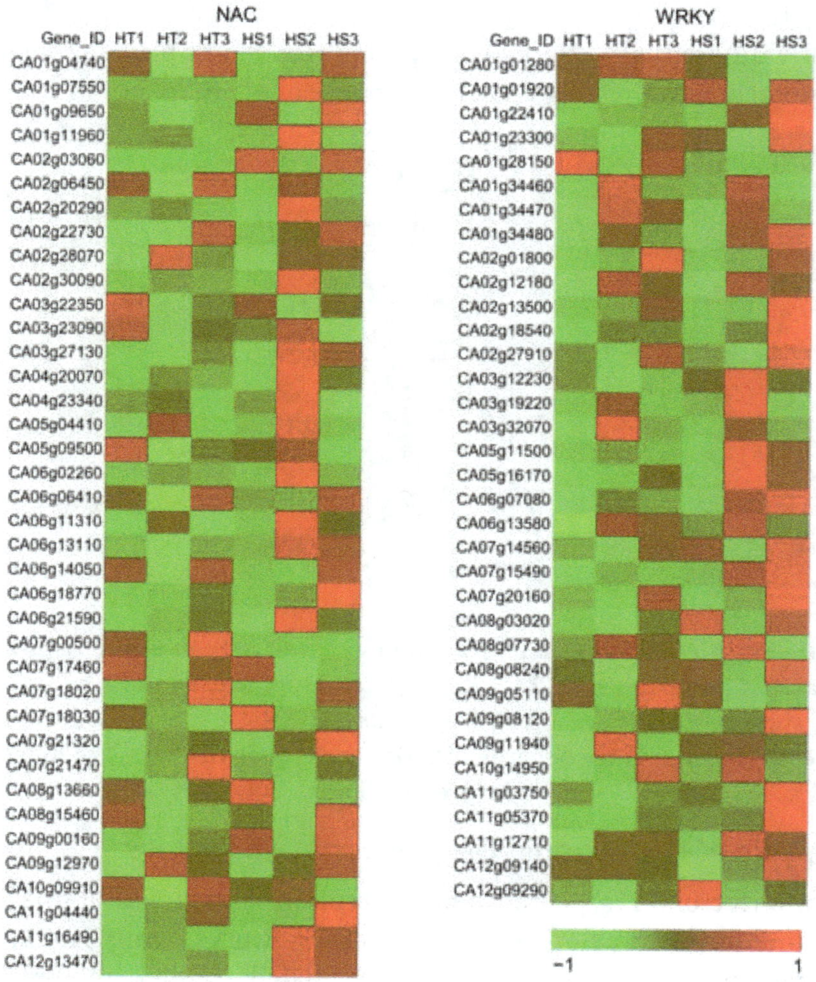

Figure 7. Heatmap of differentially expressed genes (DEGs) encoding NAC and WRKY proteins in the comparison groups CK1_vs_T1, CK2_vs_T2, CK1_vs_M1, and CK2_vs_M2. The color gradient represents the normalized fragments per kilobase of transcript per million mapped reads (FPKM) value (Z-score) of DEGs; the redder the bars, the higher the gene expression level.

4. DISCUSSION

High temperature is one of the key climatic parameters affecting plant growth and development, resulting in crop yield losses [47]. HS can restrict photosynthesis, increase photorespiration and transpiration rate through stomatal regulation, and reduce plant biomass [7]. Pepper is a highly temperature-sensitive crop [3]. Although the physiological effects of HS on pepper have been widely studied, changes in pepper at the molecular level in response to HS and subsequent recovery are poorly understood. Therefore, to better understand the HS response in pepper, it is necessary to uncover the mechanisms underlying it. In the present study, we investigated the phenotypic and physiological changes in pepper seedlings of two varieties during and after recovery from HS. Furthermore, we comparatively analyzed heat-induced transcriptomic changes to obtain a global view of HS responses in pepper.

Under high temperature stress, the structure and function of the cell protoplasmic membrane are initially damaged, resulting in an increase in cell membrane permeability and intracellular electrolyte leakage and finally leading to an increase in electrolyte leakage of tissue leachate [48]. Therefore, the degree of electrolyte extravasation and high temperature injury can be determined by measuring the relative electrical conductivity of the tissue extract. In this study, 17-03 and H1023 were verified as high temperature-resistant and high temperature-sensitive varieties, respectively (Figure 1b,c,f,g), and were used to explore the responses and recovery patterns of pepper to HS and the possible mechanisms of the different heat resistances. There was a remarkable increase in relative electrolyte leakage in both 17-03 and H1023 after HS, but the levels were significantly lower in 17-03 than in H1023 (Figure 1h), indicating that 17-03 could alleviate damage to cellular membranes during HS. However, there was little change in the two varieties from HS to recovery (Figure 1h), indicating that the damage to the cell membrane caused by HS is irreversible. Proline, an amino acid and a compatible solute, has been widely reported to accumulate in response to various abiotic stresses, such as high temperatures [37]. After 3 d of heat treatment and 1 d of recovery, proline levels were significantly increased in both 17-03 and H1023. While there was no significant difference in proline content between 17-03 and H1023 during heat treatment (Figure 1h), the proline content increase in 17-03 during recovery was higher than that in H1023 (Figure 1h), indicating that the self-repairing ability of 17-03 after HS was stronger than that of H1023. Based on these data, we conclude that 17-03 is more heat-tolerant, as it protects cell membranes from damage and alleviates osmotic stress by increasing the proline levels.

Moreover, we obtained accurate data from transcriptome analyses based on RNA-seq and analyzed the genes of metabolic pathways that were significantly affected by HS and participated in the process of plant restoration. In 17-03 and H1023, there were more DEGs in the HS stage than in the recovery stage (**Figure 2**a,b), indicating that the regulatory mechanism of HS response was more active at a transcriptional level. There were significantly more upregulated DEGs than downregulated DEGs in CK1_vs_M1 (**Figure 2**b). However, in the other groups, the number of up- and downregulated DEGs was almost the same (**Figure 2**b). Moreover, the DEGs after heat treatment were mostly different from those during recovery in 17-03 and H1023 (**Figure 2**a). These results indicate that the defense and recovery mechanisms of pepper may have common regulatory pathways, and that there are different pathways for response, resistance, and repair.

Out of a total of 11,633 DEGs, 5133 were assigned a GO classification (**Table S3**). GO enrichment analysis showed that DEGs after heat treatment were commonly enriched in "photosynthesis, light harvesting" (GO:0009765), "cellular glucan metabolic process" (GO:0006073), and "nucleosome assembly" (GO:0006334) in 17-03 and H1023. These findings are similar to the HS response in sweet maize (*Zea mays* L.) [**49**]. DEGs were also significantly enriched in "DNA replication" (GO:0006260), "DNA replication initiation" (GO:0006270), "protein folding" (GO:0006457), "protein-DNA complex assembly" (GO:0065004), "protein-DNA complex" (GO:0032993), and "nucleosomal DNA binding" (GO:0031492) in 17-03 after heat treatment, which may confer increased resistance to high temperatures. The DEGs of H1023 and 17-03 during recovery were enriched with different GO terms and KEGG pathways (**Figure 4** and **Figure 5**), indicating that the repair pathways were different, which is likely due to the different degrees of high-temperature damage. In the KEGG pathway analysis, DEGs involved in HS response were predicted to function in metabolic pathways and the biosynthesis of secondary metabolites in 17-03 and H1023, which is similar to the results of previous studies [**50,51**].

Hsps, which are involved in multiple biological processes, such as signal transduction during HS, and have deduced functions, such as being chaperones, the folding and unfolding of cellular proteins, and the protection of functional sites from the adverse effects of high temperature, range in molecular mass from 10 to 200 kDa [**52**]. Hsps have functions as molecular chaperones that affect protein quality and were initially identified as proteins that were upregulated during heat treatment [**51**]. Many Hsps have been detected as heat response factors in tomato [**53**] and grape [**54**] plants. In this study, a total of 47 Hsps were significantly differentially expressed in the four groups. Among them, 45 Hsps were common differentially expressed in 17-03 and H1023 after heat treatment (**Figure 6; Table S6**). Hsps can accumulate rapidly in sensitive organs and play important roles in protecting the metabolic apparatus of cells, thus acting as a key factor in the adaptation of plants to high temperatures [**55**]. In this study, almost all DEGs encoding Hsps were upregulated in the four groups, the expression of which was the highest after heat treatment (**Figure 6**), which may play a role in protection under and after HS in 17-03 and H1023.

In addition to Hsps, various other TF genes, such as genes of Hsfs, NAC, and WRKY TFs, were also affected by HS [**56**]. Hsfs combine with cis-acting Hsps to play important roles in both basal and acquired thermotolerance [**26,51**]. Here, 17 DEGs encoding Hsfs were identified and most were upregulated in the four groups (**Figure 6; Table S6**). Moreover, some Hsfs were significantly highly expressed in H1023 after recovery from HS; such as CA02g11030, CA03g06850, CA05g00840, CA06g08710, CA07g15920, and CA10g20440 (**Figure 6**). These significantly expressed Hsfs could play important roles in the long-term HS response of pepper, by combining with the cis-acting regulatory elements, called heat shock elements, in the promoter regions of Hsps.

Plant NAC transcription factors have been documented to have a significant impact on regulating heat stress responses. Overexpression of Arabidopsis ANAC042 in transgenic plants results in a notable increase in thermotolerance, as demonstrated by the study [57]. Our data showed that the activity of 38 transcription factors encoded by NAC domain-containing genes was likewise influenced by heat (Figure 7; Table S7). Notably, the NAC transcription factors that were elevated showed higher levels of expression in H1023 compared to 17-03, both during and after the recovery from heat stress (Figure 7). WRKY transcription factors (TFs) are a prominent family of TFs in plants and have been found to play a role in regulating the plant's response to heat stress [58]. Within this investigation, a total of 35 WRKY transcription factors (TFs) exhibited a response to heat stress (HS) across the four groups. Remarkably, nearly half of these TFs were found to have a positive regulatory effect on thermotolerance, as depicted in Figure 7. In addition, several WRKY transcription factors (TFs) were found to have a negative regulatory effect on thermotolerance, including CA01g01920 and CA01g23300 (Figure 7).

5. CONCISE SUMMARY

We confirmed that the pepper cultivars 17-03 and H1023 exhibit heat resistance and heat sensitivity, respectively. To understand the impact of heat stress (HS) and the subsequent recovery on the expression of genes involved in the heat stress response and thermotolerance pathways, we employed RNA-seq analysis. A grand total of 11,633 differentially expressed genes (DEGs) were detected across the four groups, with 1229 DEGs being common to all four groups. Functional enrichment analysis revealed that in samples 17-03 and H1023, the differentially expressed genes (DEGs) were predominantly enriched in metabolic activities under stress, as well as in photosynthesis and light harvesting during heat stress (HS) and after recovery from HS. The pathways that showed the greatest significant enrichment in both 17-03 and H1023 were same under HS conditions, but diverged throughout the recovery period. In addition, a total of 38 Heat Shock Proteins (Hsps), 17 Heat Shock Factors (Hsfs), 38 NAC Transcription Factors (TFs), and 35 WRKY Transcription Factors (TFs) were shown to be involved in the Heat Shock (HS) or recovery reactions. These findings enhance comprehension of the molecular pathways that contribute to HS and the subsequent recovery in various pepper genotypes.

AT A GLANCE

Pepper, scientifically known as *Capsicum annuum* L., holds significant economic importance as a crop on a global scale. Pepper output and quality can be severely diminished by heat stress (HS). Nevertheless, the mechanisms underlying molecular alterations in response to heat stress (HS) and the subsequent restoration remain inadequately comprehended. The investigation identified 17-03 and H1023 as heat-tolerant and heat-sensitive variants, respectively. The transcript abundance of their leaves was assessed using RNA sequencing to investigate the impact of heat stress (HS) and subsequent recovery on gene expression. There were 11,633 genes that showed differential expression, and we confirmed the differential expression of 14 randomly chosen genes using reverse-transcription polymerase chain reaction. The functional enrichment analysis showed that the metabolic processes under stress and photosynthesis and light harvesting were the pathways that were most enriched under heat stress (HS) and after recovery from HS. The pathways that showed the greatest significant enrichment in both 17-03 and H1023 were same under HS conditions, but exhibited differences during the recovery phase. In addition, we have discovered 38 heat shock factors (Hsps), 17 HS transcription factors (Hsfs), 38 NAC (NAM, ATAF1/2, and CUC2) proteins, and 35 WRKY proteins that exhibit a response to heat shock or recovery. These findings enhance comprehension of the molecular pathways that contribute to HS and the process of recovery in various pepper genotypes.

REFERENCES

1. Truong, H.T.H.; Kim, K.T.; Kim, S.; Cho, M.C.; Kim, H.R.; Woo, J.G. Development of gene-based markers for the *Bs2* bacterial spot resistance gene for marker-assisted selection in pepper (*Capsicum* spp.). *Hortic. Environ. Biotechnol.* 2011, *52*, 65–73.

2. Wang, J.; Liang, C.; Yang, S.; Song, J.; Li, X.; Dai, X.; Wang, F.; Juntawong, N.; Tan, F.; Zhang, X.; et al. iTRAQ-based quantitative proteomic analysis of heat stress-induced mechanisms in pepper seedlings. *PeerJ* 2021, *9*, e11509.

3. Dang, F.; Lin, J.; Xue, B.; Chen, Y.; Guan, D.; Wang, Y.; He, S. CaWRKY27 negatively regulates H2O2-mediated thermotolerance in pepper (*Capsicum annuum*). *Front. Plant Sci.* 2018, *9*, 1633.

4. Belehradek, J. Physiological aspects of heat and cold. *Annu. Rev. Physiol.* 1957, *19*, 59–82.

5. Schiermeier, Q. Droughts, heatwaves and floods: How to tell when climate change is to blame. *Nature* 2018, *560*, 20–22.

6. Mathur, S.; Agrawal, D.; Jajoo, A. Photosynthesis: Response to high temperature stress. *J. Photochem. Photobiol. B* 2014, *137*, 116–126.

7. Wang, Q.; Chen, J.; He, N.; Guo, F. Metabolic reprogramming in chloroplasts under heat stress in plants. *Int. J. Mol. Sci.* 2018, *19*, 849.

8. Lu, T.; Meng, Z.; Zhang, G.; Qi, M.; Sun, Z.; Liu, Y.; Li, T. Sub-high Temperature and high light intensity induced irreversible inhibition on photosynthesis system of tomato plant (*Solanum lycopersicum* L.). *Front. Plant Sci.* 2017, *8*, 365.

9. Pan, C.; Ahammed, G.J.; Li, X.; Shi, K. Elevated CO_2 improves photosynthesis under high temperature by attenuating the functional limitations to energy fluxes, electron transport and redox homeostasis in tomato leaves. *Front. Plant Sci.* 2018, *9*, 1739.

10. Rogers, H.; Munné-Bosch, S. Production and scavenging of reactive oxygen species and redox signaling during leaf and flower senescence: Similar but different. *Plant Physiol.* 2016, *171*, 1560–1568.

11. Zhang, S.; Zhang, A.; Wu, X.; Zhu, Z.; Yang, Z.; Zhu, Y.; Zha, D. Transcriptome analysis revealed expression of genes related to anthocyanin biosynthesis in eggplant (*Solanum melongena* L.) under high-temperature stress. *BMC Plant Biol.* 2019, *19*, 387.

12. Kang, S.G.; Price, J.; Lin, P.C.; Hong, J.C.; Jang, J.C. The arabidopsis bZIP1 transcription factor is involved in sugar signaling, protein networking, and DNA binding. *Mol. Plant* 2010, 3, 361–373.

13. Obertello, M.; Krouk, G.; Katari, M.S.; Runko, S.J.; Coruzzi, G.M. Modeling the global effect of the basic-leucine zipper transcription factor 1 (*bZIP1*) on nitrogen and light regulation in Arabidopsis. *BMC Syst. Biol.* 2010, *4*, 111.

14. Osuna, D.; Usadel, B.; Morcuende, R.; Gibon, Y.; Bläsing, O.E.; Höhne, M.; Günter, M.; Kamlage, B.; Trethewey, R.; Scheible, W.R.; et al. Temporal responses of transcripts, enzyme activities and metabolites after adding sucrose to carbon-deprived Arabidopsis seedlings. *Plant J.* 2007, *49*, 463–491.

15. Jagodzik, P.; Tajdel-Zielinska, M.; Ciesla, A.; Marczak, M.; Ludwikow, A. Mitogen-activated protein kinase cascades in plant hormone signaling. *Front. Plant Sci.* 2018, *9*, 1387.

16. Kong, F.; Wang, J.; Cheng, L.; Liu, S.; Wu, J.; Peng, Z.; Lu, G. Genome-wide analysis of the mitogen-activated protein kinase gene family in Solanum lycopersicum. *Gene* 2012, *499*, 108–120.

17. Pitzschke, A.; Schikora, A.; Hirt, H. MAPK cascade signalling networks in plant defence. *Curr. Opin. Plant Biol.* 2009, *12*, 421–426.

18. Sangwan, V.; Orvar, B.L.; Beyerly, J.; Hirt, H.; Dhindsa, R.S. Opposite changes in membrane fluidity mimic cold and heat stress activation of distinct plant MAP kinase pathways. *Plant J.* 2002, *31*, 629–638.

19. Xu, J.; Zhang, S. Mitogen-activated protein kinase cascades in signaling plant growth and development. *Trends Plant Sci.* 2015, *20*, 56–64.

20. Yu, W.; Wang, L.; Zhao, R.; Sheng, J.; Zhang, S.; Li, R.; Shen, L. Knockout of *SlMAPK3* enhances tolerance to heat stress involving ROS homeostasis in tomato plants. *BMC Plant Biol.* 2019, *19*, 354.

21. Ding, H.; He, J.; Wu, Y.; Wu, X.; Ge, C.; Wang, Y.; Zhong, S.; Peiter, E.; Liang, J.; Xu, W. The tomato mitogen-activated protein kinase SlMPK1 is as a negative regulator of the high-temperature stress response. *Plant Physiol.* 2018, *177*, 633–651.

22. Vacca, R.A.; de Pinto, M.C.; Valenti, D.; Passarella, S.; Marra, E.; De Gara, L. Production of reactive oxygen species, alteration of cytosolic ascorbate peroxidase, and impairment of mitochondrial metabolism are early events in heat shock-induced programmed cell death in tobacco Bright-Yellow 2 cells. *Plant Physiol.* 2004, *134*, 1100–1112.

23. Volkov, R.A.; Panchuk, I.I.; Mullineaux, P.M.; Schöffl, F. Heat stress-induced H_2O_2 is required for effective expression of heat shock genes in Arabidopsis. *Plant Mol. Biol.* 2006, *61*, 733–746.

24. Larkindale, J.; Huang, B. Thermotolerance and antioxidant systems in Agrostis stolonifera: Involvement of salicylic acid, abscisic acid, calcium, hydrogen peroxide, and ethylene. *J. Plant Physiol.* 2004, *161*, 405–413.
25. Yin, Y.; Qin, K.; Song, X.; Zhang, Q.; Zhou, Y.; Xia, X.; Yu, J. BZR1 transcription factor regulates heat stress tolerance through feronia receptor-like kinase-mediated reactive oxygen species signaling in tomato. *Plant Cell Physiol.* 2018, *59*, 2239–2254.
26. Kotak, S.; Larkindale, J.; Lee, U.; von Koskull-Döring, P.; Vierling, E.; Scharf, K.D. Complexity of the heat stress response in plants. *Curr. Opin. Plant Biol.* 2007, *10*, 310–316.
27. Fragkostefanakis, S.; Simm, S.; Paul, P.; Bublak, D.; Scharf, K.D.; Schleiff, E. Chaperone network composition in Solanum lycopersicum explored by transcriptome profiling and microarray meta-analysis. *Plant Cell Environ.* 2015, *38*, 693–709.
28. Hahn, A.; Bublak, D.; Schleiff, E.; Scharf, K.D. Crosstalk between Hsp90 and Hsp70 chaperones and heat stress transcription factors in tomato. *Plant Cell.* 2011, *23*, 741–755.
29. Mishra, S.K.; Tripp, J.; Winkelhaus, S.; Tschiersch, B.; Theres, K.; Nover, L.; Scharf, K.D. In the complex family of heat stress transcription factors, *HsfA1* has a unique role as master regulator of thermotolerance in tomato. *Genes Dev.* 2002, *16*, 1555–1567.
30. Fragkostefanakis, S.; Simm, S.; El-Shershaby, A.; Hu, Y.; Bublak, D.; Mesihovic, A.; Darm, K.; Mishra, S.K.; Tschiersch, B.; Theres, K.; et al. The repressor and co-activator HsfB1 regulates the major heat stress transcription factors in tomato. *Plant Cell Environ.* 2019, *42*, 874–890.
31. Bharti, K.; Von Koskull-Döring, P.; Bharti, S.; Kumar, P.; Tintschl-Körbitzer, A.; Treuter, E.; Nover, L. Tomato heat stress transcription factor HsfB1 represents a novel type of general transcription coactivator with a histone-like motif interacting with the plant CREB binding protein ortholog HAC1. *Plant Cell.* 2004, *16*, 1521–1535.
32. Waters, E.R.; Vierling, E. Plant small heat shock proteins—Evolutionary and functional diversity. *New Phytol.* 2020, *227*, 24–37.
33. Yang, J.; Sun, Y.; Sun, Q.; Yi, Y.; Qin, J.; Li, H.; Liu, J. The involvement of chloroplast HSP100/ClpB in the acquired thermotolerance in tomato. *Plant Mol. Biol.* 2006, *62*, 385–395.
34. Shen, L.; Liu, Z.; Yang, S.; Yang, T.; Liang, J.; Wen, J.; Liu, Y.; Li, J.; Shi, L.; Tang, Q.; et al. Pepper CabZIP63 acts as a positive regulator during *Ralstonia solanacearum* or high temperature-high humidity challenge in a positive feedback loop with CaWRKY40. *J. Exp. Bot.* 2016, *67*, 2439–2451.
35. Meng, X.; Wang, J.; Wang, G.; Liang, X.; Li, X.; Meng, Q. An R2R3-MYB gene, *LeAN2*, positively regulated the thermo-tolerance in transgenic tomato. *J. Plant Physiol.* 2015, *175*, 1–8.
36. Campos, P.S.; Quartin, V.; Ramalho, J.C.; Nunes, M.A. Electrolyte leakage and lipid degradation account for cold sensitivity in leaves of Coffea sp. plants. *J. Plant Physiol.* 2003, *160*, 283–292.
37. Ben Rejeb, K.; Abdelly, C.; Savouré, A. Proline, a multifunctional amino-acid involved in plant adaptation to environmental constraints. *Biol. Aujourdhui* 2012, *206*, 291–299.
38. Chen, Y.; Chen, Y.; Shi, C.; Huang, Z.; Zhang, Y.; Li, S.; Li, Y.; Ye, J.; Yu, C.; Li, Z.; et al. SOAPnuke: A MapReduce acceleration-supported software for integrated quality control and preprocessing of high-throughput sequencing data. *Gigascience* 2018, *7*, 1–6.
39. Anders, S.; Pyl, P.T.; Huber, W. HTSeq—A Python framework to work with high-throughput sequencing data. *Bioinformatics* 2015, *31*, 166–169.
40. Trapnell, C.; Williams, B.A.; Pertea, G.; Mortazavi, A.; Kwan, G.; van Baren, M.J.; Salzberg, S.L.; Wold, B.J.; Pachter, L. Transcript assembly and quantification by RNA-Seq reveals unannotated transcripts and isoform switching during cell differentiation. *Nat. Biotechnol.* 2010, *28*, 511–515.
41. Love, M.I.; Huber, W.; Anders, S. Moderated estimation of fold change and dispersion for RNA-seq data with DESeq2. *Genome Biol.* 2014, *15*, 550.

42. Götz, S.; García-Gómez, J.M.; Terol, J.; Williams, T.D.; Nagaraj, S.H.; Nueda, M.J.; Robles, M.; Talón, M.; Dopazo, J.; Conesa, A. High-throughput functional annotation and data mining with the Blast2GO suite. *Nucleic Acids Res.* 2008, *36*, 3420–3435.

43. Bindea, G.; Mlecnik, B.; Hackl, H.; Charoentong, P.; Tosolini, M.; Kirilovsky, A.; Fridman, W.H.; Pagès, F.; Trajanoski, Z.; Galon, J. ClueGO: A Cytoscape plug-in to decipher functionally grouped gene ontology and pathway annotation networks. *Bioinformatics* 2009, *25*, 1091–1093.

44. Jin, J.; Tian, F.; Yang, D.; Meng, Y.; Kong, L.; Luo, J.; Gao, G. PlantTFDB 4.0: Toward a central hub for transcription factors and regulatory interactions in plants. *Nucleic Acids Res.* 2017, *45*, D1040–D1045.

45. Wan, H.; Yuan, W.; Ruan, M.; Ye, Q.; Wang, R.; Li, Z.; Zhou, G.; Yao, Z.; Zhao, J.; Liu, S.; et al. Identification of reference genes for reverse transcription quantitative real-time PCR normalization in pepper (*Capsicum annuum* L.). *Biochem. Biophys. Res. Commun.* 2011, *416*, 24–30.

46. Schmittgen, T.D.; Livak, K.J. Analyzing real-time PCR data by the comparative C(T) method. *Nat. Protoc.* 2008, 3, 1101–1108.

47. Chaudhry, S.; Sidhu, G.P.S. Climate change regulated abiotic stress mechanisms in plants: A comprehensive review. *Plant Cell Rep.* 2021.

48. Demidchik, V.; Straltsova, D.; Medvedev, S.S.; Pozhvanov, G.A.; Sokolik, A.; Yurin, V. Stress-induced electrolyte leakage: The role of K+-permeable channels and involvement in programmed cell death and metabolic adjustment. *J. Exp. Bot.* 2014, *65*, 1259–1270.

49. Shi, J.; Yan, B.; Lou, X.; Ma, H.; Ruan, S. Comparative transcriptome analysis reveals the transcriptional alterations in heat-resistant and heat-sensitive sweet maize (*Zea mays* L.) varieties under heat stress. *BMC Plant Biol.* 2017, *17*, 26.

50. Mangelsen, E.; Kilian, J.; Harter, K.; Jansson, C.; Wanke, D.; Sundberg, E. Transcriptome analysis of high-temperature stress in developing barley caryopses: Early stress responses and effects on storage compound biosynthesis. *Mol. Plant* 2011, *4*, 97–115.

51. Li, T.; Xu, X.; Li, Y.; Wang, H.; Li, Z.; Li, Z. Comparative transcriptome analysis reveals differential transcription in heat-susceptible and heat-tolerant pepper (*Capsicum annum* L.) cultivars under heat stress. *J. Plant Biol.* 2015, *58*, 411–424.

52. Huang, B.; Xu, C. Identification and characterization of proteins associated with plant tolerance to heat stress. *J. Integr. Plant Biol.* 2008, *50*, 1230–1237.

53. Frank, G.; Pressman, E.; Ophir, R.; Althan, L.; Shaked, R.; Freedman, M.; Shen, S.; Firon, N. Transcriptional profiling of maturing tomato (*Solanum lycopersicum* L.) microspores reveals the involvement of heat shock proteins, ROS scavengers, hormones, and sugars in the heat stress response. *J. Exp. Bot.* 2009, *60*, 3891–3908.

54. Liu, G.; Wang, J.; Cramer, G.; Dai, Z.; Duan, W.; Xu, H.; Wu, B.; Fan, P.; Wang, L.; Li, S. Transcriptomic analysis of grape (*Vitis vinifera* L.) leaves during and after recovery from heat stress. *BMC Plant Biol.* 2012, *12*, 174.

55. Wahid, A.; Gelani, S.; Ashraf, M.; Foolad, M.R. Heat tolerance in plants: An overview. *Environ. Exp. Bot.* 2007, *61*, 199–223.

56. Qin, D.; Wu, H.; Peng, H.; Yao, Y.; Ni, Z.; Li, Z.; Zhou, C.; Sun, Q. Heat stress-responsive transcriptome analysis in heat susceptible and tolerant wheat (*Triticum aestivum* L.) by using wheat genome array. *BMC Genom.* 2008, *9*, 432.

57. Shahnejat-Bushehri, S.; Mueller-Roeber, B.; Balazadeh, S. Arabidopsis NAC transcription factor JUNGBRUNNEN1 affects thermomemory-associated genes and enhances heat stress tolerance in primed and unprimed conditions. *Plant Signal. Behav.* 2012, *7*, 1518–1521.

58. Zhang, C.; Wang, D.; Yang, C.; Kong, N.; Shi, Z.; Zhao, P.; Nan, Y.; Nie, T.; Wang, R.; Ma, H.; et al. Genome-wide identification of the potato WRKY transcription factor family. *PLoS ONE* 2017, *12*, e0181573.

INDEX

R

S

T

V

W

Z

www.ingramcontent.com/pod-product-compliance
Lightning Source LLC
LaVergne TN
LVHW062221200926
848484LV00006B/1449